Transport and Chemical Transformation of Pollutants in the Troposphere

Series editors: Peter Borrell, Patricia M. Borrell, Tomislav Cvitaš, Kerry Kelly and Wolfgang Seiler

Springer-Verlag Berlin Heidelberg GmbH

Transport and Chemical Transformation
of Pollutants in the Troposphere

Volume 8

Instrument Development for Atmospheric Research and Monitoring

Lidar Profiling, DOAS and Tunable Diode Laser Spectroscopy

Jens Bösenberg (Hamburg), **David Brassington** (London)
and **Paul C. Simon** (Brussels)
Editors

Springer

Dr. Jens Bösenberg
Max-Planck-Institut für Meteorologie
Bundesstraße 55
D-20146 Hamburg
Germany

Dr. David J. Brassington
Atmospheric Chemistry Research Unit
Silwood Park, Burkhurst Road
Ascot, Berkshire SL5 7PW
Great Britain

Dr. Paul C. Simon
Institut d'Aeronomie Spatiale Belgique
3 Avenue Circulaire
B-1180 Brusselles
Belgium

The cover picture shows a frequency-modulated tunable-diode-laser spectrometer at the Fraunhofer Institute for Atmospheric Environmental Research in Garmisch-Partenkirchen. The photograph was taken by Peter Werle.

ISBN 978-3-642-08291-7 ISBN 978-3-662-03405-7 (eBook)
DOI 10.1007/978-3-662-03405-7

Library of Congress Cataloging-in-Publication Data
Instrument development for atmospheric research and monitoring: Lidar profiling. DOAS and tunable diode laser spectroscopy/Jens Bösenberg, David Brassington, and Paul C. Simon, editors. p. cm. – (Transport and chemical transformation of pollutants in the troposphere: v. 8) Includes bibliographical references and index. ISBN 3-540-62516-X (hc). 1. Troposphere – Environmental aspects – Remote sensing. 2. Atmospheric ozone – Instruments – Design and construction. 3. Environmental monitoring – Technique. 4. Atmosphere – Laser observation. 5. Optical radar. I. Bösenberg, Jens. II. Brassington, David. III. Simon, Paul C. IV. Series. OC881.2.T75I54 1997 561.51'12–dc2' 97-368 CIP

Originally published by Springer-Verlag Berlin Heidelberg New York in 1997.
Softcover reprint of the hardcover 1st edition 1997

Cover Design: Struve & Partner, Heidelberg

SPIN 10514768 30/3136-5 4 3 2 1 0 – Printed on acid-free paper

Transport and Chemical Transformation of Pollutants in the Troposphere

Series editors: Peter Borrell, Patricia M. Borrell, Tomislav Cvitaš, Kerry Kelly and Wolfgang Seiler

Volume 1	Transport and Chemical Transformation of Pollutants in the Troposphere *An overview of the work of EUROTRAC*
Volume 2	Heterogeneous and Liquid Phase Processes *Laboratory studies related to aerosols and clouds*
Volume 3	Chemical Processes in Atmospheric Oxidation *Laboratory studies of chemistry related to troposphere ozone*
Volume 4	Biosphere-Atmosphere Exchange of Pollutants and Trace Substances *Experimental and theoretical studies of biogenic emission and of pollutant deposition*
Volume 5	Cloud Multi Phase Processes and High Alpine Air and Snow Chemistry *Ground-based cloud experiments and pollutant deposition in the high Alps*
Volume 6	Tropospheric Ozone Research *Tropospheric ozone in the regional and sub-regional context*
Volume 7	Tropospheric Modelling and Emission Estimation *Chemical transport and emission modelling on regional, global and urban scales*
Volume 8	Instrument Development for Atmospheric Research and Monitoring *Lidar profiling, DOAS and tunable diode laser spectroscopy*
Volume 9	Exchange and Transport of Air Pollutants over Complex Terrain and the Sea *Field measurements and numerical modelling; aircraft, ship, ocean platform and laboratory measurements*
Volume 10	Photo-oxidants, Acidification and Tools: Policy Applications of EUROTRAC Results

Foreword by the Series Editors

EUROTRAC is the European co-ordinated research project, within the EUREKA initiative, studying the transport and chemical transformation of pollutants in the troposphere. The project has achieved a remarkable scientific success since its start in 1988, contributing substantially both to the scientific progress in this field and to the improvement of the scientific basis for environmental management in Europe. EUROTRAC, which at its peak comprised some 250 research groups organised into 14 subprojects, brought together international groups of scientists to work on problems directly related to the transport and chemical transformation of trace substances in the troposphere. In doing so, it helped to harness the resources of the participating countries to gain a better understanding of the trans-boundary, interdisciplinary environmental problems which beset us in Europe.

The scientific results of EUROTRAC are summarised in this report which consists of ten volumes.

Volume 1 provides a general overview of the scientific results, prepared by the Scientific Steering Committee (SSC) and the International Scientific Secretariat (ISS) of EUROTRAC, together with brief summaries of the work of the fourteen individual subprojects prepared by the respective subproject coordinators.

Volumes 2 to 9 comprise detailed overviews of the subproject achievements, each prepared by the respective subproject coordinator and steering group, together with summaries of the work of the participating research groups prepared by the principal investigators. Each volume also includes a full list of the scientific publications from the subproject.

The final volume, 10, is the complete report of the Application Project, which was set up in 1993 to assimilate the scientific results from EUROTRAC and present them in a condensed form so that they are suitable for use by those responsible for environmental planning and management in Europe. It illustrates how a scientific project such as EUROTRAC can contribute practically to providing the scientific consensus necessary for the development of a coherent atmospheric environmental policy for Europe.

A multi-volume work such as this has many contributors and we, as general editors, would like to express our thanks to all of them: to the subproject coordinators who have borne the brunt of the scientific co-ordination and who have contributed so much to the success of the project and the quality of this report; to the principal investigators who have carried out so much high-quality scientific work; to the members of the International Executive Committee (IEC) and the SSC for their enthusiastic encouragement and support of EUROTRAC; to the participating governments in EUROTRAC, and in particular the German Government (BMBF) for funding, not only the research, but also the ISS publication activities; and finally to Mr. Christian Witschell and his colleagues at Springer Verlag for providing the opportunity to publish the results in a way which will bring them to the notice of a large audience.

Peter Borrell (Scientific Secretary, ISS)
Patricia May Borrell
Tomislav Cvitaš
Kerry Kelly
Wolfgang Seiler (Director, ISS)

EUROTRAC ISS
Fraunhofer Institute (IFU)
Garmisch-Partenkirchen

Table of Contents

List of Contributors

Ms. B. Allan
School of Environmental Sciences
University of East Anglia
Norwich NR4 7TJ
Great Britain

Dr. Francesco d' Amato
ENEA
INN-FIS-SPET
P.O. Box 65
Via E. Fermi 27
I-00044 Frascati (RM)
Italy

Dr. Gérard Ancellet
CNRS Service d'Aéronomie
Université P. et M. Curie
T. 15 - E.5, 4 Place Jussieu
F-75230 Paris Cedex 05
France

Mr. J. Anders
Fraunhofer-Institut für Physikalische
Meßtechnik
Heidenhofstraße 8
D-79110 Freiburg
Germany

Dr. Arnoud Apituley
National Institute of Public Health
and Environmental Hygiene (RIVM)
Postbus 1
NL-3720 BA Bilthoven
The Netherlands

Mr. H. Axelsson
Swedish Environmental Research
Institute (IVL)
P.O. Box 47086
S-402 58 Göteborg
Sweden

Dr. Giuseppe Baldacchini
ENEA - C.R.E. Frascati
C.P. 65
I-00044 Frascati
Italy

Dr. A. Bargagli
E.N.E.A., C.R.E. Casaccia
SP Anguillarese 301
I-00100 Roma
Italy

Mr. B. Bergqvist
Swedish Environmental Research
Institute (IVL)
P.O. Box 47086
S-402 58 Göteborg
Sweden

Mr. J. Bonifer
Max-Planck-Institut für Chemie
Abt. Luftchemie, Postfach 3060
Saarstraße 23
D-55020 Mainz
Germany

Dr. Jens Bösenberg
Max-Planck-Institut für Meteorologie
Bundesstraße 55
D-20146 Hamburg
Germany

Dr. Harald Böttner
Fraunhofer-Institut für Physikalische
Meßtechnik
Heidenhofstraße 8
D-79110 Freiburg
Germany

Dr. D.J. Brassington
Atmospheric Chemistry Research Unit
Silwood Park, Burkhurst Road
Ascot, Berkshire SL5 7PW
Great Britain

Mr. V.D. Burlakov
USSR Academy of Sciences
Institute of Atmospheric Optics
1 Akademicheskii Avenue
R-Tomsk 634055
Russia

Dr. J.P. Burrows
Universität Bremen, Fachbereich Physik
Institut für Fernerkundung
Postfach 33 04 40
D-28334 Bremen
Germany

Dr. Michel Carleer
Laboratoire de Chimie Physique
Moleculaire
Faculté des Sciences
Université Libre de Bruxelles
50, avenue F.D. Roosevelt
B-1050 Bruxelles
Belgium

Mr. W. Carnuth
Fraunhofer Institut für Atmosphärische
Umweltforschung
Kreuzeckbahnstraße 19
D-82467 Garmisch-Partenkirchen
Germany

Mr. I. Carrasco
C.N.R.S. Service d'Aéronomie
4, place Jussieu, Tour 15, Univ. P 6
B.P. 102
F-75252 Paris Cedex 05
France

Dr. Hugh Coe
School of Environmental Sciences
University of East Anglia
Norwich, Norfolk NR4 7T5
Great Britain

Prof. Reginald Colin
Université Libre de Bruxelles
Laboratoire de Photophysique
Moleculaire
50, avenue F.D. Roosevelt
B-1050 Bruxelles
Belgium

Dr. P. Dufour
Laboratoire d'Informatique
Université de Mons-Hainaut
Mons
Belgium

Dr. Eric Durieux
EPFL
DGR - LPAS
34, Ch. de Bellerive
CH-1007 Lausanne
Switzerland

Mr. A. Eilard
Swedish Environmental Research
Institute (IVL)
P.O. Box 47086
S-402 58 Göteborg
Sweden

Dr. C. Fayt
Laboratoire d'Informatique
Université de Mons-Hainaut
Mons
Belgium

Prof. Giovanna Finzi
Università di Brescia
Dipartimento di Elettronica per
Automazione
I-25123 Brescia
Italy

Dr. G. Fiocco
(Cattedra di Fiscica Terrestre)
Dipartimento di Fisica
Universita di Roma
I-00185 Roma
Italy

Mr. Luca Fiorani
EPFL
DGR - LPAS
34, Ch. de Bellerive
CH-1007 Lausanne
Switzerland

Dr. Horst Fischer
Max-Planck-Institut für Chemie
Abt. Luftchemie,
Postfach 3060
Saarstraße 23
D-55020 Mainz
Germany

Dr. Bo Galle
Swedish Environmental Research
Institute (IVL)
P.O. Box 47086
S-402 58 Göteborg
Sweden

Dr. Florence Goutail
Service d'Aéronomie (SA)
C.N.R.S., B.P.3
F-91371 Verrières Le Buisson
France

Dr. R. Grisar
Fraunhofer-Institut fürPhysikalische
Meßtechnik
Heidenhofstraße 8
D-79110 Freiburg
Germany

Dr. Jean-Michel Guilmot
Laboratoire de Chimie Physique
Moleculaire
Faculté des Sciences
Université Libre de Bruxelles
50, avenue F.D. Roosevelt
B-1050 Bruxelles
Belgium

Dr. Geoffrey W. Harris
Department of Chemistry
York University
4700 Keele Street
Downsview
Ontario, M3J 1P3
Canada

Mr. C. Hermans
Institut d'Aeronomie Spatiale Belgique
3, Avenue Circulaire
B-1180 Bruxelles
Belgium

Dr. F. Huard
Société BERTIN
BP 22000
13791 Aix-en-Provence cedex 3
France

Mr. U. Kempfer
Fraunhofer Institut für Atmosphärische
Umweltforschung
Kreuzeckbahnstraße 19
D-82467 Garmisch-Partenkirchen
Germany

Dr. Dieter Klemp
Forschungszentrum Jülich
ICG 2, P.O. Box 1913
D-52425 Jülich
Germany

Mr. M. Knothe
Fraunhofer-Institut für Physikalische
Meßtechnik
Heidenhofstraße 8
D-79110 Freiburg
Germany

Mr. B.S. Kostin
USSR Academy of Sciences
Institute of Atmospheric Optics
1 Akademicheskii Avenue
R-Tomsk 634055
Russia

Dr. Armin Lambrecht
Fraunhofer Institut für Physikalische
Meßtechnik
Heidenhofstraße 8
D-79110 Freiburg
Germany

Ms. Patricia Laville
Service d'Aéronomie (SA)
C.N.R.S., B.P.3
F-91371 Verrières Le Buisson
France

Mr. O. Lubin
C.N.R.S. Service d'Aéronomie
4, place Jussieu, Tour 15, Univ. P 6
B.P. 102
F-75252 Paris Cedex 05
France

Dr. Hélène Mac Leod
C.N.R.S. Service d'Aéronomie
4, place Jussieu, Tour 15, Univ. P 6
B.P. 102
F-75252 Paris Cedex 05
France

Dr. Gérard Mégie
C.N.R.S. Service d'Aeronomie
B.P. 3
F-91371 Verrières Le Buisson Cedex
France

Mr. J. Mellqvist
Swedish Environmental Research
Institute (IVL)
P.O. Box 47086
S-402 58 Göteborg
Sweden

Dr. Robert Mücke
Fraunhofer Institut für Atmosphärische Umweltforschung
Kreuzeckbahnstraße 19
D-82467 Garmisch-Partenkirchen
Germany

Mr. M. Nunes-Pinharanda
Service d'Aéronomie
C.N.R.S., B.P. 3
F-91371 Verrières Le Buisson Cedex
France

Dr. Alexandros Papayannis
Nat. Tech. University of Athens
Department of Physics
Zografou Campus
GR-157 73 Athens
Greece

Mr. U. Parchatka
Max-Planck-Institut für Chemie
Abt. Luftchemie,
Postfach 3060
Saarstraße 23
D-55020 Mainz
Germany

Dr. John M. C. Plane
School of Environmental Sciences
University of East Anglia
Norwich NR4 7TJ
Great Britain

Dr. Ulrich Platt
Universität Heidelberg
Institut für Umweltphysik
Im Neuenheimer Feld 366
D-69120 Heidelberg
Germany

Dr. Jean-Pierre Pommereau
Service d'Aéronomie
C.N.R.S., B.P. 3
F-91371 Verrières Le Buisson Cedex
France

Mr. H. Poncet
C.N.R.S. Service d'Aéronomie
4, place Jussieu, Tour 15, Univ. P 6
B.P. 102
F-75252 Paris Cedex 05
France

Dr. W.J. Riedel
Fraunhofer-Institut für Physikalische Meßtechnik
Heidenhofstraße 8
D-79110 Freiburg
Germany

Dr. J. Roths
Max-Planck-Institut für Chemie
Abt. Luftchemie, Postfach 3060
Saarstraße 23
D-55020 Mainz
Germany

Mr. Thorsten Schaberl
Max-Planck-Institut für Meteorologie
Bundesstraße 55
D-20146 Hamburg
Germany

Mr. C. Schiller
Department of Chemistry
York University
4700 Keele Street
Downsview
Ontario, M3J 1P3
Canada

Mr. C. Senff
Max-Planck-Institut für Meteorologie
Bundesstraße 55
D-20146 Hamburg
Germany

Dr. Paul C. Simon
Institut d'Aeronomie Spatiale Belgique
3, Avenue Circulaire
B-1180 Bruxelles
Belgium

Dr. Franz Slemr
Fraunhofer-Institut für Atmosphärische Umweltforschung
Kreuzeckbahnstraße 19
D-82467 Garmisch-Partenkirchen
Germany

Ms. N. Smith
School of Environmental Sciences
University of East Anglia
Norwich NR4 7TJ
Great Britain

Mr. Jochen Stutz
Universität Heidelberg
Institut für Umweltphysik
Im Neuenheimer Feld 366
D-69120 Heidelberg
Germany

Prof. Dr. Maurus Tacke
Fraunhofer-Institut für Physikalische Messtechnik
Heidenhofstraße 8
D-79110 Freiburg
Germany

Dr. Thomas Trickl
Fraunhofer Institut für Atmosphärische Umweltsforschung
Kreuzeckbahnstrasse 19
D-82467 Garmisch-Partenkirchen
Germany

Dr. Ann Carine Vandaele
Institut d'Aeronomie Spatiale
3, Avenue Circulaire
B-1180 Brussels
Belgium

Dr. Peter Werle
Fraunhofer Institut für Atmosphärische Umweltforschung
Kreuzeckbahnstraße 19
D-82467 Garmisch-Partenkirchen
Germany

Mr. T. Zenker
Max-Planck-Institut für Chemie
Abt. Luftchemie, Postfach 3060
Saarstraße 23
D-55020 Mainz
Germany

Mr. L. Zetterberg
Swedish Environmental Research Institute (IVL)
P.O. Box 47086
S-402 58 Göteborg
Sweden

Mr. R. Zitzelsberger
Max-Planck-Institut für Chemie
Abt. Luftchemie, Postfach 3060
Saarstraße 23
D-55020 Mainz
Germany

Dr. V. V. Zuev
USSR Academy of Sciences
Institute of Atmospheric Optics
1 Akademicheskii Avenue
R-Tomsk 634055
Russia

Transport and Chemical Transformation of Pollutants in the Troposphere

Volume 8, Part I

Tropospheric Environmental Studies by Laser Sounding

Development and Application of Lidar Instruments in Atmospheric Profiling

Jens Bösenberg (Hamburg)
Coordinator

TESLAS Steering Group

Gerard Ancellet	Paris
Roberto Barbini	Frascati
Martin J. T. Milton	Teddington

Participating Institutions

BAT:	Battelle Institut e.V., Am Römerhof 35, Frankfurt am Main, Germany V. Klein, M. Endemann
ENEA:	ENEA, Dip. INN-SVIL, P.O. Box 65, I-00044 Frascati, Italy R. Barbini, F. Coalo, A. Palucci, S. Ribezzo
EPFL:	EPFL-LIDAR Group, Bellerive 34, CH-1007 Lausanne, Switzerland E. Durieux, B. Calpini, L. Fiorani, L. Jaquet, H. van den Bergh
IAO:	Institute of Atmospheric Optics, SB RAS, 1, Akademicheskii Ave.,Tomsk, 634055, Russia V.V. Zuev, A.A. Mitzel, G.S. Evtushenko, Ju.N. Ponomarev, J.S. Tyryshkin, V.D. Burlakov, M.Ju. Kataev, V.O. Troitskii
IFU:	Fraunhofer-Intitut für Atmosphärische Umweltforschung, D-82467 Garmisch Partenkirchen, Germany T. Trickl
IROE:	I.R.O.E.-C.N.R., Firenze, Italy L. Stefanutti, F. Castagnoli, M. Del Guasta, M. Morandi, V.M. Sacco, L. Zuccagnoli
LIT:	Lund Institute of Technology, dep. of Physics, P.O. Box 118, S-22100 Lund, Sweden H. Edner, S. Svanberg, E. Wallinder
MPI:	Max-Planck-Institut für Meteorologie, Bundesstr. 55, D-20146 Hamburg, Germany J. Bösenberg, V. Matthias, T. Schaberl
NPL:	National Physical Laboratory, Teddington, Middlesex, United Kingdom M.J.T. Milton, P.T. Woods
RIVM:	Rijksinstituut voor Volksgezondheid en Milieu, Lab. f. Air Research, P.O. Box 1, NL- 3720 BA Bilthoven, The Netherlands A. Apituley, D.P.J. Swart, E.P. Visser
SA/CNRS:	Service d'Aéronomie du CNRS, Université Paris 6, Boite 102 4 Place Jussieu, 75230 Paris Cedex 05, France G. Ancellet, M. Beekmann, G. Mégie
GKSS:	GKSS-Forschungszentrum Geesthacht GmbH, D-2054 Geesthacht, Germany C. Weitkamp, U.B. Goers, J. Glauer, W. Lahmann, W. Michaelis

Chapter 1

Introduction

Jens Bösenberg

Max-Planck-Institut für Meteorologie, Bundesstr. 55, D-20146 Hamburg, Germany

TESLAS, which stands for **T**ropospheric **E**nvironmental **S**tudies by **La**ser **S**ounding, was formed in November 1987 as a subproject of EUROTRAC to enhance the measurement capabilities for vertical profiling of ozone in the troposphere by means of laser remote sensing. For studies of several atmospheric processes related to the formation and redistribution of photo-oxidants there was a clear need for measuring extended time series with appropriate vertical and temporal resolution. These could not be obtained by conventional in situ techniques, at least not with affordable effort, so remote sensing appeared to be the best way to obtain the required information.

At the beginning of the subproject, some **Di**fferential **A**bsorption **L**idar (DIAL) systems for measuring the vertical distribution of ozone already existed, but their use was restricted to very few laboratories and very few measurement campaigns, since the instruments were highly complex, rather unreliable, and required extensive efforts for maintenance and operation by skilled scientists. In addition, the accuracy of these measurements under a variety of meteorological conditions was not really well established. The main tasks within TESLAS therefore were to develop fully the DIAL-methodology for remote sensing of tropospheric ozone, and to develop instruments which are accurate, reliable, easy to operate, and suitable for field deployment or airborne operation.

For studies of the role of ozone in the chemistry of the polluted atmosphere processes involving very different scales, both temporally and spatially, are important. In principle, the scales range from centimeter-sized eddies associated with turbulent transport, over mesoscale processes like convective transport, synoptic scale phenomena as tropopause folding events, to global distribution and secular trends. While the very small scales are still the domain of *in situ* measurements, the resolution of laser remote sensing has been pushed to several tens of meters in the vertical and several seconds temporally, so that a large variety of processes can actually be addressed with this technique. Also, the reliability, accuracy, and ease of operation which have been achieved provide for an excellent tool to address the processes associated with medium or large scales. The main advantage of using a remote sensing technique for such studies is, that they can be performed continuously (or at least quasi-continuously) over extended periods of time.

A total of twelve groups from seven European countries were involved in TESLAS during its lifetime:

BAT:	Battelle Institut e.V., Frankfurt, Germany
ENEA:	Ente per le Nuove Tecnologie, l'Energia e l'Ambiente, Frascati, Italy
EPFL:	Ecole Polytechnique Federale, Lausanne, Switzerland
GKSS:	GKSS Forschungszentrum, Geesthacht, Germany
IAO:	Institute for Atmospheric Optics, Tomsk, Russia
IFU:	Fraunhofer Institut für Atmosphärische Umweltforschung, Garmisch-Partenkirchen, Germany
IROE:	Institute di Ricerca sullo Onde Elettromagnetiche del CNR, Firenze, Italy
LIT:	Lund Institute for Technology, Lund, Sweden
MPI:	Max-Planck-Institut für Meteorologie, Hamburg, Germany
NPL:	National Physical Laboratory, Teddington, Great Britain
RIVM:	Rijksinstituut voor Volksgezondheid en Milieu, Bilthoven, The Netherlands
SA/CNRS:	Service d'Aeronomie du CNRS, Paris, France

The TESLAS subproject started with nine groups from five European countries: IFU, GKSS, MPI, Batelle, IROE, ENEA, SA/CNRS, LIT, NPL, coordinated by Jacques Pelon, SA/CNRS Paris. During the years, some groups had to leave the project (BAT) or reduce their activities (IROE, LIT, ENEA, NPL) due to funding problems, but new groups were joining in: RIVM, EPFL, IAO. In 1990 the coordination was shifted from Jacques Pelon to Jens Bösenberg, MPI Hamburg. A regular exchange of information was established, formally through twelve mostly well attended workshops, which turned out to be very efficient for the dissemination of new results, and for the discussion of different approaches to the solutions of a large variety of technical and methodological problems. In addition, of course, close personal connections between the group members provided for additonal informal exchange of information, which was at least as important as the official links.

Regarding the main methodology, two distinct approaches were taken. Most of the groups had decided to operate in the Hartley-Huggins band of ozone absorption, which is located in the UV part of the spectrum, but two groups were exploring the possibility of using absorption lines of the vibrational-rotational spectrum in the infrared around 10 μm, the emission region of CO_2 lasers (BAT, ENEA). Of these two, only ENEA was in a position to build and successfully test a system, Battelle had to give up due to lack of financial support for this project. For the UV systems, different approaches regarding the laser technology were used: three groups followed the traditional design using tunable dye lasers (EPFL, GKSS, LIT), four groups were using Nd:YAG lasers with subsequent Raman shifting (IROE, NPL, RIVM, SA/CNRS), and three groups employed KrF-excimer lasers with subsequent Raman shifting (IFU, LIT, MPI). The use of metal vapor lasers with frequency mixing was also explored (IAO). So a wealth of different techniques was investigated with good success, each of them having advantages for special applications. Of the twelve groups finally nine succeeded in setting up at least one complete system for use in atmospheric studies, compared to just one system which existed in Europe before TESLAS was started. One group (SA/CNRS) also finished the construction of an airborne system.

The development of DIAL-systems within TESLAS was not performed as a pure engineering task, mainly because not all required properties of the instrument could be specified in the beginning. It was rather performed as a feedback process. The atmospheric studies carried out with prototype versions of the instruments yielded experience for the next step of improvement and adaption to the specific needs of a large variety of applications. The feedback from actual experiments to the instrument design, and vice versa, was found to be essential in the development of a complex system to be applied for studies of processes which were not well known at the outset.

For a new system of considerable complexity with a number of potential error sources, to be applied under a large variety of conditions, a detailed assessment of accuracy of the measurements was considered essential. This was performed by a combination of several methods. First, a detailed understanding of the methodology and the physics of the processes involved in the measurement is mandatory for estimating the error margins. The impact of different approaches to the data processing on the accuracy of the results has been studied by an intercomparison of the algorithms used by different groups, applying them to some selected data sets. Overall tests of the system accuracy were performed through intercomparison experiments.

It has already been mentioned, that several systems have been applied for studies of atmospheric processes or climatology involving tropospheric ozone. It is these applications which best assess the usefulness of lidar remote sensing of tropospheric ozone for atmospheric research. It should also be noted, that several of these applications were performed in cooperation with other subprojects of EUROTRAC, in particular with TOR, TRACT, and EURAD. In addition TESLAS instruments were involved in European campaigns addressing transport (TOASTE) or chemical processes (OCTA, POLLUMET), as well as in other major measurement campaigns, such as ICE, EUCREX, GOME, and LITE.

Since for a DIAL system which is to be used in a specific application the specifications can vary considerably, depending basically on the scales of the process to be studied, no DIAL system for general use has been developed. The different instruments are more or less specialized for a specific range of applications. Although this range of applications has been broadened considerably during the developments within TESLAS, the necessary specialization is probably one of the reasons, why the development of a commercially available instrument has not been fully successful, although complete systems can now be purchased from the industry.

This final report of the TESLAS subproject is organized somewhat differently from the reports of other EUROTRAC groups, in that a homogeneous description of the important aspects of the tropospheric ozone lidar technique is attempted rather than a collection of reports from the individual groups. The main reason for this is to provide a better overview over the capabilities of this method, and to avoid duplication of material which is common for all the different developments. So the report will start with a general description of the methodology, Chapter 2, followed by a description of the systems, Chapter 3. Chapter 4 is devoted to the data processing and evaluation schemes, and Chapter 5 to the assessment of accuracy for the different systems. Finally, in Chapter 6 a number of applications will be described. While Chapters two to five have been prepared by single authors, based on the relevant contributions from the other groups, Chapter six is a collection of contributions from the individual groups. Chapter 7 gives a list of publications from the subproject.

Chapter 2

Methodology

Gerard Ancellet[1] and Jens Bösenberg[2]

[1] Service d'Aeronomie du CNRS, Université Paris 6, Boite 102, 4 Place Jussieu, 75230 Paris Cedex 05, France
[2] Max-Planck-Institut für Meteorologie, Bundesstr. 55, D-20146 Hamburg, Germany

2.1 Introduction

In this section, it is intented to present the work performed by the TESLAS participants on the definition of the methodology of the lidar measurements to fulfill the general objectives outlined in the introduction. Previous studies using lidar systems to monitor ozone in the atmosphere have shown that the DIfferential Absorption Lidar (DIAL) technique is the most efficient way to derive an ozone profile with good range and sensitivity [1–3]. Fluorescence or Raman-scattering measurements of pollutants are restricted to source monitoring where range and sensitivity requirements are not critical. In this subproject, the DIAL technique is then considered as the basic method for monitoring the vertical distribution of tropospheric ozone, but two spectral domains (UV and IR) were investigated, and various schemes are proposed for atmospheric interference corrections. The selection of the appropriate spectral domain and of the optimum correction algorithms are defined from numerical simulation studies using an atmospheric model and realistic lidar parameters. Before reporting the main findings of these calculations, the spatial and temporal scales and the concentration levels are described for tropospheric ozone monitoring. Since a lidar measures a molecular density and not directly a mixing ratio (this requires the knowledge of the pressure and temperature profile which are generally given with sufficient accuracy by the nearest meteorological radiosounding station), the approximate conversion factors between the various units used in this report shall be given here for selected altitudes in the troposphere:

Table 2.1: Conversion factors between ozone number concentration, mass concentration, and mixing ratio (by volume) at selected altitudes.

Altitude	Volume mixing ratio / ppb	Mass concentration / $\mu g\ m^{-3}$	Number concentration / molecules cm^{-3}
0	100	199	2.5×10^{12}
5 km	100	119	1.5×10^{12}
15 km	100	68	8.6×10^{11}

2.2 Concentration levels, spatial and temporal scales for ozone monitoring

The scientific studies which require ozone profiling address questions as different as transport of minor constituents across atmospheric boundaries, or photochemical production during a pollution episode over Europe. As a result, almost all the spatial and temporal scales are addressed by field measurements of ozone. Three main topics have been dealing with ozone profiling:

1. climatology and trend analysis
2. photochemical production/destruction study near a source of large amount of ozone precursors (NO_x, Hydrocarbons)
3. transport studies when ozone is considered as a quasi-passive tracer (mass flux in the planetary boundary layer, stratosphere/troposphere mass exchanges).

2.2.1 Climatology and trend analysis

Climatology of ozone aims at the assessment of the seasonal variations at a given site and will provide the ozone horizontal distribution on a regional or global scale when using a network of measuring stations. Typical seasonal variation at some European sites is presented in Fig. 2.1 for a mid tropospheric level showing that the ozone concentrations change from 40 to 65 ppb (corresponding to an ozone density of 48–77 mg m^{-3} [4].

For studies of this kind the lidar instrument should have an accuracy at least 5 times less than this ozone change, corresponding to 5 ppb. The temporal and spatial scales needed are the same as the ones required for a balloon-borne ozone

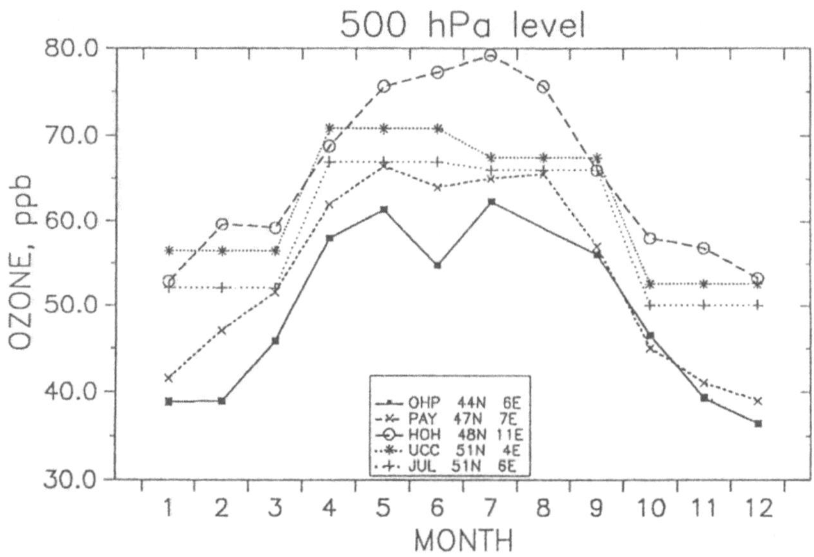

Fig. 2.1: Ozone seasonal variations deduced from balloon-borne ozone soundings performed in Europe: OHP = Observatoire de Haute Provence, PAY = Payerne Observatory, HOH = Hohenpeissenberg Observatory, UCC = Uccle Observatory, JUL = Julich Observatory [4].

sounding program: 3 profiles per week with an averaging over more than one hour if needed, vertical resolution better than 500 m. It should be stressed that a meteorological bias (+40 % in the lower troposphere in summer, –40 % near the tropopause in spring) may exist for a lidar instrument running during clear days only. This bias is only removed when data acquisition is performed also during partially cloudy days (up to $N = 6/8$ on the octas scale) using only short periods of clear sky (of the order of 15 min.) to derive the ozone profile [5]. Trend analysis requires the assessment of a concentration change of 0.5 to 1 pbb per year implying a data set of at least 500 independent profiles (equivalent to a minimum record of 5 years) and a precision less than the average ozone residual variability (*i.e.*, corrected for the seasonal variation) which is of the order of 25 % at 500 hPa, again a 5 ppb precision is good enough for the ozone measurement. The main limitation for trend estimates is the possibility of a trend in the systematic errors associated to the lidar data (instrument improvement/aging leading to smaller/larger electronic noise or alignment sensitivity, aerosol influences).

2.2.2 Polluted-area studies

In order to study the photochemical production/destruction of ozone, one must generally perform budget studies during 2 to 4 days episodes in continental areas near the pollution sources. This can be done during campaigns involving intensive field measurements and modeling activities (*e.g.*, the TRACT experiment). The concentration variations to be resolved are of the order of 1–3 ppb/h as illustrated by a diurnal cycle observed at European remote stations (Fig. 2.2).

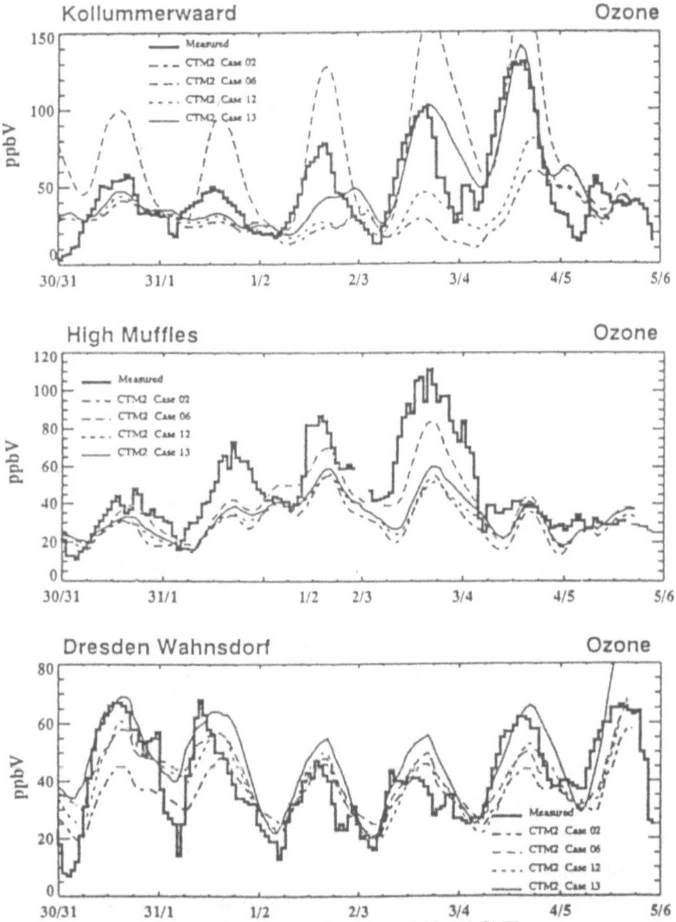

Fig. 2.2: Ozone diurnal cycles measured at several ground based stations in NL, UK, and FRG (thick lines) compared to model calculations (EUROTRAC/TOR report, 1993).

This rate of increase is mainly driven by fast photochemistry involving the NO oxydation by RO_2 radicals, with a production rate as high as 10–20 ppb/h, balanced by an approximate 5 ppb/h destruction rate due to chemical destruction of O_3 and dry deposition (of the order of 2 ppb/h) [6]. To obtain the production rate to an accuracy on the order of 1–3 ppb/h, one has to account for transport processes removing part of the locally produced ozone by advection or downward mixing of air with smaller ozone concentrations. Since a large horizontal gradient may exist in the emission rates, a horizontal ozone gradient larger than 10 ppb/ 20 km associated to advection of the order of 5 m/s can balance a 10 ppb/h photochemical production. In a similar way the entrainment of air from above at a speed of 14 cm/s has the same effect for a vertical ozone gradient of –10 ppb/500 m. The ozone measurement alone is generally insufficient for estimating the various processes independently, but increasing the spatial and temporal coverage of the measurements is a step forward towards the understanding of the magnitude of the process. Lidar instruments are well suited for this task provided that they can have an accuracy better than 2 ppb, a vertical resolution of the order of 100 m, and an integration time less than 10 min.

Since the ozone profiles are also used either for model initialization or model evaluation, the vertical resolution and the integration time of the ozone measurement should be comparable to the vertical resolution of regional models (*e.g.* the EURAD model), which is of the order of 40 hPa for a 40 × 40 km horizontal grid [7]. This will correspond to a vertical resolution of 400 m in the lower layer and 600 m at 500 hPa, while the integration should be less than 30 min. assuming an average velocity of 20 m/s for the advection of a meteorological system. Although the range should ideally span the whole troposphere, episode studies generally concentrate separately on the lower (0–6 km) and upper (5–15 km) troposphere. These resolution values should however be divided by at least a factor of 5 when the ozone values are used by non hydrostatic models (*e.g.* EUMAC Zooming Model).

2.2.3 Transport studies

Tranport studies across the top of the Planetary Boundary Layer (PBL) near 1000 m or across the tropopause can be conducted using ozone as a quasi-passive tracer. Ozone profiles can then be used for process studies like ozone flux profile in the PBL or tropopause fold evolution between 4 and 12 km. This implies that the transport time constant is smaller than the chemical time constant. This corresponds to less than 3 days at the tropopause level, but to a much shorter time scale in the range 5–50 min for boundary layer studies depending on the ozone precursor concentrations and the solar radiation. The measurements of ozone flux profiles in the PBL require very high spatial and temporal resolution to resolve the turbulent eddies relevant for the ozone flux. The vertical resolution should be less than 100 m, and the temporal resolution should be < 60 s for moderate horizontal wind speeds of 5–10 m/s [8, 9]. For application under a broader range of meteorological conditions even higher resolution is probably required. The ozone fluctuations typically range from 2–5 ppb in the middle of the PBL and can reach values of 10–15 ppb near the top of the PBL [10]. Therefore, the ozone measurements need to have an accuracy of 1 ppb or better.

For tropopause-fold studies, one should be able to resolve an ozone-rich layer which is approximately 1 km deep and 200 km wide (Fig. 2.3) [11–13].

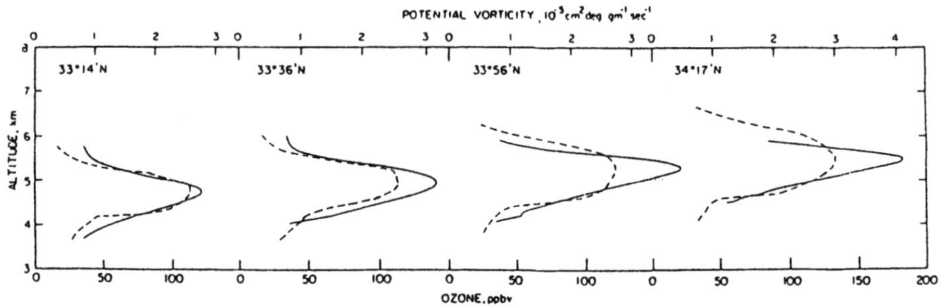

Fig. 2.3: Tropopause fold derived from an airborne lidar measurement (thick line) and potential vorticity calculations (dotted line) [12].

This implies to perform an ozone profile with an integration time less than 30 min and a vertical resolution better than 500 m. The accuracy of the ozone measurement is not a strong issue and should only be better than 10 ppb.

2.3 The DIAL technique

During the course of the TESLAS project, range-resolved ozone measurements have been performed using the DIAL technique where signals at two wavelengths λ_1 and λ_2 are simultaneously recorded. Light at λ_1 is more strongly absorbed than at λ_2, the differential ozone absorption cross section $\Delta\sigma(O_3) = \sigma_1 - \sigma_2$ should be larger than 10^{-18} cm^2 in order to achieve sufficient sensitivity at tropospheric ozone concentration levels of the order of 50 ppb. The simplified equation for the single wavelength lidar return from an atmospheric layer of thickness ΔR at range R is given by:

$$P(R,\lambda) = C\frac{\Delta R}{R^2}E\beta(R,\lambda)\exp\left\{-2\int_0^R\left[n_{O3}(r)\sigma_{O3}(\lambda)+\alpha_{mol}(r,\lambda)+\alpha_{aer}(r,\lambda)\right]dr\right\} \quad (2.1)$$

where $P(R,\lambda)$ is the received power from range R and wavelength λ, C is the system efficiency independent of range, E the transmitted power, $\beta(R,\lambda)$ is the total volume backscatter coefficient (molecular plus aerosol) in the layer ΔR, $n_{O3}(r)$ is the ozone molecular density at range r, α_{mol} and α_{aer} are the molecular and aerosol extinctions, respectively. In equation (2.1) one has considered only the elastic scattering process with no wavelength changes, Raman scattering also occurs and the signal received at the Raman shifted wavelength follows the same equation. Of course, $\beta(R,\lambda)$ then is the Raman scattering coefficient corresponding to a particular gas and the argument of the exponential term should be split in two since the signal wavelength changes along a round-trip absorption path. For the elastic scattering case, the concentration can be derived from:

$$\begin{aligned} n_{O3}(R) = & -\frac{1}{2\Delta\sigma_{O3}}\frac{\mathrm{d}}{\mathrm{d}R}\ln\frac{P_1(R)}{P_2(R)} && \{\mathrm{i}\} \\ & +\frac{1}{2\Delta\sigma_{O3}}\frac{\mathrm{d}}{\mathrm{d}R}\ln\frac{\beta_1(R)}{\beta_2(R)} && \{\mathrm{ii}\} \\ & -\frac{1}{\Delta\sigma_{O3}}\Delta\alpha_{\mathrm{aer}}(R) && \{\mathrm{iii}\} \\ & -\frac{1}{\Delta\sigma_{O3}}\Delta\alpha_{\mathrm{mol}}(R) && \{\mathrm{iv}\} \end{aligned} \tag{2.2}$$

Term {i} on the right-hand side of equation (2.2) is determined directly from the measurements, terms {ii}-{iv} are interference terms due to differential backscatter, differential aerosol extinction, and differential molecular extinction, respectively. The best sensitivity of the technique is obtained when the optical depth (*i.e.*, the integral term in equation (2.1)) is close to unity over the measurement range. For tropospheric ozone this requirement is fulfilled only for wavelengths in the range 260–300 nm. This wavelength region is also convenient as it permits solar blind detection and hence daytime measurements. The disadvantage of this spectral domain is the lack of narrow structure in the ozone absorption spectrum (Fig. 2.4) which requires a minimum wavelength separation of 5 nm to fulfill the initial condition of $\Delta\sigma_{O3} > 10^{-18}$ cm^2 [14].

Alternatively the IR domain near 9.6 μm may be used for ozone measurements, although the largest $\Delta\sigma_{O3}$ is of the order of 5×10^{-19} cm^2.

In the following sections, an error analysis is presented which on one hand is based on simulation studies to estimate the statistical error on $n_{O3}(r)$ due to the random nature of the measured signal, and on the other hand on an estimate of the amplitude of the neglected terms in equation (2.2).

2.4 Simulation studies

Simulation studies have been conducted to calculate the expected random error on a concentration measurement as a function of the range [15]. The signal-to-noise ratio (SNR) can be approximated by assuming that the detected signal follows a Poisson distribution when the received number of photons is small enough. The SNR can then be written as:

$$SNR(R) = \frac{P(R)}{\sqrt{p(R) + P_b}} \tag{2.3}$$

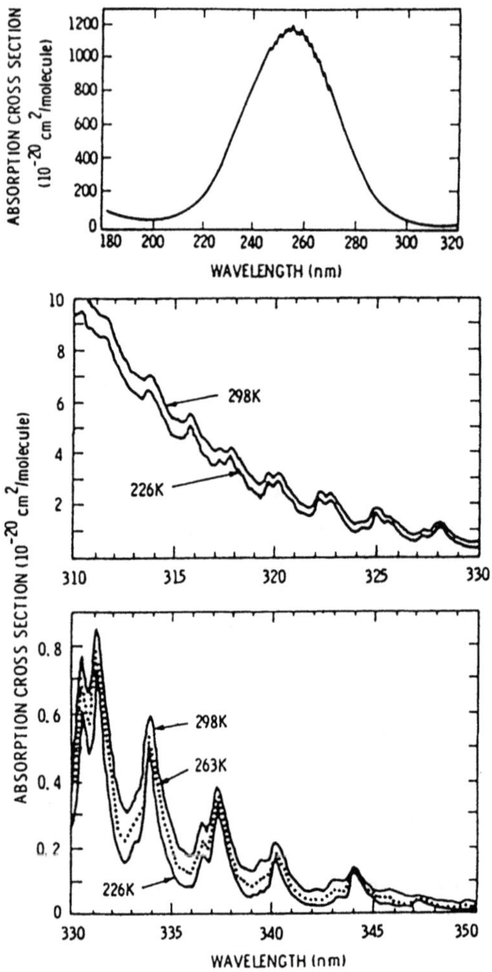

Fig. 2.4: Ozone absorption coss sections in the region 180-350 nm [14].

where P_b is the measured signal due to the background skylight. Assuming that the signals are not correlated over the range ΔR, the corresponding statistical error on the ozone concentration is given by:

$$\varepsilon(R) = \frac{1}{2n_{O3}\,\Delta\sigma_{O3}\,\Delta R}\sqrt{\frac{1}{N}\left(\frac{2}{SNR_1^{\,2}}+\frac{2}{SNR_2^{\,2}}\right)} \tag{2.4}$$

where N is the number of laser shots used to provide an average lidar signal $P(R)$. Although the signal statistics may not follow exactly a Poisson distribution this formulation allows the error ε to be calculated with sufficient accuracy using typical system parameters and an atmospheric model.

The error ε has been calculated for a lidar system operated in the UV using either a Nd-YAG laser or an KrF exiplex laser as a light source (see chapter 3). Both a ground-based and an airborne lidar were considered, and the system parameters necessary to compute the different terms of equation (2.1) are indicated in Table 2.2.

Table 2.2: Ozone lidar receiver parameters.

Telescope Diameter	$D = 80$ cm[(a)] or 30 cm[(b)]
Receiver spectral bandwith	$BW = 2$ nm
Telescope field of view	$FOV = 1.5$ mrad
PMT quantum efficiency	$q = 0.20$
Optical efficiency	$k = 0.30$

(a) Ground-based lidar
(b) Airborne lidar

The laser wavelengths and the associated energies E are indicated in Table 2.3, assuming that the laser wavelength can be upshifted by stimulated Raman scattering in hydrogen or deuterium (see Chapter 3).

Table 2.3: Laser wavelengths and output energies

		Wavelength	Conversion efficiency	Output energy
KrF laser (200 mJ)		248.0 nm	25 %	30 mJ
Hydrogen	S1	277.1 nm	30 %	24 mJ
	S2	313.2 nm	20 %	16 mJ
Deuterium	S1	268.4 nm	30 %	36 mJ
	S2	291.8 nm	15 %	18 mJ
Nd-YAG laser (100 mJ)		266.0 nm	25 %	15 mJ
Hydrogen	S1	299.1 nm	30 %	12 mJ
	S2	341.5 nm	20 %	8 mJ
Deuterium	S1	289.0 nm	30 %	18 mJ
	S2	316.2 nm	15 %	9 mJ

The wavelength of 248 nm will not be considered in the following study as the strong extinction at this wavelength reduces the range to much less than 1 km for ozone measurements.

The atmospheric model used for the backscatter and extinction terms includes:

- the U.S. Standard Atmosphere density model [16],
- the mean ozone profile of Krueger and Minzer [17],
- the LOWTRAN 6 code assuming a uniform PBL particle concentration up to 1.5 km, which corresponds to a 10 km visibility range [18].

An example of the calculated lidar received signal given by equation (2.1) is plotted in Fig. 2.5 for the 3 wavelengths of the Nd-YAG laser ground-based system, showing that the 266 nm signal decreases faster than the signals at the longer wavelength.

The error ε is plotted as a function of height in Fig. 2.6 for daytime conditions, using wavelengths lower than 300 nm to minimize the background sky light.

The number N of laser shots averaged together is of the order of 1000, corresponding to an integration time of 1.66 min. for a 10 Hz laser repetition frequency. The spatial resolution ΔR is kept constant and equal to 1 km at each altitude. Errors corresponding to other system parameter values could be calculated from the results of Fig. 2.6 since ε is proportional to $\Delta R^{-3/2} \cdot E^{-1/2} \cdot D^{-1} \cdot N^{-1/2}$. For the wavelength pair 289/299 nm and 277/292 nm, ε remains lower than 5 % for altitudes lower than 10 km while for the wavelengths 266/289 nm the 5 % value is already exceeded at an altitude of 7 km. For night-time measurements the wavelength pair 292/313 nm was also considered and

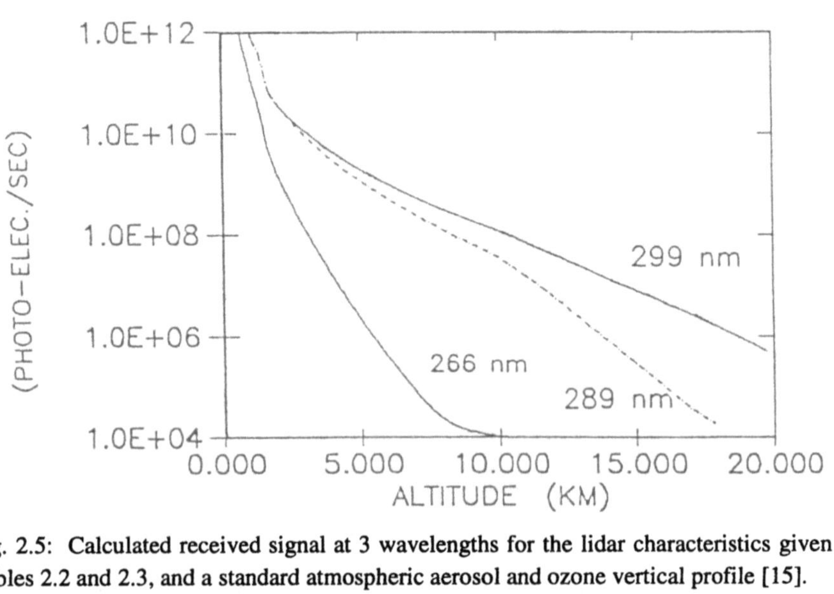

Fig. 2.5: Calculated received signal at 3 wavelengths for the lidar characteristics given in Tables 2.2 and 2.3, and a standard atmospheric aerosol and ozone vertical profile [15].

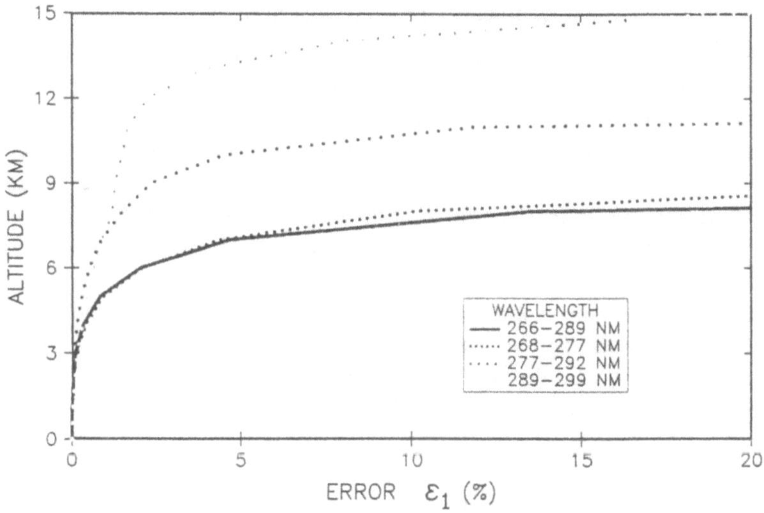

Fig. 2.6: Daytime statistical error ε *vs* altitude for various wavelength pairs used by a ground based lidar, and a vertical resolution of 1 km [15].

ozone measurements could be performed up to 15 km using both wavelength pair 289/299 nm and 292/313 nm. So both laser sources are equivalent for ozone measurements in the free troposphere. The vertical resolution can be increased to 300 m provided that the altitude range does not exceed 6 km.

For an airborne nadir-looking lidar flying at a 5 km altitude, the error ε is calculated using a 250 m vertical resolution (Fig. 2.7).

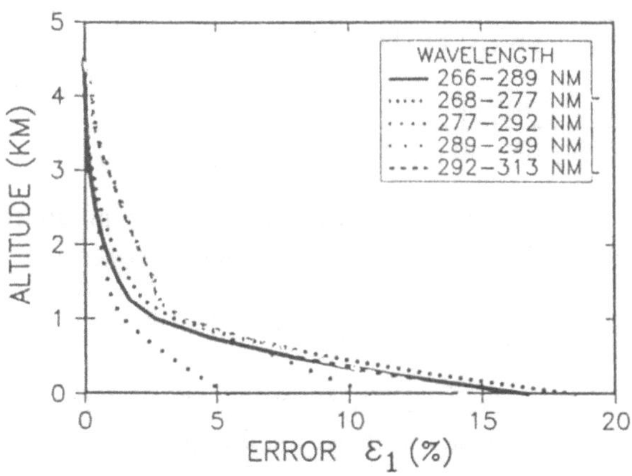

Fig. 2.7: Daytime statistical error ε *vs* altitude for various wavelength pairs used by an airborne lidar flying at 5 km, and a vertical resolution of 250 m [15].

It shows that the error is as large as 15 % for the Nd-YAG laser system while the wavelength pair 292 / 313 nm gives the best performances. The KrF laser is then the best candidate for an airborne system, although it is easier to operate a Nd-YAG laser onboard an aircraft. The previous results indicate the optimum achievable performance only, because additional instrumental errors (laser beam alignment with the telescope axis, time coherent electronic noise in the receiver components, memory effects in the light detector) will further reduce the accuracy of the signal acquisition. As it will be shown in chapter 3, the maximum achievable dynamic range is of the order of 1000. From the curves shown in Fig. 2.5, the real maximum range will be 5 km for the wavelength 289 / 299 nm and 2.5 km for 266 / 289 nm, assuming the starting altitude for signal processing is 0.5 km. Shifting the starting altitude to 3.5 km allows a maximum range of 12 km for the wavelength 289 / 299 nm and 6 km for the wavelength 266 / 289 nm.

2.5 Aerosol interferences

The influence of aerosol scattering properties on DIAL measurements of tropospheric ozone has been described previously [3, 15, 19]. Following mainly the derivation of Browell *et al.* [19], the terms {ii}, {iii} and {iv} in equation (2.2) are corrections due to the wavelength dependence of the total backscatter, the aerosol extinction, and the molecular extinction, respectively.

2.5.1 Differential aerosol backscatter

Most important is the differential aerosol backscatter correction, term {ii} in eq. (2.2). Introducing the inverse S_K of the widely used scattering ratio K_S, $S_K = \beta_{mol}/(\beta_{mol} + \beta_{aer})$ term {ii} can be written as

$$\{ii\} = \frac{1}{2\Delta\sigma}\frac{d}{dR}\ln\frac{\beta_1}{\beta_2} = \frac{1}{2\Delta\sigma}\frac{d}{dR}\ln\left(\frac{\beta_{mol,1}\cdot S_{K,2}}{\beta_{mol,2}\cdot S_{K,1}}\right) \qquad (2.5)$$

For abbreviation set $\Lambda = \lambda_1/\lambda_2$. If we assume an Ångström law for the particle backscatter coefficient, $\beta_{aer,1} = \beta_{aer,2}\,\Lambda^{k_\beta}$ then term {ii} reads

$$\{ii\} = \frac{1}{2\Delta\sigma}\frac{d}{dR}\ln\left[S_K + (1-S_K)\cdot\Lambda^{4+k_\beta}\right] \qquad (2.6)$$

To get a better overview over the factors which determine the magnitude of the differential backscatter correction term, we use the approximation

$$\Lambda^{4+k_\beta} \approx 1 + (4+k_\beta)\cdot\frac{\Delta\lambda}{\lambda_2}$$

which holds for typical wavelengths separations $\Delta\lambda = \lambda_1 - \lambda_2 << \lambda_2$. After some transformations one arrives at

$$\{ii\} \approx \frac{\Delta\lambda}{2\,\Delta\sigma \cdot \lambda_2} \cdot (4 + k_\beta) \cdot \frac{dS_K}{dR} \qquad (2.7)$$

Hence the correction term is determined by three factors:

1. the relative wavelength difference divided by the differential absorption cross section, which is fixed by the selection of wavelengths for a particular system,
2. the factor $(4 + k_\beta)$, which is determined by the composition and size distribution of the aerosol,
3. the vertical gradient of the inverse scattering ratio dS_K/dR, which is determined by the vertical distribution of the aerosol.

The term $\Delta\lambda / 2\lambda\ \Delta\sigma_{O3}$ is a somewhat complex function of the two wavelengths used, it is plotted in Fig. 2.8 as a function of the online wavelength and the wavelength difference.

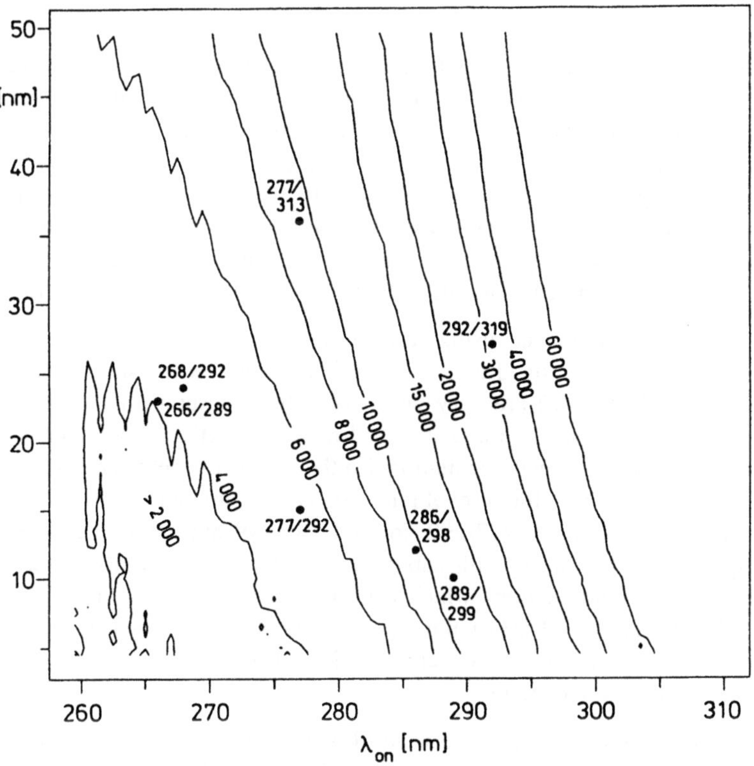

Fig. 2.8: Differential aerosol backscatter sensitivity.

The coordinates for some widely used tropospheric ozone lidars are marked, Table 2.4 gives the numerical values of the sensitivity to differential backscatter gradients for selected wavelength pairs.

Table 2.4: Sensitivity to gradients in differential aerosol backscatter.

λ_{on} / λ_{off} nm	$\Delta\lambda$ / 2 $\Delta\sigma$ λ_{off} µg m^{-2}
266 / 289	4038
268 / 292	4485
277 / 292	5118
286 / 298.3	9452
277 / 313	9531
289 / 299	11965
292 / 319	31640

It is obvious, that the minimum is found for $\lambda_{on} \approx 270$ nm, the region of the largest wavelength dependence of the ozone absorption cross section. Overall, the sensitivity to aerosol interference is primarily influenced by the choice of the on-line wavelength, which also determines the maximum achievable range, and to a lesser extent only by the choice of the off-line wavelength, as long as the wavelength separation is kept reasonably small. Obviously, not much can be gained by making $\Delta\lambda$ very small, in particular when considering that other errors in the measurement increase rapidly with decreasing differential absorption.

Apparently, measurements using the wavelength pairs of 268/292 nm or 266/289 nm (KrF or quadrupled Nd:YAG lasers with subsequent Raman conversion in deuterium) are very well suited for measurements in regions with high aerosol load. However, these short online wavelengths cannot always be used, because of too strong signal attenuation. So the tradeoff is between extended range and minimum sensitivity for aerosol interference. For measurements with extended range the combinations of 277/292 nm and 289/299 nm are well suited (KrF or quadrupled Nd:YAG lasers with subsequent Raman conversion in hydrogen and deuterium). The combinations which can be achieved using a single Raman cell only, 277/313 nm or 292/319 nm, are considerably more sensitive to aerosol interference, about a factor of two compared to the optimum which can be reached at the same online wavelength.

The proportionality of the correction to $(4 + k_\beta)$ clearly shows that it is important to know rather exactly the exponent k_β of the wavelength dependence of the aerosol backscatter coefficient.

The last factor, dS_K/dR makes this correction important only in regions of spatially inhomogeneous aerosol. It also shows clearly, that the correction depends largely on the resolution of the DIAL measurement, both in space and time, since gradients are substantially decreased by averaging. This is important, since wavelengths with rather high sensitivity to aerosol interference have to be used for upper tropospheric measurements from ground based systems. Since the requirements regarding vertical resolution are much reduced in the upper troposphere, sufficiently accurate measurements are still possible.

2.5.2 Wavelength dependence of aerosol backscatter

This has been the subject of a model study [20], where a few typical aerosol size distributions and compositions were used for calculations of the backscatter and extinction coefficients as a function of wavelength, using Mie theory and literature values for the complex refractive index. The main result was, that for the backscatter coefficient the commonly assumed λ^{-1} dependence is a suitable approximation for wavelengths longer than ≈ 350 nm only. Due to the onset of absorption at shorter wavelengths the backscatter is considerably reduced, in most cases resulting in a decrease of the backscatter coefficient with decreasing wavelength rather than the expected increase. Since the relative humidity modifies the refractive index of the particles, it plays an important role in the determination of the scattering properties. As a result, in the UV the wavelength dependence of the aerosol backscatter coefficient is better characterized by λ_0 in the moist case (90 % humidity), and λ^{+1} or $(\lambda - \lambda_0)^{+2}$ in the standard case (70 % humidity), where λ_0 is the wavelength of a possible maximum in the backscatter coefficient. For the wavelength dependence of the aerosol extinction, it is still safe to assume λ^{-1}, although deviations may occur.

2.5.3 Differential aerosol and molecular extinction

Considering the differential aerosol extinction correction, term {iii} in equation (2.2), the wavelength dependence of the aerosol extinction coefficient is usually also approximated by a power law, with the Ångström coefficient k_α

$$\alpha_1 = \alpha_2 \left(\frac{\lambda_1}{\lambda_2}\right)^{-k_\alpha} \tag{2.8}$$

For small wavelength differences $\Delta\lambda$ this can be approximated by

$$\Delta\alpha \approx -\alpha_2 \cdot k_\alpha \frac{\Delta\lambda}{\lambda_2} \tag{2.9}$$

Hence the accuracy of the determination of $\Delta\alpha_M$ is dependent on the accuracy of the determination of the aerosol extinction coefficient itself at the reference wavelength, and on the knowledge of k_α. The model calculations for several aerosol types showed that $-1.0 < k_{\alpha} < -0.5$ is generally observed, so about ±25 % is about the error margin resulting from this factor. The accuracy of the determination of the aerosol extinction coefficient using some variant of the Klett or Fernald method has been discussed extensively in the literature. It depends strongly on the particular aerosol distribution and a priori knowledge about the aerosol. This shall not be discussed here in more detail.

The differential molecular extinction due to Rayleigh scattering can be calculated with high accuracy from known air density and the λ^{-4} dependence. Its magnitude for a wavelength pair with rather large separation, 277–313 nm, is 7.5 ppb or about 13 $\mu g/m^3$ (at ground level). Since the air density can generally be estimated to better than 3 % [21], the accuracy of this term can easily be kept to better than 1 $\mu g/m^3$.

A detailed analysis of many measurements shows, that good accuracy can generally be achieved over major portions of the troposphere. However, a careful analysis of the aerosol load is mandatory, and small regions at the boundaries of aerosol layers remain where the uncertainty in the retrieved ozone density can still be large. The main source of uncertainty is the incomplete knowledge of the aerosol backscatter wavelength dependence.

2.6 Interference with other gases

If another gas with strong absorption bands in the near UV (250 - 320 nm) is present, its absorption should be considered in the ozone calculation from the DIAL equation (2.2). At least four molecules must be considered: SO_2, NO_2, CH_2O and O_2. Other gases such as many hydrocarbons absorb in this spectral region, too, but have to be considered only in highly polluted areas. In Table 2.5, the absorption coefficients at 298 K are presented for the various UV wavelengths obtained with a Nd-YAG or KrF laser source. We have assumed the same concentration of 38 ppb (10^{12} molecules cm^{-3}) at ground level for O_3, NO_2 and SO_2, while the formaldehyde concentration is set to a value of 3.8 ppb (10^{11} molecules cm^{-3}) corresponding to an average urban level for this species.

Oxygen absorption should be corrected for wavelengths shorter than 270 nm, since the interference for an ozone concentration of 10^{12} cm^{-3} is of the order of 7 % using a wavelength pair 266/289 nm. This effect is in principle easy to account for as the oxygen mixing ratio is altitude independent, and the density profile used for the Rayleigh scattering interferences can be used to correct for this error. The main limitation is probably the uncertainty about the oxygen absorption in this spectral domain, where the values of Cann *et al.* [26] are deduced from theoretical calculations. The SO_2, NO_2, and CH_2O interferences are only relevant for measurements in the polluted boundary layer, considering the concentrations which have been used to produce the figures in Table 2.5. Furthermore, for this part of the atmosphere, typical wavelength pairs used for ozone measurements are the ones with large O_3 differential absorption cross sections. The CH_2O interferences could be as large as the SO_2 ones, especially when using the wavelength pair 266/289 nm or 266/299 nm (error of 5 % on the ozone estimate in the polluted boundary layer). The SO_2 interference can always be reduced to a value less than 5 % for example using the 266/299 nm or 289/316 nm wavelength pair for a Nd-YAG system and using the 268/292 nm pair for the KrF system. In most of the cases the NO_2 interferences can be neglected.

Table 2.5: Absorption coefficient in 10^{-6} m^{-1} of chemical species versus wavelength.

Species	Wavelength in nm									References
	248	266	268.4	277.1	289	291.8	299	313.2	316.2	
O_3 (38 ppb)	1071	947	838	502	159	113	45.4	6.8	4.6	[14]
SO_2 (38 ppb)	8	48	58	77	9	80	41	20	7	[22, 23]
NO_2 (38 ppb)	1	2	2.1	4.6	8.6	11	14	22	24	[24]
CH_2O (3.8 ppb)	3	3	11	12	35	6	38	10	25	[25]
O_2	1000	60	60	7	3	3	1	–	–	[26]

2.7 IR measurements

DIAL measurements can be performed in the ozone IR absorption band at 9.6 μm using a TEA CO_2 laser as a laser source and heterodyne detection to achieve a 3 km range. In the IR the backscatter coefficient in eq. (2.1) is only related to the scattering by aerosol particles, while the optical depth is only a function of the molecular absorption due to O_3, CO_2, and H_2O (see Table 2.4). This is an important advantage of using this spectral domain when aerosol interference is a major problem.

For ozone measurements the selected CO_2 laser lines are the P_{14} (1052.195 cm^{-1}) and P_{24} (1043.163 cm^{-1}) in the 9 μm CO_2 absorption band. This choice corresponds to a 11.7 cm^{-1} atm^{-1} (4.6×10^{-19} cm^{-2}) differential absorption cross section coefficient at a pressure of 1000 hPa and a temperature of 290 K . The temperature T and pressure P dependence of the absorption cross section a can be calculated using the following equation:

$$\alpha(T,p)=\alpha_0(T_0,p_0)\frac{\gamma(T,p)\left[\pi\Delta\nu^2+\gamma(T_0,p_0)^2\right]}{\gamma(T_0,p_0)^2\left[\pi\Delta\nu^2+\gamma(T,p)^2\right]}\cdot\exp\left[cE\left(\frac{1}{T_0}-\frac{1}{T}\right)\right] \quad (2.10)$$

where $\Delta\nu$ is the laser linewidth, E the ground state energy, c a constant value, and γ is the collisional halfwidth of the O_3 absorption. This halfwidth is generally proportional to P/P_0 and to $(T/T_0)^{-1/2}$.

Therefore the main drawback of this wavelength domain for ozone measurements is the low sensitivity as the O_3 differential absorption coefficient for an ozone concentration of 10^{12} cm^{-3} is less than 50×10^{-3} km^{-1}. This value is 2 to 15 times less than the ozone differential absorption values in the UV which are indicated in Table 2.3. Although the IR absorption could increase by a factor of 2 for measurements in the upper troposphere ($p < 500$ hPa), the sharp aerosol concentration decrease above the first 3 km makes an IR measurement unlikely in this part of the atmosphere. The interferences of the H_2O and CO_2 on the ozone measurements are 0.175 ppb/torr(H_2O) and 0.035 ppb/ppm(CO_2). Only the CO_2 interference is large enough to cause problems (of the order of 10 ppb of ozone), but can be corrected quite easily considering the low variability of the CO_2 mixing ratio. For a typical TEA CO_2 coherent lidar system the expected sensitivity in the urban planetary boundary layer is of the order of 50 ppb with a 200 m range resolution. When using a quasi-horizontal path within the boundary layer the maximum range will be of the order of 4 km.

Table 2.6: Absorption and scattering coefficients for the R_{18} and R_{20} lines of the CO_2 10 μm band for a moderate aerosol load (visibility of 10 km).

CO_2 line	Mol. absorpt. km^{-1}	Aer. absorpt. km^{-1}	Aer. scatt. km^{-1}	Mol. backsc. km^{-1} sr^{-1}	Aer. backsc. km^{-1} sr^{-1}
10R18	0.23	0.01	0.02	10^{-8}	5×10^{-4}
10R20	1.46	0.01	0.02	10^{-8}	5×10^{-4}

2.8 Multiwavelength techniques

A major problem of the DIAL technique as described above is that there are more unknowns than measured quantities. To be specific, the aerosol backscatter and extinction at both wavelengths, and the ozone density are 5 independent sets of unknowns, which of course cannot be retrieved rigorously from just two sets of measurements. The term „set“ indicates, that all quantities are height dependent.

The problem is circumvented in the DIAL technique with aerosol correction by prescribing the wavelength dependence of both aerosol extinction and backscatter (generally with the same Ångström law for both parameters), and the lidar ratio

α_{aer}/β_{aer} as a function of height (generally taken as constant, due to lack of better information).

In their description of the aerosol correction method, Browell *et al.* [19] have shown, that the correction may be rather sensitive to the assumed lidar ratio for the aerosol involved. Therefore, an independent measurement of the aerosol extinction is most welcome. This has first been discussed by Papayannis *et al.* [15], suggesting the additional measurement of the Raman-scattered signal of molecular nitrogen. Since the Raman backscatter coefficient is determined by the density of the scattering molecules, and since this is well known, the measurement of the Raman signal directly yields the total extinction coefficient, including molecular and particle scattering and absorption in gases [27]. By combination with the DIAL technique, the aerosol extinction can be separated.

Papayannis *et al.* [15] proposed an iterative solution to deduce the ozone concentration and aerosol backscatter and extinction. Matthias *et al.* [28] introduced an analytic solution for the same case, solving the three lidar equations (two for elastic scattering, one for Raman scattering) simultaneously. It should be noted here, that the wavelength dependence of the aerosol backscatter and extinction still have to be prescribed when using this technique, hence three equations are sufficient for an analytic solution. The analysis of the system of equations shows, that generally there can be more than one solution, but, depending on the conditions imposed by the measured quantities and the actual wavelengths chosen for the measurement, this can reduce to one or even no solution. In the case of two solutions, which appears to be quite common, it is generally possible to exclude one of them because of unrealistic results. A clear advantage of the analytical solution is, that the most probable solution can be selected, in contrast to the iteration scheme, where there is little posibility to influence the actual convergence to either of the solutions. Else, of course, both approaches are equivalent.

It should be noted here that there are other variants of multi-wavelength techniques. First, it is possible to determine the ozone distribution from Raman measurements at two different wavelengths. This has the clear advantage, that only molecular backscattering properties are used, which are very well known. Differential aerosol extinction still has to be corrected for in that case, which is unlikely to be a major problem. The main disadvantage of course is the small Raman-scattering cross section, which leads to rather long integration times, which are generally not acceptable for tropospheric studies. For stratospheric measurements, this technique has already been applied successfully [29].

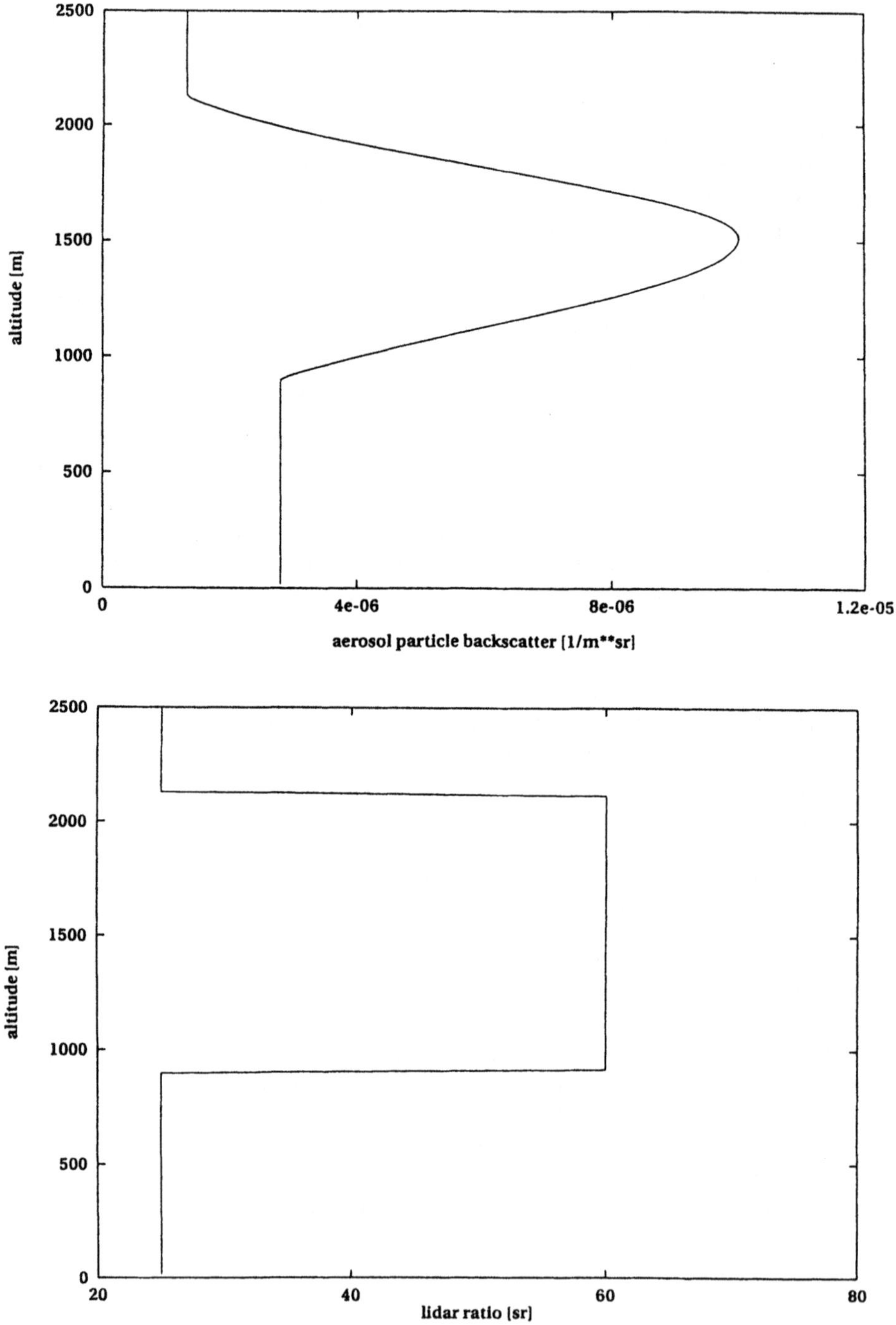

Fig. 2.9. a) Prescribed aerosol backscatter profile. b) Prescribed aerosol lidar ratio profile.

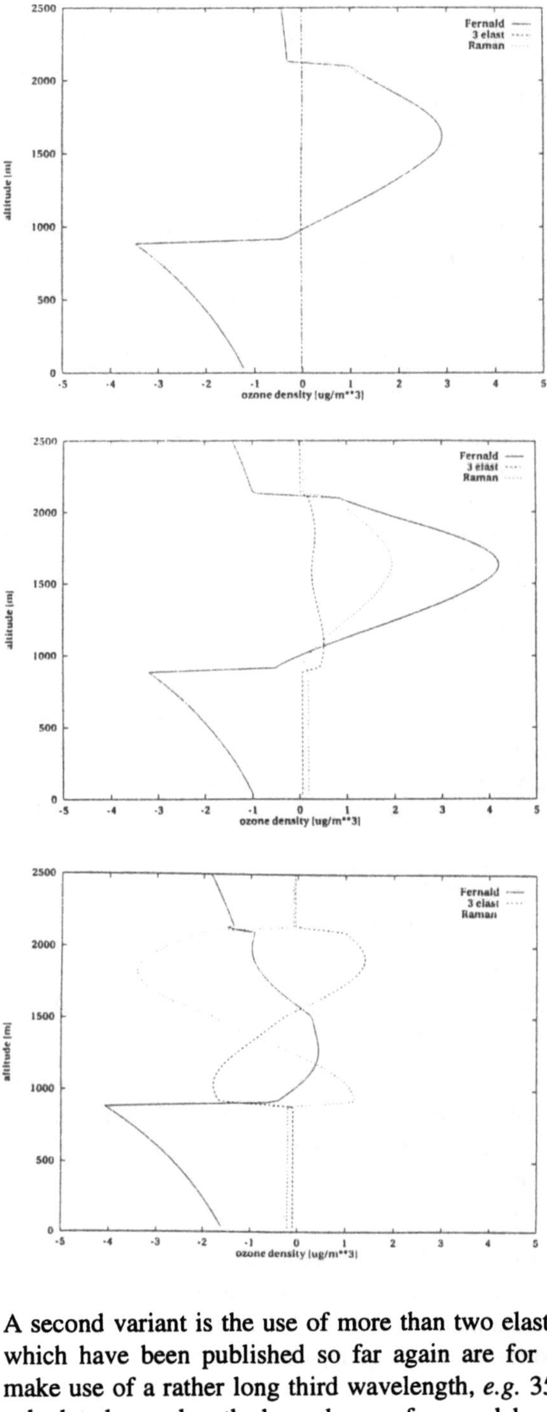

Fig. 2.10. Intercomparison of DIAL with aerosol correction according to Fernald and three-wavelength-technique with or without Raman channel for deuterium wavelengths. Ångström coefficients and calibration have been chosen differently.

a) $k_\alpha = -0.6(-0.6)$, $k_\beta = -0.2(-0.2)$, $S = 40$ sr, correct calibration;

b) $k_\alpha = -0.3(-0.6)$, $k_\beta = +0.3(-0.2)$, $S = 40$ sr, $\beta(R_c = 2700\text{ m}) = 2 \cdot \beta_{\text{correct}}$;

c) $k_\alpha = -0.9(-0.6)$, $k_\beta = -0.7(-0.2)$, $S = 40$ sr, $\beta(R_c = 2700\text{ m}) = 2 \cdot \beta_{\text{correct}}$.

A second variant is the use of more than two elastically scattered signals. Results which have been published so far again are for stratospheric measurements, or make use of a rather long third wavelength, *e.g.* 355 nm or longer. In view of the calculated wavelength dependence of aerosol backscatter in the UV [20], these long wavelengths probably are not very useful. An analytical method similar to the decribed Raman-assisted method has been developed [28]. The detailed properties of the solutions obtained with this technique are still under investigation.

Figs. 2.9 and 2.10 show a synthetic example to demonstrate the main features of the different approaches.

A backscatter profile has been prescribed at 313 nm, which exhibits typical levels and gradients of aerosol backscatter. The lidar ratio was assumed to be height dependent, with a value of S = 60 sr betweeen 900 m and 2200 m, and S=25 sr elsewhere. The Ångström coefficients were selected as $k_\alpha = -0.6$ and $k_\beta = -0.2$, according to typical values as calculated by [20]. Fig. 2.10 shows the error in the retrieved ozone density, using the standard correction technique.

In our case this is based on the algorithm as described by Fernald *et al.* [30], considering both aerosol and molecular scattering. For the retrieval in Fig. 2.10a, the lidar ratio has been assumed constant with S = 40 sr. With either of the three-wavelength-techniques, the lidar ratio is retrieved rather than prescribed, so the ozone density is calculated correctly in this case. In Fig. 2.10b,c the wavelength dependencies and the calibration factor are varied in a reasonable range, $k_\alpha = \pm 0.3$, $k_\beta = \pm 0.5$, and $\beta(R_c = 2700\text{ m}) = 2 \cdot \beta_{correct}$.

Generally, the errors of the three-wavelength-techniques are lower than for the conventional correction method, mostly due to the retrieval of the correct lidar ratio. In particular, the errors are restricted to regions of inhomogeneous aerosol distribution, while for the standard correction rather large errors can also occur in homogeneous regions, due to potential errors in the calculated backscatter profile.

Apparently, the different correction techniques individually have their strong and weak points. Further analysis is needed to find the solution which is least sensitive to insufficient information about the aerosol optical properties.

Experimental studies using the Raman-assisted technique have been performed with MPI system, where a Raman channel at 296 nm has been installed. As an example, Fig. 2.11 shows an actual measurement.

In this particular case the aerosol backscatter distribution as determined with the conventional Fernald algorithm differs substantially from that retrieved with the Raman method, no matter which values are chosen for the calibration and the lidar ratio, see Fig. 2.12. The reason for this is the strong dependence of the lidar ratio on height (Fig. 2.13), which certainly could not be guessed *a priori*.

As a result, the ozone concentration calculated with the conventional DIAL technique including aerosol correction, and with the Raman-assisted method, show considerable differences, with a maximum of 30 $\mu g\ m^{-3}$ at 2300 m height. But also in the region 1500 - 2000 m the differences are quite substantial.

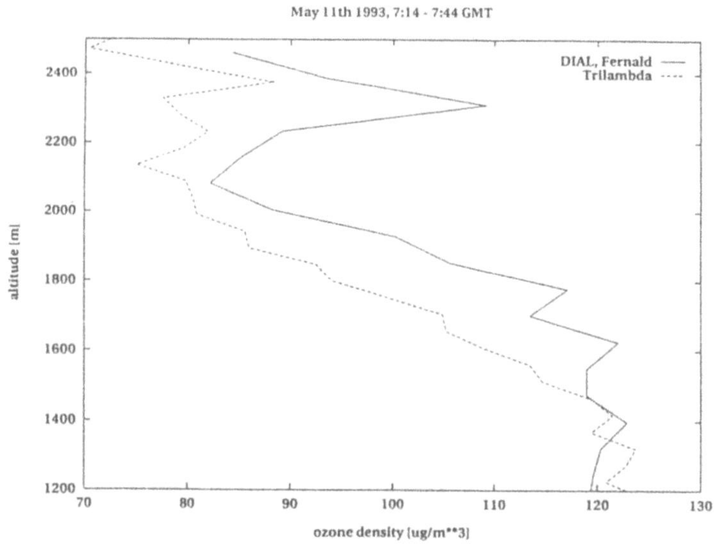

Fig. 2.11: Retrieved ozone profiles, standard and Raman assisted corrections. May 11, 1993, Itzehoe.

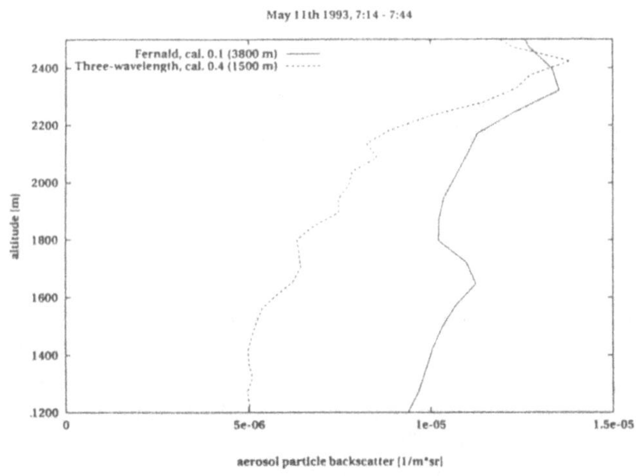

Fig. 2.12: Retrieved backscatter profiles, using the three-wavelength-method and the Fernald algorithm, respectively. Data are from May 11, 1993, Itzehoe.

In this example it cannot be decided which of the values are more correct, since independent information on the actual ozone concentration is lacking. However, the differences are clearly demonstrated , and the results of the Raman assisted retrieval appear more reasonable (*e.g.* the peak at 2300 m, which is rather unlikely to be real, is much reduced).

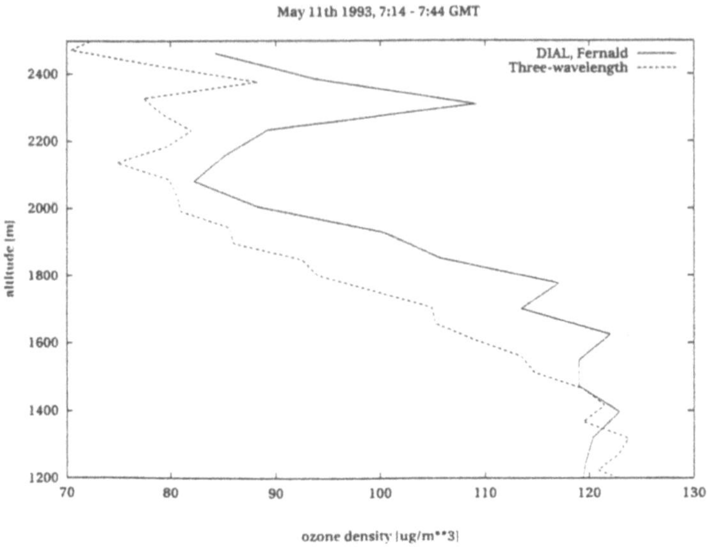

Fig. 2.13: Retrieved lidar ratio profile, May 11, 1993, Itzehoe.

2.9 References

1. O. Uchino, M. Tokunaga, M. Maeda, Y. Miyazoe; Differential Absorption Lidar measurement of tropospheric ozone with Excimer-Raman hybrid laser, *Opt. Lett.* **8** (1983) 347-349.
2. E. Browell; Lidar measurements of tropospheric gases, *Opt. Eng.* **21** (1982) 128.
3. J. Pelon, G. Mégie, Ozone monitoring in the troposphere and the lower stratosphere: evaluation and operation of a ground based lidar station, *J. Geophys. Res.* **87C** (1982) 4947-4955.
4. M. Beekmann, G. Ancellet, G. Mégie; Climatology of Ozone in Southern Europe and its relation to potential vorticity, *J. Geophys. Res.* **99D** (1994) 12841-12853.
5. H. DeBacker, E. Visser, D. DeMuer, D. Swart; Potential for meteorological bias in lidar ozone data sets resulting from the restricted frequency of measurement due to cloud cover, *J. Geophys. Res.* **99D** (1994) 1395-1401.
6. A. Volz, D. Mihelcic, P. Müsgen, H. Pätz, G. Pilwat, H. Geiss, D. Kley; Ozone production in the Black Forest: Direct measurements of RO_2, NO_x, and other relevant parameters, in: I. Isaksen (ed.), *Tropospheric Ozone*, Reidel Publ.Company, Dordrecht 1988.
7. *EUROTRAC Annual Report '93*, Part 5, EUMAC, EUROTRAC ISS, Garmisch-Partenkirchen 1994.
8. D.H. Lenschow; Micrometeorological techniques for measuring biosphere-atmosphere trace gas exchange, in: *Biogenic Trace Gas Emissions from Soil and Labor*, Blackwell Scientific Publications, Oxford 1994.
9. C. Senff, J. Bösenberg, G. Peters; Measurement of water vapor flux profiles in the convective boundary layer with lidar and Radar-RASS, *J. Atmos. Oceanic Technol.* **11** (1994) 85-93.

10. T. Schaberl; Messung des Ozonflusses in der unteren Troposphäre mit einem neuen Ozon-DIAL-System und einem Radar-Rass, *Ph.D. Thesis*, Universität Hamburg, 1995.

11. E. Danielsen; Stratospheric-tropospheric exchange based on radioactivity, ozone and potential vorticity, *J. Atmos. Sci.* **25** (1968) 502-518.

12. E. Browell, E. Danielsen, S. Ismail, G.L. Gregory, S.M. Beck; Tropopause fold structure determined from airborne lidar and in situ measurements, *J. Geophys. Res.* **92D** (1987) 2112-2120.

13. G. Ancellet, J. Pelon, M. Beekmann, A. Papayannis, G. Mégie; Ground based lidar studies of ozone exchanges between the stratosphere and the troposphere, *J. Geophys. Res.* **96** (1991) 22401-22421.

14. L. Molina, M. Molina, Absolute absorption cross sections of ozone in the 185 to 350 nm wavelength range, *J.Geophys.Res.* **D91** (1986) 14501-14508.

15. A. Papayannis, G. Ancellet, J. Pelon, G. Mégie; Multiwavelength Lidar for ozone measurements in the troposphere and the lower stratosphere, *Appl. Opt.* **29** (1990) 467-476.

16. *U.S. Standard Atmosphere*, NOAA, NASA, USAF, US Government printing Office, Washington, D.C. 1976, p. 227.

17. A. Krueger, R. Minzer; A mid latitude ozone model for the US Standard Atmosphere, *J. Geophys. Res.* **81** (1976) 4477-4481.

18. F.X. Kneizys, *et al.*; Atmospheric transmittance/radiance: Computer code Lowtran 6, AFGL-TR-80-0067, Air Force Geophysics Laboratory, Bedford, Ma. 1980.

19. E. Browell, S. Ismail, S. Shipley; Ultraviolet DIAL measurements of ozone profiles in regions of spatially inhomogeneous aerosols, *Appl. Opt.* **24** (1985) 2827-2836.

20. P. Völger; Optische Eigenschaften von Aerosolpartikeln und deren Auswirkungen auf Ozonmessungen mit einem Lidar, *Diplomarbeit*, Universität Hamburg 1993.

21. P.B. Russell, B.M. Morley, J.M. Livingston, G.W. Grams, E.M. Patterson; Orbiting lidar simulations. 1: Aerosol and cloud measurements by an independent-wavelength technique, *Appl. Opt.* **21** (1982) 1541-1553.

22. O. Thomsen; Messung des Absorptionswirkungsquerschnitts von Schwefeldioxid im Wellenlängenbereich von 265 bis 298 nm, *M.Sc. Thesis*, University of Hamburg, 1990, GKSS-Forschungszentrum Geesthacht GmbH, GKSS-90/E/36.

23. D.J. Brassington; Sulfur dioxide absorption cross-section measurements from 290 nm to 317 nm, *Appl. Opt.* **20** (1981) 3774-3779.

24. A.M. Bass, A.E. Ledford, A.H. Laufer; Extinction coefficients of NO_2 and N_2O_4, *J. Res. Nat. Bur. Stand.* **80A** (1976) 143.

25. A. Bass, L. Glasgow, C. Miller, J. Jesson, D. Filkin; Temperature dependent absorption cross sections for formaldehyde (CH_2O): the effect of formaldehyde on stratospheric chemistry, *Planet. Space. Sci.* **28** (1980) 675-679.

26. M.W.P. Cann, *et al.*; Oxygen absorption in the spectral range 180 - 300 nm for temperatures to 3000 K and pressures to 50 atm, *Can. J. Phys.* **62** (1984) 1738.

27. A. Ansmann, U. Wandinger, M. Riebesell, C. Weitkamp, W. Michaelis; Independent measurement of extinction and backscatter profiles in cirrus clouds by using, *Appl. Opt.* **31** (1992) 7113-7131.

28. V. Matthias, J. Bösenberg, V. Wulfmeyer; Improvement of ozone measurements with DIAL by using an additional Raman channel, in: P.M. Borrell, P. Borrell, T. Cvitas, W. Seiler (eds.), *Proc. EUROTRAC Symp. '94*, SPB Academic Publishing, The Hague 1994, pp. 326-329.

29. T.J. McGee, M.R. Gross, R.A. Ferrare, W.S. Heaps, U.N. Singh; Raman lidar measurements of stratospheric ozone in the presence of volcanic aerosols, *Geophys.Res.Lett.* **20** (1993) 955-958.

30. F.G. Fernald; Analysis of atmospheric lidar observations: some comments, *Appl. Opt.* **23** (1984) 652-653.

Chapter 3

Instruments

Alexandros Papayannis

Department of Physics, National Technical University of Athens,
Zografou Campus, 157 73 Athens, Greece

3.1 Laser sources

Ozone DIAL measurements in the PBL and the free troposphere can be performed in both the UV and the IR spectral regions, as already discussed in Chapter 2. For the UV region, it has also been shown that a fairly broad range of wavelengths can be used, and that rather large spacing between the on- and off-resonance wavelengths can be admitted without considerably increasing the sensitivity to aerosol interference. Since all performance characteristics of an UV DIAL system depend on the choice of the wavelengths, with partially contradicting requirements, any choice of transmitter wavelengths must be a compromise.

When selecting a laser source for ozone DIAL measurements, the most important parameters to be considered are:

- the altitude range and concentration of ozone to be detected, as well as the spatial and temporal resolution required,
- the concentration and spectral characteristics of other gases that may interfere with the ozone measurements,
- the type of aerosols present in the atmosphere and its scattering characteristics,
- the ease of operation in monitoring applications.

The choice of a laser source for use in a DIAL system is crucial. It depends not only upon the wavelength, energy and repetition rate that are required to reduce the systematic and statistical errors, but also on the reliability, ease of use and operating costs, which are very important characteristics mainly for ozone monitoring purposes. In the past, the most widespread choice has been that of pulsed tunable dye lasers pumped by Nd:YAG or excimer lasers [1-7]. The main disadvantage of these dye-laser-based systems has been the large effort required for maintenance and operation. Therefore, these systems have been used for campaign studies rather than for routine operation or monitoring. Mainly this drawback led to the consideration of other laser sources for routine ozone monitoring stations.

In addition to the pulsed tunable dye lasers, three other types of pulsed laser sources have demonstrated their ability of performing accurate ozone DIAL measurements in the troposphere: the Raman-shifted excimer laser [8-11], the Nd:YAG-laser with frequency quadrupling and subsequent Raman-shifting [12-17], both of these for measurements up to the lower stratosphere, and the CO_2 lasers, with demonstrated performance up to 2500 m altitude range [18-22]. A fourth type of lasers are the metal-vapour lasers, which are able to produce high laser output in the UV [23, 24]. However, their ability to perform accurate ozone measurements in the troposphere needs to be verified.

Of these lasers, only the KrF-excimer laser and the frequency-quadrupled Nd:YG laser, both with subsequent Raman-shifting, offer the advantage that only a single laser with no external wavelength control has to be operated. The simple operation of these lasers with fixed output wavelengths makes them particularly suitable for routine operation.

In the next paragraphs we will briefly describe the main characteristics of the four types of pulsed laser sources used for ozone DIAL measurements.

3.1.1 Tunable dye lasers

The main part of the DIAL transmitter comprises two independent pulsed lasers that optically pump two dye lasers, which produce the two widely tunable DIAL wavelengths separately. This allows to fire one laser source after the other. The time delay between two successive on- and off-resonance DIAL wavelengths is set as short as possible to probe a "frozen" atmosphere, but long enough to avoid any interference between two successively received lidar signals (typically > 100 μs). The laser pumping sources are based either on a Nd:YAG or on an excimer (XeCl) laser. The output of the dye lasers is then frequency doubled using potassium-dihydrogen phosphate (KD*P) crystals to produce the desired wavelengths in the UV spectral region.

Typical pulse energies of the emitted DIAL wavelengths range from a few mJ up to 20-25 mJ and the repetition rates range from 10 up to 100 Hz. The divergence (full angle) of the emitted DIAL wavelengths is of the order of 0.1 mrad.

The main advantage of these lasers are full tunability for both on- and offline wavelengths. The price which has to be paid for this is a quite complex system, which has to be adjusted by an expert, performance degradation due to aging of dyes, and need for precise wavelength control. This makes these lasers less desirable for operation in routine monitoring systems.

3.1.2 KrF lasers

Rare-gas halide lasers yield high power emission directly in the ultraviolet spectral region. Pulse energies of several Joules and repetition rates of more than 500 Hz have been demonstrated. This makes these lasers ideal candidates for sensitive ozone measurements with high temporal resolution. Two different types of excimer lasers have been used within TESLAS. Mostly, a master oscillator-power amplifier (MOPA) design is used, consisting of a narrow-band oscillator (0.001 nm) with a tuning range of about 0.8 nm injection seeding a power amplifier with unstable resonator optics. Typical values of the output energy are 250 mJ per pulse at repetition rates up to 80-100 Hz. The unstable-resonator design of the KrF laser amplifier yields a beam divergence (full angle) lower than 0.2 mrad. Typical pulse widths (FWHM) are 17-22 ns and the beam shape ($v \times h$) is 21 × 5 mm. The laser line is horizontally polarised (90:10 or 70:30). The standard design, which is operated in the broadband mode, unpolarized, has also successfully been used for tropospheric ozone measurements, further reducing the effort needed for operation.

The KrF laser wavelength itself (248.5 nm) is not used for sensitive ozone measurements in the free troposphere due to the high absorption of the laser radiation by ozone. Instead, the use of the stimulated-Raman-scattering technique (section 3.2) permits the simultaneous generation of laser wavelengths in the 270-320 nm spectral region, which are very well suited for ozone measurements from ground up to the lower stratosphere. The conversion efficiencies for the different Raman orders are different for the broadband and the narrowband versions. This will be further discussed in the corresponding section.

An important problem arising from the use of KrF lasers is the deposition of impurities on the laser cavity mirrors. This problem can be reduced if two (due to the different chamber pressure in the oscillator and amplifier chambers) cryogenic gas purifiers are used in the oscillator-amplifier laser cavities. Typical values of the windows lifetimes are 10^6 laser shots without gas purifiers and 10^7 laser shots with gas purifiers.

3.1.3 Nd:YAG lasers

The main part of the DIAL transmitter as used within TESLAS comprises two pulsed Nd:YAG lasers emitting typical energies of 600-1000 mJ/pulse at 1064.52 nm with a repetition frequency of 10-20 Hz. Two KD*P (Type II) crystals are mounted in a serial arrangement to generate the second and the fourth harmonic of the Nd:YAG laser at 532 nm and 266 nm, respectively. The average total conversion efficiency is 10 %, which corresponds to a 70-110 mJ available energy per pulse at each laser at 266 nm. To prevent drift of the harmonic output energy, each harmonic generation (HG) crystal is supplied with an electronically controlled temperature stabilizer to maintain the crystal at a constant temperature (± 0.10 K). Generally, the temperature stabilisation is sufficient if one needs to operate the Nd:YAG laser at 266 nm, for small time scales (around 10-20 minutes). If larger time scales are needed (*i.e.*, several hours), as in the case of routine ozone DIAL monitoring, then the temperature stabilisation of the crystal is not sufficient (especially when important temperature gradients are present outside the crystal housing). In this case, an active electronic servo-loop, constantly keeping the KD*P crystal in a phase-matched condition, is needed [25]. To overcome the temperature-related problems of the KD*P crystal, one is offered the possibility of using the beta-barium borate (BBO) crystal to frequency double from 532 to 266 nm. The BBO crystals, although less sensitive to temperature fluctuations, offer only slight advantages versus the KD*P crystals.

The divergence of the laser radiation emitted at 266 nm is generaly lower than 0.5 mrad and the pulse width (FWHM) has a duration of 3-5 ns. The laser beam spatial distribution has a Gaussian profile in the far field while the beam diameter is 8-10 mm. High energy dichroic beamsplitters separate the 266 nm wavelength from the simultaneously emitted wavelengths at 532 nm and at 1064.52 nm. Using the 266 nm laser radiation as the pump, additional wavelengths in the ultraviolet spectral region are generated using the stimulated-Raman-scattering (SRS) effect (Section 3.2 and Table 2.3). The unused energy at 1064.52 nm and 532 nm is disposed of by using beam dumps.

3.2 The stimulated Raman-scattering technique

The stimulated Raman-scattering (SRS) technique has extensively been used for "shifting" the wavelength of pulsed laser light, mainly into longer wavelengths [14, 26-30]. During the TESLAS project two different gases were used as Raman media to generate the first (S1) and second Stokes (S2) orders from both KrF and frequency-quadrupled Nd-YAG lasers: hydrogen and deuterium. The frequency shifts available from the vibrational transitions Q(1) in hydrogen and Q(2) in deuterium are 4155.25 cm^{-1} and 2987.23 cm^{-1} respectively [31]. The wavelengths available from these combinations have been presented in Table 2.3. Additional wavelengths can be generated by the use of sequential shifts (third or higher Stokes orders) in deuterium and hydrogen. The use of hydrogen deuteride (HD), although it offers an interesting Raman shift frequency (3606 cm^{-1}), has not been investigated because of its low gain at room temperature [29].

3.2.1 Basic principles

When an intense laser beam is focused into a cell containing a compressed molecular gas, a Stokes beam may be initiated from spontaneous Raman scattering and subsequently amplified by SRS, at the expense of the pump beam. If the intensity of the first-Stokes beam (S1) becomes sufficiently high, it stimulates a second-Stokes (S2) beam and so on, to produce higher-order Stokes (Sn) frequencies. However, the production of second- and higher-order Stokes beams in a gaseous medium is often dominated by the competitive parametric four-wave mixing (4WM) processes [32].

For monochromatic pump and Stokes plane waves, both propagating along the z axis and in a low output Stokes signal regime, the intensity $I_s(x,y,L)$ of the Stokes beam in the small signal regime evolves with the interaction length (L) following the exponential relationship [33]:

$$I_s(x,y,L) = I_s(x,y,0)\exp(RG) = I_s(\mathrm{x,y,0})\exp(RGC \cdot I_p(x,y) \cdot L) \qquad (3.1)$$

where $I_p(x,y)$ is the intensity of the pump beam and *RGC* is the Raman Gain Coefficient.

In the following paragraph we will describe very briefly the most important parameters and processes which govern the efficient generation of the Stokes beams: the Raman gain (*RG*) and the four-wave mixing processes.

3.2.2 Raman gain

The Raman gain ($RG = RGC \cdot I_p(x,y) \cdot L$) depends on several important parameters [33, 34]. First, and of main importance, it is proportional to the frequency of the Stokes frequency, hence it decreases for successively higher orders. In the pressure broadened regime (pressure $\geq$ 15 bar for H_2) the *RGC* is independent of pressure, while in the Doppler broadened limit at low pressure, the *RGC* is proportional to the pressure. The *RG* is also proportional to $I_p(x,y) \cdot L$, which is directly connected with the divergence of the pump beam and the spatial pump beam quality. Finally, the *RG* is directly connected to the spontaneous Raman-scattering cross-section. The addition of a foreign gas to a Raman-active medium increases the linewidth without increasing the number density of the active gas, therefore the gain decreases as the foreign gas pressure increases. This phenomenon provides a useful method for varying the gain to the value required for a particular application.

3.2.3 Four-wave mixing

The efficient use of SRS to generate new wavelengths is conciderably complicated by the existence of third-order interactions (*i.e.* four-wave mixing processes). The four-wave mixing processes can take place involving a Stokes wave, a pump wave and an anti-Stokes wave. This interaction, unlike the pure Raman cascade, is sensitive to the phase mismatch between the waves. Four-wave mixing processes also contribute to the generation of higher-order Stokes radiation, often to a greater extent than the cascade processes [35]. In the presence of four-wave mixing, the divergence of each order will be different because of the different wavelengths involved. This effect, which has been experimentally confirmed within TESLAS and studied in some detail by [30], is undesirable in laser remote sensing applications because it gives rise to a number of overlapping beams with different divergences. Increasing the focal length of the lens used for focussing decreases the angle of the pump beam. This is one way to inhibit the build-up of four-wave mixing processes, thus minimising the divergençe of the generated Raman radiation. Higher SRS conversion efficiencies for the lowest Stokes orders may be achieved.

The other factor that must be considered when varying the total pressure inside the Raman cell, is the coherence length of the interaction [36] which decreases as the total pressure increases. Consequently, as the pressure is increased the extent of four-wave mixing processes decreases because of the dependence of the phase mismatch on the coherence length. A reduction in the extent of four-wave mixing processes reduces the amount of energy transfered to higher Stokes orders.

3.2.4 Experimental results

Stimulated Raman scattering in hydrogen and deuterium

The Q(1) transition of hydrogen has been used extensively for Raman shifting because of its high gain and large frequency shift. The Raman gain coefficient of deuterium is lower by nearly a factor of 6 as compared to hydrogen [34]. Consequently, we expect to need a greater partial pressure to produce results comparable to hydrogen.

Results with Nd:YAG lasers

Extensive investigations have been performed into the Raman shifting of radiation from a quadrupled Nd:YAG laser in hydrogen, deuterium and hydrogen deuteride [14]. Their results apply to a system used as part of a large fixed station for monitoring ozone in the troposphere and stratosphere. The objective of the work described here has been to achieve comparable results using smaller lasers and gas cells suitable for mobile or transportable DIAL systems.

Experiments performed at CNRS, France (*G. Ancellet*). The laser radiation at 266 nm (40 mJ) was focussed into a stainless-steel Raman cell (length 1.5 m) using Suprasil lenses (of 1.5 m focal length). The entrance and exit windows are made of quartz (Suprasil) and are tilted by 1 degree with respect to each other in order to avoid back-reflections into the laser. The output laser beams were then recollimated using the same type of lenses. The SRS efficiency was measured using different mixtures of H_2 or D_2 with He as buffer gas. The gas purity was 99.7 % for deuterium, 99.999 % for hydrogen and 99.999 % for helium.

Results are presented in Fig. 3.1 where the output energies are corrected for the absorption of the prism used for separating the various output wavelengths.

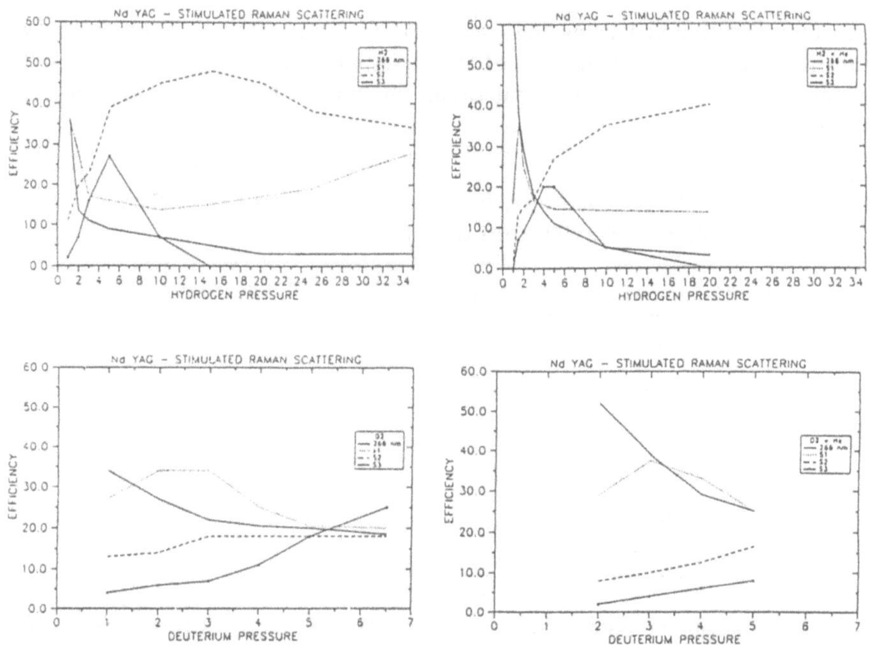

Fig. 3.1: Efficiency of the stimulated Raman scattering in H_2 and D_2 using pure gas and a gas mixture with He (H_2 + 10 bar of He, D_2 + 5 bar of He).

Efficiencies of 35 % are possible in hydrogen and deuterium at low pressures (1-3 bar), addition of He only slightly improves these values given the above pump energy at 266 nm. The main advantage of using a buffer gas is an increase of the transmitted residual energy at 266 nm up to 10 mJ in 1 bar of hydrogen while the 299 nm output energy remains at 7 mJ. Table 3.1 presents the gas mixtures used and pulse energies achieved with the DIAL system of the Service d'Aeronomie du CNRS.

Table 3.1: Gas mixtures and pulse energies for the DIAL system of the Service d'Aeronomie du CNRS.

Cell	Raman gas	Buffer gas (He)	E_{266}/pulse	E_{S1}/pulse
D_2	3 bar	5 bar	15 mJ	10 mJ
H_2	1 bar	10 bar	10 mJ	7 mJ

Experiments performed at NPL, UK. (*M. Milton, B. Robinson*). The laser radiation at 266 nm (60-70 mJ) was focussed into a stainless-steel Raman cell (length 1.5 m) using Suprasil lenses (of 1.1 m focal length). The entrance and exit windows are made of quartz (Suprasil) and are tilted by 1 degree with respect to each other in order to avoid back-reflections into the laser. The output laser beams were then recollimated using the same type of lenses. The SRS efficiency was measured using different mixtures of H_2 or D_2 with Ar and N_2 as buffer gas. The gas purity was 99.7 % for deuterium, 99.999 % for hydrogen, argon and 99.9999 % for nitrogen.

Table 3.2: Gas mixtures, focusing lens and energy conversion efficiencies for the NPL DIAL system.

Conditions	Focusing lens	Efficiency (%)
H_2, 35 bar	1 m	18
H_2, 35 bar	1.1 m	25
50 % H_2, 50 % N_2 40 bar total	1.1 m	65
50 % H_2, 50 % Ar 40 bar total	1.1 m	61

Table 3.2 presents the output efficiency at the 1st Stokes, using two different focal lengths and varying the total pressure inside the Raman cell (Raman active gas: hydrogen). This Table shows the effect of using equal partial pressures of hydrogen and nitrogen at a total pressure of 40 bar. The efficiency is much higher (60-65 %) and the output beam has a better profile.

Experiments performed at RIVM, Netherlands (*A. Apituley, A. Sunesson*). The laser beam at 266 nm (80-90 mJ) was focussed into a stainless-steel Raman cell (length 2 m) using high-energy anti-reflection coated quartz lenses (of 1.1 m focal length). The entrance and exit windows are made of quartz (Suprasil) and are tilted by 1 degree with respect to each other in order to avoid backreflections into the laser. The output laser beams were then recollimated using the same type of lenses.

The SRS efficiency was measured using different mixtures of H_2 or D_2 with Ar as buffer gas. The buffer gas was used to enhance energy transfer to the 1st Stokes line through lowering of the Raman gain, as discussed in section 3.2.2. The gas purity was 99.7 % for deuterium, 99.999 % for hydrogen and 99.999 % for argon. In this experimental set-up, quantum efficiencies as high as 70 % for H_2 and 40 % for D_2 have been achieved [17]. Using N_2 as buffer gas comparable conversion efficiences were achieved. However, SRS also occurs in N_2 which yields extra output lines in addition to the desired H_2 and D_2 lines. This complication is avoided by using buffer gases that are not Raman active. Next to Ar, He was used as a buffer gas, but lower efficiences were found. The results of the experiments described above are shown in Figs. 3.2 and 3.3, for the hydrogen and deuterium, respectively.

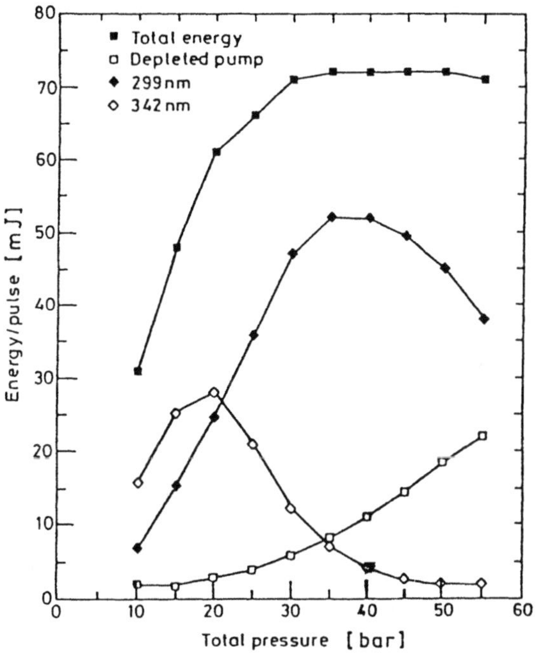

Fig. 3.2: Pulse energy generated in forward Raman orders from a quadrupled Nd:YAG laser versus pressure of H_2 using a 1.1-m focusing lens and a 2-m Raman cell. The pump energy was 85 mJ.

In Fig. 3.2 a fixed partial pressure of hydrogen was used and the partial pressure of nitrogen was varied. In an experiment performed in this way, the density of Raman-active molecules is constant, with only the linewidth and the coherence length varying as the partial pressure of the buffer gas is varied. In this case it can be seen that at a partial pressure of 17 bar of nitrogen the energies generated in the first and second Stokes are equal. As the partial pressure is increased, the ratio of first to second Stokes increases significantly. This is a clear example of the suppression of second Stokes by the introduction of a buffer gas which causes a reduction in the coherence length and a lower Raman gain.

Fig. 3.3 shows results obtained with a partial pressure of 15 bar of deuterium with nitrogen as a buffer gas. At low pressures considerably more second Stokes than first is generated. The second Stokes is effectively reduced by introducing the buffer gas. Despite the low Raman gain of deuterium, the peak conversion efficiencies obtained are not significantly lower than those for hydrogen.

The optimum gas mixtures and pulse energies used for the ozone measurements are listed in Table 3.3. Identical results could be obtained for many months with the same gas fill.

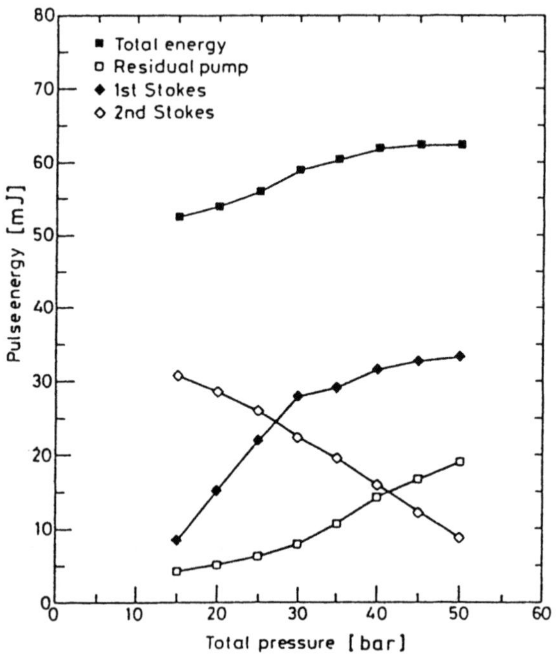

Fig. 3.3: Pulse energy generated in forward Raman orders from a quadrupled Nd:YAG laser versus pressure of D_2 using a 1.1-m focusing lens and a 2-m Raman cell. The pump energy was 85 mJ.

Table 3.3: Gas mixtures and pulse energies for the RIVM DIAL system.

Cell	Raman gas	Buffer gas (Ar)	E_{266}/pulse	E_{S1}/pulse
D_2	10 bar	25 bar	15 mJ	25 mJ
H_2	5 bar	23 bar	10 mJ	40 mJ

Results with KrF lasers

The energy available from a KrF laser is generally higher than that from a frequency-quadrupled Nd:YAG laser (by a factor of three to eight). This higher energy combined with the higher Raman gain at higher frequencies, enables the interaction to proceed by cascade to the second and higher-order Stokes wavelengths, even without the use of tight focusing. Previous workers have observed such an orderly cascade between Raman orders in a collimated pump beam [37]. The need to reduce this gain in order to generate just the first and second Stokes with high efficiency has led to the use of buffer gases to control the gain.

High conversion efficiencies have been achieved using injection-seeded KrF lasers at low gas pressures. These observations have not been reported before and represent an efficient and simple method for generating Raman-shifted radiation.

Experiments performed at GKSS, Germany *(C. Weitkamp, M. Krause, W. Lahmann, W. Michaelis).* The SRS experiments were performed with an injection-seeded KrF laser (150 mJ energy/pulse, 250 Hz max. rep. rate), a Raman cell, 920 mm long withstanding pressures up to 100 bar. The focusing lens (focal length of 1350 mm) and the cell windows are made of uncoated quartz. The output of the Raman cell is then recollimated by using a same type of lens and the wavelegths generated are dispersed with the use of a Pellin-Broca prism. The SRS efficiency was measured (for forward SRS scattering) using hydrogen and deuterium as active Raman gases. The optimum results obtained are summarised in Table 3.4.

Table 3.4: Gas mixtures, pulse energies and energy conversion efficiencies for the GKSS DIAL system.

Cell pressure / bar	Input energy / mJ	Output energy / mJ	Energy conversion efficiency
1.5, H_2	145	44.0	46.9 %
40, H_2	163	39.6	41.2 %
4, D_2	145	29.5	37.8 %
29 D_2, 10 He	140	19.2	28.1 %

Table 3.5: Divergence of the output laser beams versus the gas mixtures for the GKSS DIAL system.

Gas	Line	Raman gas (bar)	Buffer gas He (bar)	Divergence (mrad)
-	Pump	-	-	0.5
H_2	S_1	1.5	-	0.44
H_2	S_1	1.5	3	0.2
H_2	S_1	40	-	2.0
H_2	S_2	1.5	3	0.58
H_2	S_2	40	-	1.6
D_2	S_1	4	-	0.22
D_2	S_1	4	1	0.5
D_2	S_1	11.5	10	0.44
D_2	S_2	4	-	0.68
D_2	S_2	4	1	0.68
D_2	S_2	11.5	10	0.40

Additional measurements concerning the divergence of the output laser beams have also been performed and are presented in Table 3.5. The divergence of the 1st Stokes line is usually on the same order as the divergence of the pump beam, whereas the diverence of the 2nd Stokes line tends to be wider. The divergence increases when the pressure inside the Raman cell increases. The addition of buffer gas (He) tends to decrease the divergence of the output beams as seen in Table 3.5.

Experiments performed at IFU, Germany (*T. Trickl, U. Kempfer, W. Carnuth, R. Lotz*). The laser beam at 248.538 nm (250 mJ/pulse) is focused into a 1.8 m long stainless-steel Raman cell by using a focusing lens with a focal length of $f = 1.0$ m. Shorter f values are, in general, not recommended because of the onset of efficiency lowering processes (*i.e.* electric breakdown, multiphoton excitation, or intensity-induced refractive index changes). The focusing lens is made of quartz and is broadband dielectrically coated in the 248-313 spectral range. The Raman cell contains hydrogen at 1.5-2 bar pressure or deuterium at 7-10 bar. The cell windows are made of CaF_2 and are coated at the outer side. The output of the Raman cell is collimated by an $f = 5$ m concave spherical mirror with more than 99 % reflectance for the wavelengths under consideration. The beam expansion results in a total beam divergence of less than 0.2 mrad.

Fig. 3.4 shows the energy conversion efficiencies for the first Stokes orders as a function of the hydrogen pressure for average pulse energies of about 198 mJ inside the Raman cell.

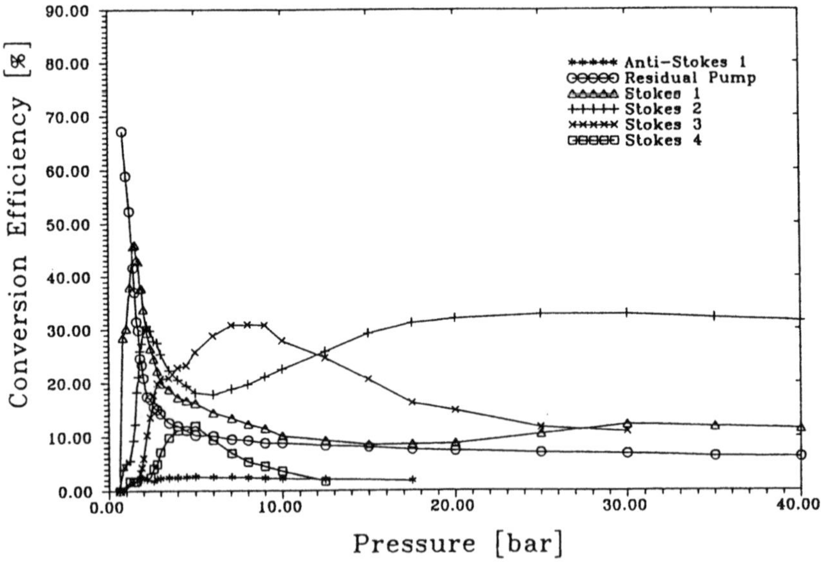

Fig. 3.4: Conversion efficiency to Raman orders in H_2 of a KrF laser as a function of the gas pressure. The pump energy of 196 mJ was focused into a 1.9 m cell with 1 m lens. Reprinted from [10] with permission of the American Insitute of Physics.

In that figure we can see that the first Stokes reaches a minimum at a pressure of approximately 12 bar and then starts to increase slowly. This minimum is caused by the transfer of energy to higher orders. At low pressures, the generation of the higher Stokes orders is dominated by four-wave mixing. At the onset of second Stokes, a 4WM cone with a maximum full angle of 14 mrad is observed. As the thresholds for the higher orders are reached, cones with greater divergence appear. The importance of four-wave mixing is also demonstrated by the rapid growth of third Stokes without a decrease in second Stokes and by the presence of some anti-Stokes radiation.

Fig. 3.5 shows the behavior of the Stokes orders at lower pressures in more detail.

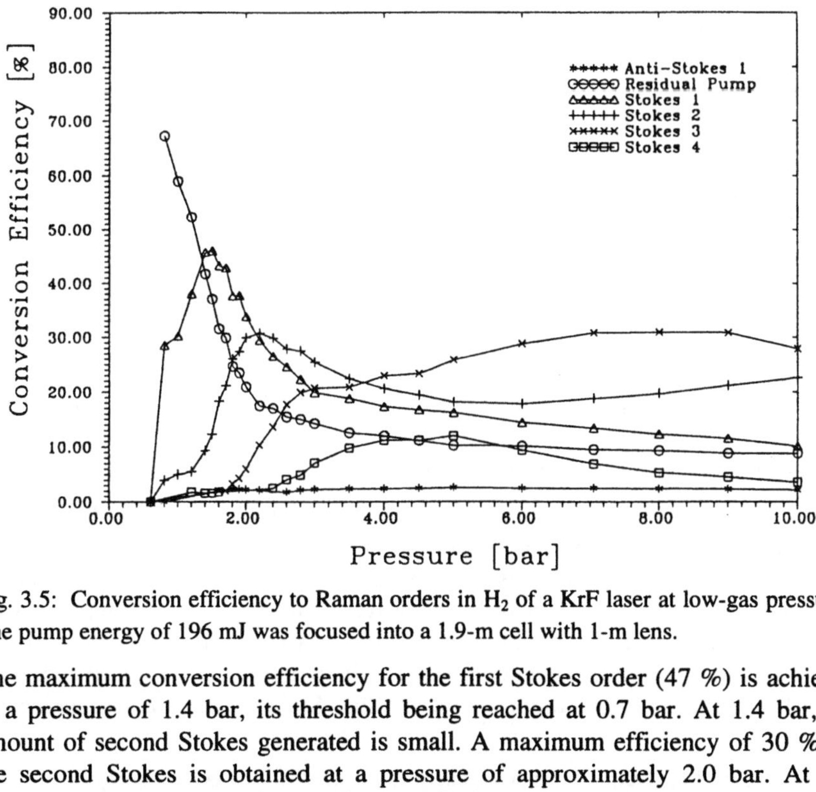

Fig. 3.5: Conversion efficiency to Raman orders in H_2 of a KrF laser at low-gas pressures. The pump energy of 196 mJ was focused into a 1.9-m cell with 1-m lens.

The maximum conversion efficiency for the first Stokes order (47 %) is achieved at a pressure of 1.4 bar, its threshold being reached at 0.7 bar. At 1.4 bar, the amount of second Stokes generated is small. A maximum efficiency of 30 % for the second Stokes is obtained at a pressure of approximately 2.0 bar. At this pressure, the efficiencies for the first two Stokes orders are the same. Therefore, this pressure is well suited to the simultaneous generation of these orders for lidar measurements.

The importance of the oscillator-amplifier configuration for producing high conversion efficiencies at pressures below 2 bar is confirmed by blocking the narrow band laser oscillator at the input to the amplifier which produces a slight reduction in the pulse energy but does not change the divergence. However, a substantial reduction of the Raman-shifted energy is observed. This observation is ascribed to a change in the polarization purity of the beam which would also affect the efficiency [10].

The results shown in Fig. 3.6 were taken under the same conditions as Fig. 3.5, but with deuterium in place of hydrogen.

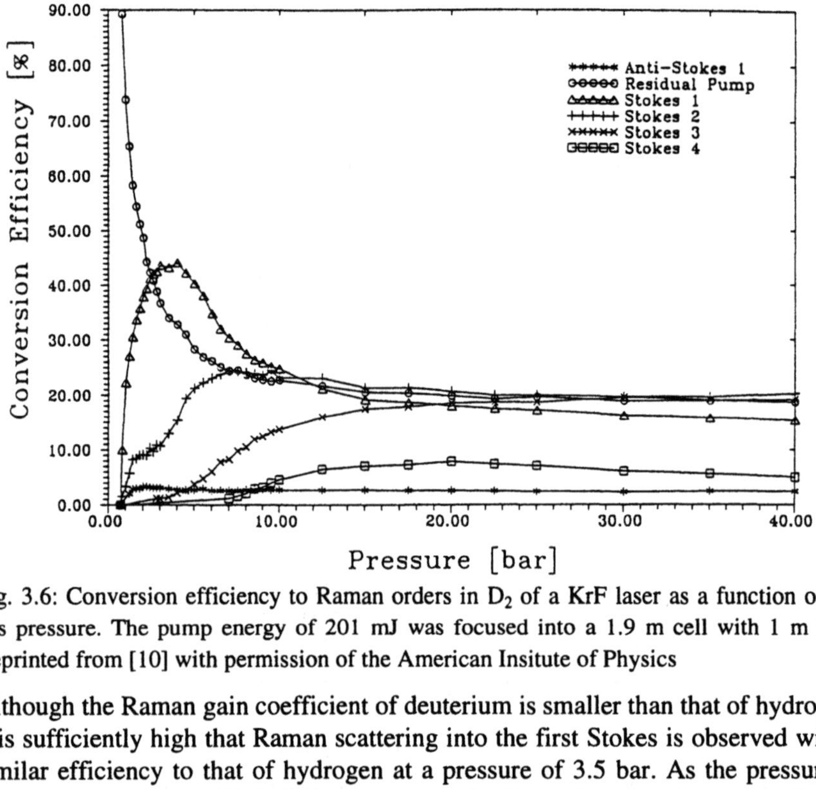

Fig. 3.6: Conversion efficiency to Raman orders in D_2 of a KrF laser as a function of the gas pressure. The pump energy of 201 mJ was focused into a 1.9 m cell with 1 m lens. Reprinted from [10] with permission of the American Insitute of Physics

Although the Raman gain coefficient of deuterium is smaller than that of hydrogen, it is sufficiently high that Raman scattering into the first Stokes is observed with a similar efficiency to that of hydrogen at a pressure of 3.5 bar. As the pressure is increased, the extent of higher-order processes is much reduced compared with hydrogen. In this case, the addition of a buffer gas is not necessary because the gain of deuterium is lower. The optimum results obtained with a pump energy of 240 mJ (simultaneous generation of 1st and 2nd Stokes lines) are summarised in Table 3.6.

Table 3.6: Output laser beam energies obtained with a pump energy of 240 mJ under optimum gas mixtures for the IFU DIAL system.

Raman order	Conditions	Energy
1st Stokes, hydrogen	H_2 2.05 bar	72 mJ
2nd Stokes, hydrogen	H_2 2.05 bar	72 mJ
1st Stokes, deuterium	D_2 7.5 bar	70 mJ
2nd Stokes, deuterium	D_2 7.5 bar	58 mJ

Experiments performed at LIT, Sweden (*H. Edner, P. Ragnarson, S. Svanberg, E. Wallinder*). The KrF laser radiation (240 mJ/pulse within 0.4 mrad divergence) is focused into a high-pressure Raman cell, using a lens of 1 m focal length. The Raman cell has a length of 1.5 m and can withstand pressures as high as 100 bar. The focusing lens and the cell windows are made of quartz and are broadband dielectrically coated for at least 248 nm to 131 nm. The output of the Raman cell is then recollimated by using a same type of lens. The SRS efficiency was measured using hydrogen and deuterium as active Raman gases. No buffer gases were considered. Fig. 3.7 shows the conversion efficiency in the case of hydrogen, for the 1st and 2nd Stokes frequencies.

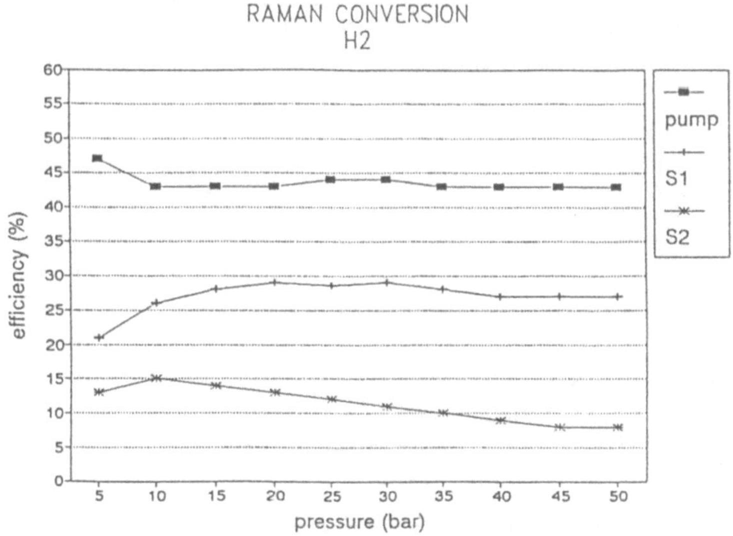

Fig. 3.7: Raman conversion efficiency of KrF laser radiation in H_2.

The maximum efficiency obtained in pure hydrogen, at S_1 is thus almost 30 % (at 20-30 bar pressure) and at S_2 (at 10 bar pressure) is 15 %. The maximum for the 1st and 2nd Stokes does not occur at the same pressure.

Experiments performed at MPI, Germany (*T. Schaberl*). The SRS experiments were performed with an injection-seeded KrF laser (250 mJ energy/pulse, rep. rate 80 Hz) and a Raman cell 1m long, withstanding pressures up to 40 bar. The focusing plano-convex lens has a focal length of 1 m. The entrance and exit windows are made of quartz (Suprasil), AR-coated on the outer surfaces, and are tilted by 1 degree with respect to each other in order to avoid backreflections into the laser. The output of the Raman cell is then recollimated by using a same type of lens. The SRS efficiency was measured (for forward SRS scattering) using hydrogen and deuterium as active Raman gases.

Figs. 3.8 and 3.9 show how operating conditions can be controlled to optimise the output at the desired wavelengths.

Fig. 3.8 shows how the output of a cell filled with hydrogen to a pressure of 4 bar varied as the input pump energy was increased. Threshold for the generation of first Stokes radiation was observed at approximately 15 mJ. Above this threshold there was a large increase in conversion efficiency together with high depletion of the pump beams. At a pump energy of about 20 mJ the threshold for the generation of second Stokes was reached. This was accompanied by some depletion of both the first Stokes and the residual pump beams. At an energy of 50 mJ the threshold for third Stokes generation was reached and at energies greater than 120 mJ anti-Stokes orders became detectable. A particular feature of the conditions used in Fig. 3.8 is that for energies higher than 70 mJ, the efficiency for first Stokes generation was approximately constant.

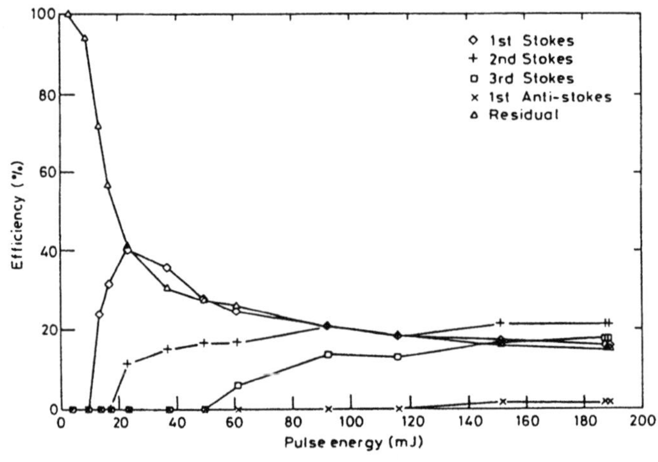

Fig. 3.8: Energy generated by SRS in H_2 from a KrF laser versus pump energy using a 0.95-m cell and a 1-m focusing lens. The pressure of H_2 was 4 bar.

Helium is suitable for use as a buffer gas and can also be used to control the divergence of the Stokes beams [30]. The results shown in Fig. 3.9 were taken with a mixture of hydrogen and helium.

At low pressures an increase in first Stokes was observed, whilst at higher pressures there was a decrease in all Stokes orders. At room temperature anti-Stokes orders can only be generated by four-wave mixing. This is further evidence for the presence of four-wave mixing at pressures less than 10 bar. Fig. 3.9 confirms that the use of a buffer gas is effective in optimising the generation of first and second Stokes by reducing four wave mixing. The optimum results are summarised in Table 3.7.

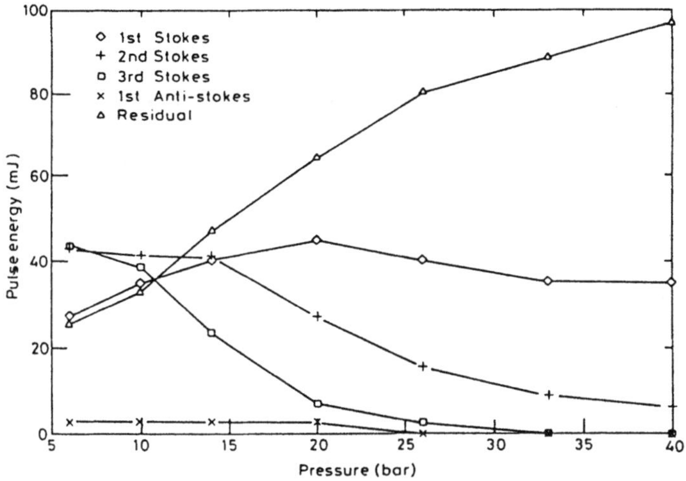

Fig. 3.9: Pulse energy generated in forward Raman orders from a KrF laser in H_2 with He buffer gas. The partial pressure of H_2 was 6 bar. He buffer gas was used to establish the total pressure indicated. The pump energy was 188 mJ.

Table 3.7: Gas mixtures and pulse energies for the MPI DIAL system.

Cell	Raman gas	Buffer gas (He)	E_{pump}/pulse	E_{S1}/pulse	E_{S2}/pulse	E_{S3}/pulse
D_2	40 bar	-	60 mJ	33 mJ	44 mJ	4 mJ
H_2	20 bar	20 bar	62 mJ	27 mJ	62 mJ	-

Stimulated rotational Raman scattering. At the 6th TESLAS workshop it had been reported by T. Schaberl, MPI, that Stimulated Rotational Raman Scattering (SRRS) had been observed in addition to the energy transferred by the vibrational SRS interactions. In principle, there should be no SRRS from linearly-polarised light [36, 37]. Hence the presence of any SRRS indicates that there is a circularly-polarized component in the incident pump radiation. This can either be due to the polarization of the pump laser itself, or can be caused by stress-induced birefringence in the windows of the pressure cell. It was therefore decided to make some survey experiments regarding this problem for all KrF-laser based systems. The results of the measurements performed at LIT, Sweden, are shown in Table 3.8. It shows the energy of the rotational components generated by a KrF pump laser under some selected operating conditions, both for hydrogen and deuterium.

These measurements are consistent with the general observation that the energy scattered into rotational orders by a KrF pump laser can be reduced when operating at higher pressures. Thus, above 1.8 bar in hydrogen SRRS is negligible. This series of measurements was performed without buffer gas because the increase in first Stokes, caused by the addition of helium buffer gas, was not more than a few per cent.

Summary. The results presented here show that there are two possible approaches to the optimisation of the Raman output when using KrF lasers. The highest energies are achievable in the low-pressure regime in which 4WM is the dominant interaction. In this mode of operation the beam divergence may be high, which can give rise to errors in differential-absorption lidar measurements. If the beam divergence cannot be reduced by a rather large beam expander the preferred operation is in the high-pressure regime or with a buffer gas, which improves the beam quality and the shot-to-shot stability at the expense of the energy.

Table 3.8: Measurement of the energy scattered by stimulated rotational Raman scattering from a KrF pump laser.

Conditions	Pump	1st Stokes	Rotational components
Hydrogen 1bar	90 %	10 %	
	0.1		2nd a-Stokes
	3.2	5	1st a-Stokes
	73.0	65.5	residual
	21.7	29.5	1st Stokes
	1.9		2nd Stokes
Hydrogen 2.5 bar	22 %	36 %	
	0.4	0.7	1st a-Stokes
	93.8	93.9	residual
	5.8	5.3	1st Stokes
Deuterium 7bar	40 %	29 %	
	0.2	0.2	1st a-Stokes
	96.7	96.2	residual
	3.1	3.5	1st Stokes
Deuterium 14 bar	34 %	25 %	
		0.1	1st a-Stokes
	99.5	98.9	residual
	0.5	1	1st Stokes

3.2.5 Infrared lasers

Pulsed CO_2 TEM_{00} Transverse Excited Atmospheric pressure (TEA) lasers yield high power emission directly in the infrared (IR) spectral region (9-12 μm). The CO_2 lasers offer tunability for many lines, when an intracavity transmission grating is introduced into the laser cavity. More than 65 different lines in the 9-12 μm can be generated. In ozone DIAL applications the on- and off-resonance wavelengths correspond to the same molecular branch. The use of a negative branch resonator provides a nearly pure TEM_{00} transverse mode and a far field divergence of less than 0.5 mrad. Typical laser output energy ranges from 1-4 J and the FWHM pulse length is of the order of 100 ns. Typical repetition rates are of the order of 1-5 Hz.

3.2.6 Metal-vapour lasers

The copper-vapour (CVL) as well as the manganese- (MVL) and the gold-vapour (AVL) lasers are referred to as pulse-periodic lasers with an easy synchronisation [23, 24]. The laser-active media are transparent for all radiation wavelengths which greatly simplifies the schemes of spatial matching of the generated laser beams.

Using non-linear optical techniques a series of additional laser wavelengths can be generated, corresponding to basic, frequency mixing and second-harmonic generation of laser lines of the aforementioned vapour lasers (Table 3.9). Since all the 15 UV lines presented in Table 3.9 are located in the 255 to 314 spectral range, these laser lines can in principle be used for tropospheric ozone measurements. Using metal-vapour lasers (Mn, Au, Cu) of average powers greater than 10 W, laser lines in the UV spectral region can easily be generated simultaneously using the second-harmonic generation (SHG) and the sum-frequency mixing (SFG) techniques in non-linear BBO crystals. The harmonic generation efficiency depends on the energy and beam divergence of the pump laser, as well as on the non-linear characteristics of the BBO crystal.

Table 3.9: Available wavelengths from metal-vapour lasers for the IOA DIAL system.

		Cu	Mn	Mn	Cu	Au
	λ / nm	510.6	534.1	542.0	578.2	627.8
Cu	510.6	255.3	261	262.9	271.2	281.6
Mn	534.1		267	269	277.6	288.6
Mn	542.0			271	279.8	290.9
Cu	578.2				289.1	301
Au	627.8					313.153

For a typical CVL (pulse power of some tens of kW) a very slow increase of conversion efficiency is observed. For the SHG of the CVL green line in a BBO crystal, the conversion efficiency is about 9 % [38].

The metal-vapour lasers provide simultaneous emission of multiple UV and visible wavelengths of low power (≈ 0.5 W) at very high repetition frequencies. This can be used to advantage in ozone DIAL systems working only in the photoncounting detection mode, where the simultaneous detection of ozone in the troposphere and the stratosphere is made possible. The problems related to the overload of the detection systems (PMTs) at the near-range are drastically reduced (see section 3.4).

3.2.7 New perspectives

In the early 1980s colour-center lasers emerged as solid-state tunable sources at near- and mid-infrared (IR) wavelengths (800 nm to 4000 nm). This type of lasers is operated at cryogenic temperatures and therefore is far from being a convenient laser source. Toward the end of the 1980s laser research was focused on the development of a new solid-state laser source: the titanium-doped sapphire (Ti:sapphire) laser. This type of laser offers a wide tuning range exceeding even that of dye lasers. Ti:sapphire is one of a range of vibronic tunable laser materials (together with the alexandrite Cr^{3+}:$BeAl_2O_4$) that are the basis of many commercially available solid-state tunable laser systems. Although flashlamp pumping of Ti:sapphire lasers is possible, most commercial systems use a blue-green laser as a pump (*i.e.* 3rd harmonic of the Nd:YAG laser).

Recently, new types of tunable pulsed lasers have become also available, such as solid-state dye lasers, rare-earth-doped UV lasers, optical parametric oscillators (OPOs), which can be frequency doubled, tripled or quadrupled to generate tunable UV radiation in the 260-340 nm spectral region.

The solid-state dye lasers seem very promising laser sources, but they are still in their developing phase and thus they often present important operational problems. In this type of lasers an organic dye is captured in a solid material. Currently, a maximum average power of 1.5 W has been demonstrated at 10 Hz repetion rates. Such lasers would be easier to use than current liquid-dye lasers, and they would still have the advantage of wide tunability through visible wavelengths. However, the main drawback of the solid-state dye lasers is that the solid dye, like the liquid dye, becomes depleted after some time, so longer lifetimes for dyes are necessary.

Rare-earth-doped ultraviolet laser sources (cerium-doped or chromium-doped lithium strontium aluminium fluoride (Ce:LiSAF or Cr:LiSAF) lasers) seem to be the most efficiently continuously-tunable materials reported to date [39]. The LiSAF host material is a uniaxial crystal from the colquirite family that was first developed in 1988. The Ce:LiSAF laser has a broad absorption centered at 266 nm, which overlaps well with the fourth harmonic output of the Nd:YAG laser. The emission band is quite broad (275 to 330 nm) with the gain peak centered at 290 nm. Using a frequency quadrupled Nd:YAG laser, which delivers 10-ns pulses at a repetion rate of 10 Hz, and using an intracavity tuning prism, continuously tunable UV output from 285 nm to 297 nm has been obtained (output energy per pulse up to 4 mJ). The Cr:LiSAF laser has absorption bands at 670 nm as well as extensive tuning ranges in the near irfrared (800-1000 nm). Direct diode pumping of Cr:LiSAF lasers has very recently been reported to produce pulsed laser light in the near-IR.

Pulsed OPOs have generated extensive interest as sources of tunable light and several commercial systems have appeared. The OPO is not a laser itself in the strict meaning, but it is a source that generates coherent light through optical amplification of noise and is usually classified as a nonconventional laser. The OPOs are powerful sources of broadly tunable coherent radiation covering the UV up to the mid-IR spectral regions. Presently, three basic nonlinear materials are available for OPO applications: the BBO, the KTP and the Lithium-tri-Borate (LBO). These materials present certain advantageous properties such as: high optical damage threshold, high nonlinearity and large birefringence. The generation of tunable laser radiation, when optically pumped by a laser source (XeCl, Nd:YAG, Er:YAG etc.) is performed either by temperature or by angle tuning of the crystal phase matching. For instance, when an OPO is optically pumped by the 2nd (532 nm), 3rd (355 nm) or the 4th harmonic (266 nm) of a pulsed Nd:YAG laser, tunable laser radiation is generated in the 307 nm–325 nm, 414 nm - 2.4 μm, or 707 nm - 2.15 μm spectral range, respectively.

The main advantages of tunable solid-state lasers versus other types of tunable laser sources (*i.e.*, liquid-dye lasers) are: compactness, long life, non-toxicity, tunability, ease of use and reliability. Their main limit has been the rather low transmitted energy, but there appears to be some potential for increasing this. Therefore, these laser sources will have to be further developed to be used for ozone DIAL measurements in the troposphere and the lower stratosphere. Additionally, other new solid-state laser sources may become more attractive for air pollution monitoring purposes than those presently used.

3.3 Receiving optics

The essential components of a lidar optical receiver system comprise a receiving telescope and an associated dual- or multi-wavelength resolving optical system (spectrometer or interference filters) in order to efficiently separate the DIAL wavelengths.

3.3.1 Receiving telescope

The backscattered radiation is collected by a telescope, with the diameter of the primary mirror typically in the range of 10 - 80 cm. The telescope design is not of major importance, Cassegrain, Newtonian, or Dall-Kirkham types have successfully been used. Also a rather simple and flexible design, using only one parabolic mirror with output coupling in the primary focus to a large core optical fiber, has been used with good success [11]. In the case of UV DIAL systems, highly reflective coatings (aluminum protected by MgF_2) have been applied to the primary and secondary mirrors (reflectivity factor of 80-90 %) of the receiving telescopes. In the case of IR DIAL systems, gold-coated copper mirrors are used.

The backscattered lidar signal has an extremely high dynamic range (typically 10^6–10^9) which cannot be handled with the required accuracy by a single detector. The dynamic range can be greatly reduced by the use of two spatially separated receiving telescopes: the first adapted for the lower altitudes (near-field telescope) and the second adapted for the higher altitudes (far-field telescope).

The receiving telescopes can be used either in a coaxial or in a non-coaxial configuration, thus offering different starting altitudes from wich ozone profiles can be measured [40]. It is important to note, that complete overlap often is achieved at rather large distance only, in particular when using the non-coaxial configuration. From the experience gained within TESLAS, it cannot be recommended to use the region of incomplete overlap for DIAL measurements. This would require a very high degree of uniformity of the spatial distribution of energy in the two beams used for a DIAL measurement, which seems to be extremely hard to achieve.

For this reason, typical starting altitudes (where full overlap between the laser beams and the field-of-view (FOV) of the receiving telescope is achieved) range from 100 m to 2000 m, depending on the FOV, the diameters of the receiving telescope and the transmitted beam, and the distance between the optical axes of the laser beam and the telescope. The telescopes' field of view can be adjusted by a diaphragm in the focal plane from a minimum value compatible with the laser divergence ($\approx$ 0.5 mrad) up to 1.5-2 mrad.

At RIVM a method has been developed to tune the overlap function using computer controlled stepping motors mounted on the beam steering mirrors. The motors are used to scan the outgoing laser beam over a rectangular grid, while the backscattered signals are monitored at a range where geometrical effects (transverse focus travel and depth of field) are negligible. With this method, excellent similarity can be achieved between overlap functions from two or more different laser beams. Also, if the field of view of the receiving system is reduced to near its angular resolution (typically a 0.15 mrad or better) this tool can be used to measure the beam divergence of the outgoing laser beams. As an example Fig. 3.10 shows a beam profile measured with this method [41].

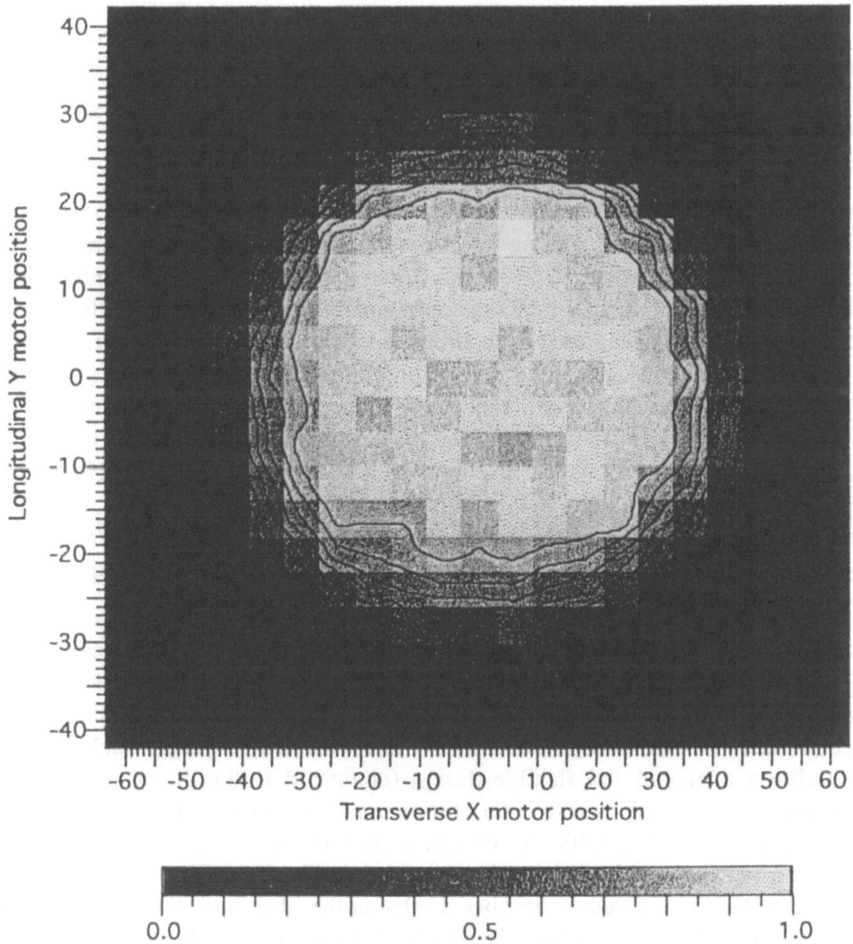

Fig. 3.10: The normalized intensity distribution of the emitted laser beam at 289 nm, integrated between 3600 and 3900 m, measured with the beam scanning technique. Lighter shades correspond to higher intensities. In the transverse (or *X*) direction one motor step corresponds to 0.02832 mrad, while in the longitudinal (or *Y*) direction one motor step corresponds to 0.04 mrad. The total grid displayed in the figure is therefore sized: (3.577 × 3.36) mrad. The field of fiew of the receiving system was set to 1.67 mrad. The center of the light area was chosen as the position yielding optimal alignment for the laser beam at 289 nm.

3.3.2 Wavelength separation

In the DIAL technique, when the necessary wavelengths are generated simultaneously (by a single laser source), the separation of wavelengths in the receiving system is of crucial importance. This separation can be performed either by using a grating spectrometer or a set of specially designed narrow-band interference filters.

Grating spectrometers

One good solution to separate closely-spaced wavelengths and to suppress straylight is the use of a grating spectrometer having optical path lengths of a typical value of 300 mm up to 1500 mm (Czerny Turner type). This configuration has the direct advantage of a high degree of flexibility in the design of the system, since additional wavelengths can be simultaneously detected in an upgrade phase (*e.g.*, Raman channels). A plane ruled grating blazed at wavelengths ranging from 250 to 290 nm, provides a sufficient linear dispersion of 0.2-3.6 nm/mm (1200-3600 grooves/mm) at the output slits. The grating spectrometer offers the advantage of a greatly reduced detection bandwidth (0.2-1.2 nm FWHM) at all wavelength channels, a high straylight reduction (of the order of 10^{-8}), while the total optical transmitance of the system is of the order of 40-50 %. Therefore, the grating spectrometer is a good choice, when daytime operation is required and where an important background skylight rejection has to be applied.

One good solution to couple the grating spectrometer to the receiving telescope is the use of a large core quartz optical fiber [42]. The optical fiber transmits the lidar signal from the telescope's focal plane to the entrance slit of the grating spectrometer (transmission greater than 80 % at wavelengths larger than 280 nm). This allows a rugged and flexible design (compared to the direct conventional coupling through a collimating lens) of the optical receiving unit. It also enables a better optical coupling with the detection system and improves the electrical insulation between the laser and the detection electronics. A further advantage is that the beam is scrambled in the fiber, so that at the fiber output the differences in beam geometry for signals from different heights are largely reduced.

Narrow-band interference filters

The use of narrow-band interference filters to discriminate the DIAL signals is cheaper and easier to install than the solution of the grating spectrometer. However, it suffers from a low total optical transmittance (lower than 20-35 %), and when closely spaced wavelengths are used the crosstalk between the channels may be too large. This latter disadvantage can easily be avoided when two lasers are used for the generation of λ_1 and λ_2, by introducing a time delay of typically 300-500 μs between the laser shots [17].

3.4 Detectors

The detection of the lidar signals in the UV spectral region is performed by Photomultiplier Tubes (PMTs), while in the mid-IR spectral region (10 μm)

Mercury-Cadmium-Tellurium (MCT) detectors are used. The output signals from the PMTs are then pre-amplified using signal amplifiers with broad bandwidths (5-300 MHz).

In the UV DIAL systems several types of PMTs have been used and their performance has been critically examined. The criterium of an optimum performance of the PMTs is their ability of performing range-gated analogue current measurements as well as photon counting measurements. Typical values of the high voltage (HV) PMT supply range from 900 V to 2500 V. The PMTs tested and subsequently recommended for an optimum performance in the DIAL technique are presented in Table 3.10.

In DIAL systems for tropospheric measurements generally the close-range signal is much stronger than the far range signals (several orders of magnitude). Considering the high accuracy requirements, special care has to be taken to avoid overexposure of the photocathode, overload of the final PMT stages, and afterpulsing associated with ion feedback in the PMT.

Table 3.10: Main characteristics of the various types of PMTs used within TESLAS.

Model	Spectral region	Type	Gain	Quantum efficiency
EMI 9813QB	250-355 nm	Linear focussed	10^7	15-20 %
EMI 9816QA	250-355 nm	Linear focussed	10^7	15-20 %
EMI 9817QA	250-355 nm	Linear focussed	3×10^6	15-20 %
EMI 9883QB	250-355 nm	Linear focussed	6.7×10^7	15-20 %
EMI 9893B	250-355 nm	Linear focussed	8.3×10^7	15-20 %
Hamamatsu R1464	185-850 nm	Head-on	10^6	25 %
Hamamatsu R928	185-900 nm	Side-on	10^7	25 %

Overexposure of the photocathode can only be avoided by correct optical layout of the system (suppression of close-range signals). When the gain of the PMT is fixed, the signal dynamic range is restricted by the maximum permissable current in the final PMT stages. Exceeding this current results in a nonlinear response curve. The usable signal range can be increased by reducing the gain of the PMT for the close range signals only, which is performed by changing the voltage difference in the dynode chain [4, 10].

Afterpulsing shows as a deformed and delayed echo of the original pulse shape, while SIB is a slowly decaying tail, continuing after the true signal has vanished. The mechanism for the occurrence of afterpulsing and SIB is not well understood. They seem to be the result of a complicated interaction between the exposure of the photocathode and the gain of the PMT (*i.e.*, the structure of the dynode chain and the electrical network). Afterpulsing and SIB are often grouped under the same name "signal-induced noise", since they have the same source but are separated here since they are dealt with differently.

The effects of non-linearity, SIB and afterpulsing, were studied using an artificial lidar signal [27], as well as using real DIAL signals [43]. This showed that the influence of non-linearity on the concentration calculation precludes the use of signals larger than 400 mV into 50 Ω for the PMTs that were used. Therefore, neutral density filters have to be used to keep the signal from exceeding this limit for low altitude measurements. Earlier publications [44-46] suggest that the afterpulsing is a function of the PMT gain alone, or that both gain and gain variations are important. So, afterpulsing can be minimized by ensuring voltage stability at each dynode, however, this does not remove SIB. SIB seems to be a function of the height of the primary pulse and the decay time constant of the signal. A way to correct for SIB is to perform a fit on the background part of a lidar signal, after subtraction of the DC-level and to subsequently subtract the new-fitted-base-line, extrapolated to $R = 0$ from the signal. Observations [45, 47] confirm that SIB can be treated as an exponentially decaying tail. Therefore, a linear fit on the logarithm of the lidar signal was performed. This method is sensitive to noise, however, and better results can be obtained by fitting an exponential function to the raw signal. An additional problem is that the effects can vary substantially between individual PMTs.

In the case of IR DIAL systems working in the 9-11.5 μm spectral region the collected lidar backscattered signals are focused onto a liquid nitrogen cooled HgCdTe detector operating at 77 K. The photoconductive alloy is optimised to have a maximum detectivity at 10 μm [2×10^{10} cm $Hz^{1/2}$ $W^{-1/2}$] and is matched with a low-noise voltage amplifier . The bandwidth integrated video NEP is of the order of 10^{-8} W.

3.5 Data acquisition

Due to the important dynamic range of the backscattered lidar signal coming from altitudes as high as 13-15 km, two different signal detection techniques have to be applied: the analogue detection technique (near field measurements) and the photon counting detection technique (far field measurements).

3.5.1 Analog detection systems

The data aquisition systems used for the digitisation of the lidar signals are based on the use of fast analogue transient digitizers. Data aquisition systems with to 2 to 6 channels for simultaneous aquisition are used, with sampling rates of 5 up to 50 MSa/s and an accuracy of 8 up to 16 bits. The resulting spatial resolution, thus, ranges from 3 to 30 m. However, the use of at least a 12-bit digitizer, with 2048 storing channels, is desirable in order to capture the backscattered lidar signals with an adequate precision. The transient digitizers used are equiped with 24-bit on-board Digital Signal Processing (DSP) systems able to perform fast on-line hardware averaging of the lidar signals.

To improve efficiently the signal-to-noise (SNR) and the dynamic range of the detectable lidar signal the averaging technique is applied. The number of the laser shots averaged (typicaly around 100-15000) is variable, since it directly depends on the spatio-temporal resolution and on the accuracy required for the ozone measurements. Therefore, laser sources with high pulse repetition rates are desirable in order to achieve a high temporal resolution on the retrieval of the ozone vertical profile. However, in practice there is a limit to this refinement procedure, due to the imperfection of the digitizers used.

Careful double-shielding of the detection electronics turned out to be very essential, in the case of excimer-based DIAL systems, because of the presence of strong local radio-frequency (rf) sources [10]. However, better shielded KrF lasers have become available which minimise this interference problem.

3.5.2 Photon counting systems

For the higher altitude levels (above *ca.* 5-10 km) where the lidar signal is low enough, fast multichannel photon counters have to be used. Photon-counting systems comprise the use of fast multi-channel scalers which are connected to the PMTs in parallel to the analogue output. Typical values of the lidar signal sampling frequency range from 0.5 MHz up to 2 MHz and the resulting altitude resolution ranges from 75 to 300 m. Special attention should be paid to problems arising from the overexposure of the PMTs photocathodes (SIB pulses), as discussed in the previous section, as well as to the matching of the impedance of the analogue and photon-counting outputs.

3.5.3 Signal acquisition and storing

After the signal digitization generally some averaging has to be performed in order to reduce the data rate to an amount which can be handled by standard storage devices like hard disks or tape drives. This is mostly done in hardware averagers, but the use of fast signal processors also allows to use software averaging, which may also include range check and calculation of signal variance (MPI system, see section 3.6.8).

For the rest of the data handling, standard components as available for small computers are used, the details need not be reported here. It is, however, essential that at least some experiment control is provided online. At least the raw signals have to be visualized in real time. The additional display of the logarithm of the background and range corrected signals offers great advantages and has been implemented in most systems. It is also worth noting that the computer can be used to set or at least record the main experimental parameters (*e.g.*, acquisition time, altitude resolution, values of the various laser wavelengths, number of laser shots, PMT high voltage input).

3.6 Operational DIAL systems for ozone measurements

In the following paragraphs a brief presentation of the experimental set-ups of the various ozone DIAL systems developed within the TESLAS Project will be performed.

3.6.1 CNRS ozone DIAL system

G. Ancellet, A. Papayannis, J. Pelon and G. Mégie

The transmitter of the CNRS ozone DIAL system is based on the use of two Raman-shifted quadrupled Nd:YAG lasers [12]. The energy available at the emitted wavelengths 266 nm, 289 nm and 299 nm is 15 mJ, 10 mJ and 7 mJ, respectively. The energy stability of the laser output at 266 nm is obtained through the use of an active feed-back loop controlling the KD*P crystal tuning angle [25].

The backscattered lidar signals are collected by a vertically pointing Cassegrain telescope of either 30 or 80 cm diameter (not simultaneously) , which inputs a four channel grating spectrometer for wavelength separation and background skylight suppression. PMTs directly mounted on the exit slit of the spectrometer detect the lidar signals at each wavelength. The DIAL system is of the non-coaxial type (20 or 50 cm distance between the laser beams and the telescope axis) to avoid overloading of the PMTs.

The data acquisition system includes four 12-bits, 10-MHz transient digitizers for near-range detection (0.5-2 km or 1.5-8 km, depending on the size of the telescope) and two 1-MHz photoncounting channels to record the far-range signals (2-5 km or 8-15 km). Data are averaged on dedicated electronic boards and then stored on a 386 PC computer, which provides a real time display of the raw signals and a colour coded display of the ozone vertical profiles as a function of time. The number of averaged laser shots can be selected by the operator (typically 600-1000 shots) corresponding to a 1-2 min. time resolution. The main characteristics of the CNRS DIAL system are summarized in Table 3.11, while the experimental set-up is presented in Fig. 3.11.

Table 3.11: The SA/CNRS DIAL system characteristics

Transmitter:	Two Nd: YAG lasers (GCR-11 Spectra Physics) 4th harmonic + Stimulated Raman Scattering in hydrogen and deuterium
Receiver:	Cassegrain Telescope: area = 30 cm or 80 cm Czerny Turner Spectrometer with a 3600 gr/mm holographic grating: bandwidth = 1.2 nm efficiency = 40 % Multialkali photomultipliers: Hamamatsu R1464 size = 14 mm risetime = 2.2 ns gain = 10^6
Data acquisition:	Analog detection (4 units): resolution = 12 bits sampling = 10 Mhz Photocounting detection (2 units): time gate = 1 μs counter frequency = 250 MHz Hardware averaging Data storage and processing on a 386 PC computer (Compaq)

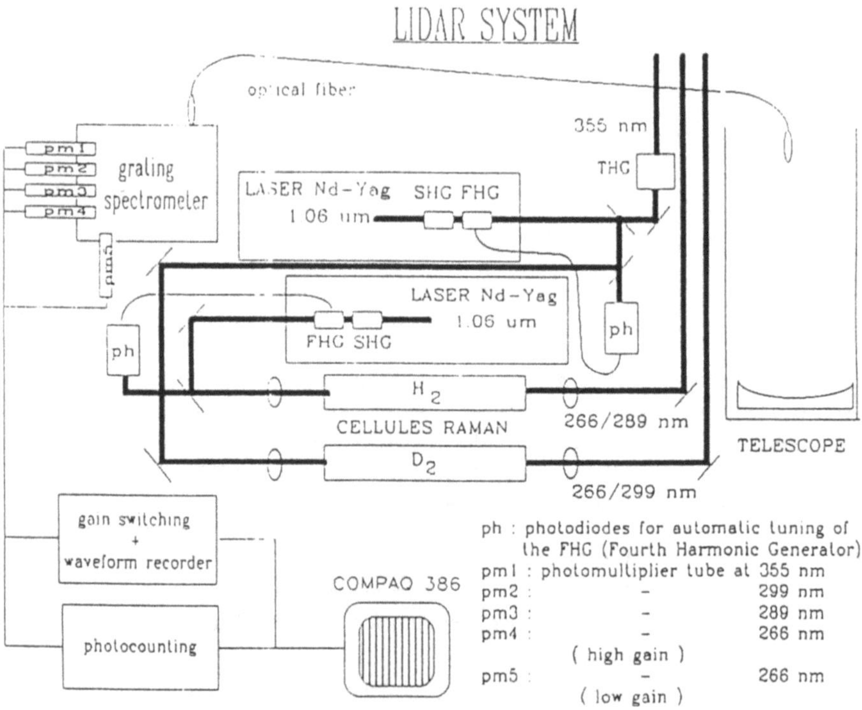

Fig. 3.11: Experimental layout of the SA/CNRS DIAL system.

3.6.2 ENEA ozone DIAL system

R. Barbini and A. Palucci

The experimental set-up of the ENEA ozone DIAL system is presented in Fig. 3.12.

Two identical pulsed TEM_{00} Transverse Excited Atmosphere (TEA) CO_2 lasers are installed inside a large container (50 m^2). A suitable negative branch resonator (SFUR: Self Filtering Unstable Resonator) has been used, providing a nearly pure TEM_{00} transverse mode and a far field divergence of 1.5 mrad. The laser output energy ranges from 1-4 J and the FWHM pulse length is 80 ns. Introducing an intracavity transmission grating 64 individual laser lines can be selected in the 9-11.5 μm spectral region [48-50]. In the case of DIAL air pollution measurements two different laser pulses, at slightly different wavelengths (of the same molecular branch), are transmitted into the atmosphere using gold-coated copper mirrors mounted symmetrically on the receiving telescope in a coaxial configuration. The main characteristics of the ENEA DIAL system are summarized in Table 3.12.

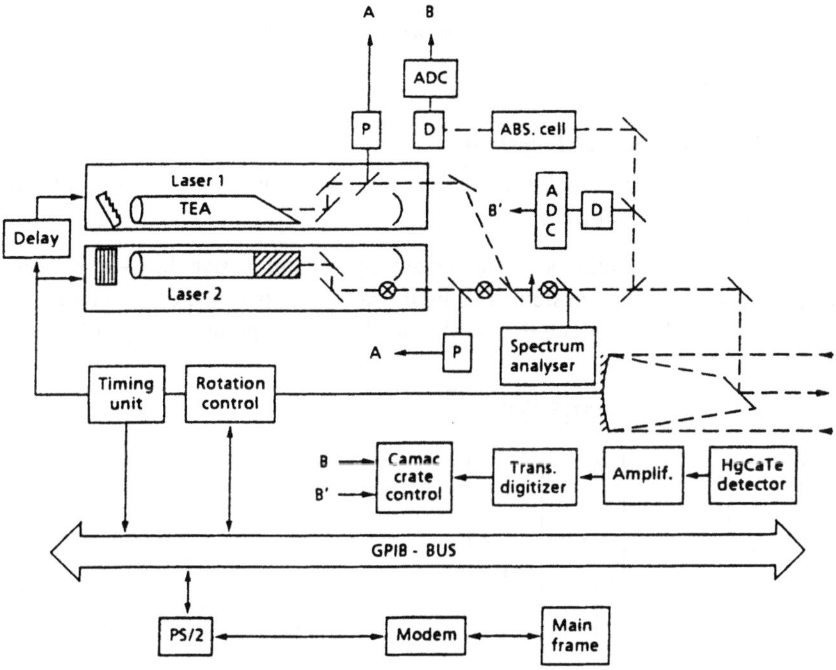

Fig. 3.12: Experimental layout of the ENEA DIAL system.

Table 3.12: The ENEA DIAL system characteristics.

Transmitter	**Receiver**	
Two TEA CO_2 lasers	Telescope	
Emission $\lambda = 9\text{-}11.5\ \mu m$	Type Newtonian	
Pulse energy = 1-4 J TEM	Mirror diameter $D = 38$ cm	
Pulse duration = 80 ns	*f*-number	$f = 3$
Beam divergence = 1 mrad (f.a.)	Field of view	FOV = 1 mrad
Transient digitizers	**Detector**	
DSP Transiac	HgCdTe at 77 K	
12 bits 10 Msamples/s	Type SBRC/PC	
8 bits 100 Msamples/s	Area $1 \times 1\ mm^2$	
	Detectivity $D^* = 2 \times 10^{10}$ cm $Hz^{1/2}\ W^{-1/2}$	

The pulsed CO_2 TEA lasers described have been completely designed and assembled at ENEA and are used for both range resolved DIAL ozone/air pollution measurements as well as for wind-velocity Doppler measurements [51, 52].

3.6.3 EPFL ozone DIAL system

E. Durieux, B. Calpini, A. Clappier, L. Fiorani, M. Flamm, L. Jaquet and H. van den Bergh

The EPFL DIAL system is characterised by an innovative high-repetition-rate shot-per-shot acquisition system. The lidar transmitter is based on two independent laser sources used to produce the DIAL wavelengths, separately (Fig. 3.13).

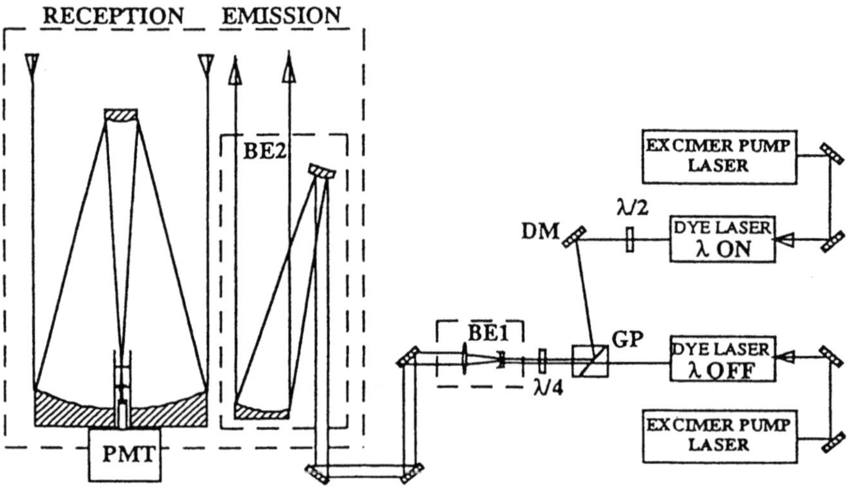

Fig. 3.13: Experimental layout of the EPFL DIAL system.

This allows to fire one laser source after the other with an time interval of ≈ 50 μs, in order to probe a 'frozen' atmosphere. Each laser source is based on an excimer (XeCl)-pumped frequency-doubled dye laser. An output energy of ≈ 1 mJ per pulse at 100 Hz repetition rate is obtained in the UV, with the on-resonance (ON) wavelength set at 272.5 nm and the off-resonance (OFF) wavelength set at 299 nm. The ON wavelength polarization is rotated by 90°, reflected by a UV mirror (DM) toward a Glan Prism (GP). In such configuration the ON and OFF wavelengths are combined in the same axis. A λ/4 plate ensures that no polarization effect will affect the data by making it circular. Two beam expanders are used (BE1 × 5, BE2 × 10) to reduce the laser energy density falling the steering mirrors and to reduce the beam divergence to values lower than 0.1 mrad [7].

The backscattered lidar signal are collected by a 60 cm diameter Cassegrain telescope (4.6 m equivalent focal length, 1 mrad field of view) which axis is parallel to the BE2 axis and at 498 mm apart. A 40 mm positive lens collimates the light bem before entering a solar blind filter. The UV light is then detected by a

PMT (EMI 9829 QB) and digitized by a transient digitizer. The sampling rate of the transient digitizer can be varied from 6.25 MHz up to 250 MHz for 2048 channels on 8-bits and up to 10 MHz for 512 channels on 12 bits. New emission-reception layout is on the way to be implemented, containing two Cassegrain telescopes, which will be movable around azimuth and elevation axis. The detection of the DIAL wavelengths will be performed by a 0.5 m spectrometer, thus allowing to increase the dynamic range of the experiment by using analog detection on the first telescope and a photon counting detection at the second telescope. The experimental set-up of the EPFL DIAL system is presented in Fig. 3.14, while the main characteristics of the system are presented in Table 3.13.

Table 3.13: The EPFL DIAL system characteristics.

Lasers	Two XeCl-pumped dye lasers
Wavelengths	272. nm 5 (ON), 299 nm (OFF)
Output energy	1 mJ
Pulse repetition frequency	100 Hz
Beam-expansion	× 50
Output beam divergence	< 0.1 mrad
Telescope diameter	600 mm
Telescope focal length	4600 mm
Receiver field of view	1 mrad
Digitizer (adjustable sampling rate)	6.25-250 MHz, 2048 channels, 8bits or 10 MHz, 512 channels, 12 bits
Detector	PMT EMI 9829 QB

A general overview of the digital electronic and data aqcuisition system is presented in Fig. 3.15 [7]. The VME bus has been chosen to support the communication between the various system modules. The detectors of the DIAL system are connected to front end modules.The system is presently capable to record shot-per-shot digitized analog signals, counts, and digitized time intervals.

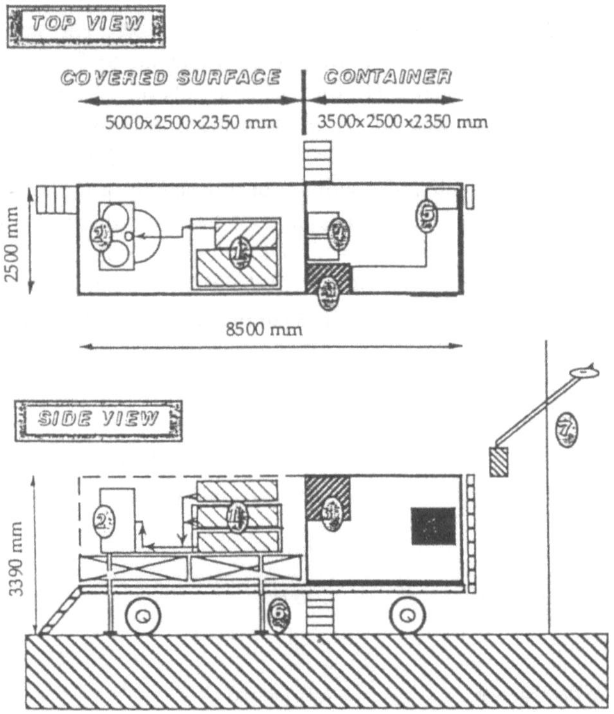

Fig. 3.14: Trailer layout of the EPFL DIAL system: 1- Laser sources; 2- Telescope; 3- Cooling system; 4- Electronic racks; 5- Ground analysers; 6- Stabilizers; 7- Meteorological mast.

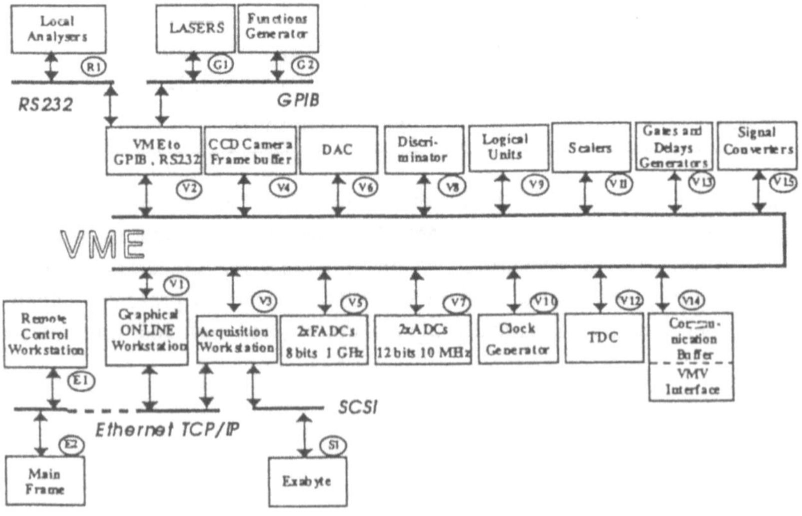

Fig. 3.15: Front end electronic and data acquisition system layout of the EPFL DIAL system.

It can generate analog signals and generate or record video signals which are interfaced to the GPIB and RS-232 buses.

3.6.4 GKSS ozone DIAL system

U.B. Goers, R,. Baumgart and C. Weitkamp

The GKSS ozone DIAL system is part of the ARGOS mobile DIAL system, which is designed for trace gas measurements in the 270-450 nm spectral region. The on- and off-resonance wavelengths are generated by two Nd:YAG-pumped dye lasers with a time lag of 50 μs in between. The possibility of a continuous wavelength tuning in the 270-450 nm spectral region was chosen to minimise the SO_2 interference to ozone measurements.

The experimental set-up of the ARGOS system is shown in Fig. 3.16 [53, 54].

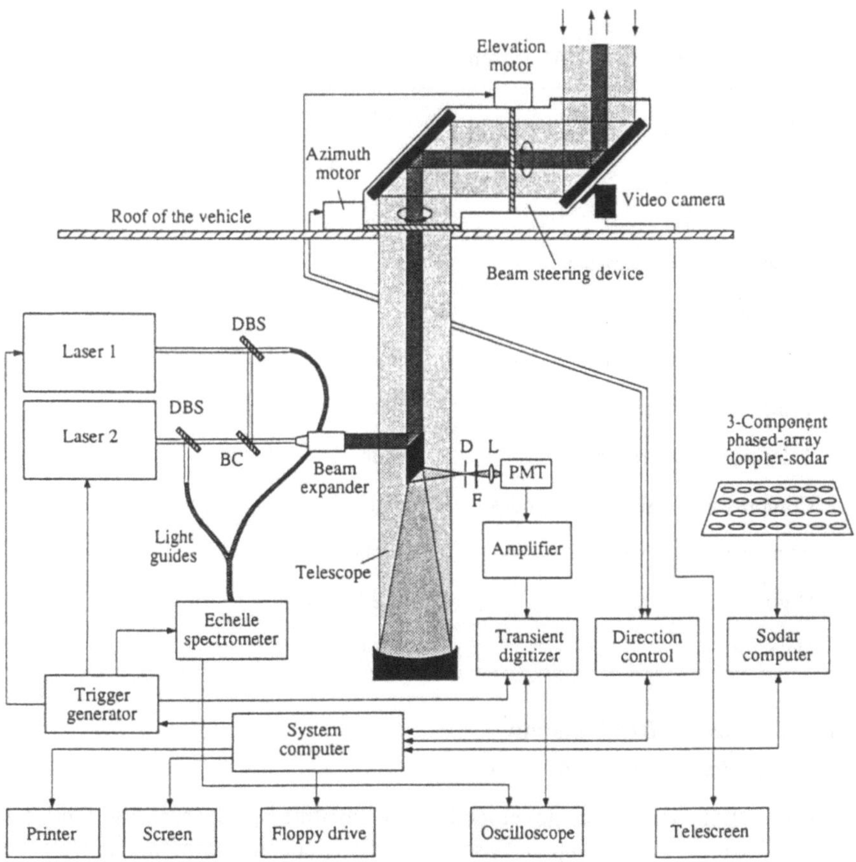

Fig. 3.16: Experimental layout of the GKSS DIAL system.

The dye lasers output is frequency doubled by two BBO crystals. Two dichroic beam splitters (DBS) transmit part of the fundamental dye laser radiation into an echelle spectrometer used for the wavelengths adjustement and for the stability detection during the measurements. The two laser beams are combined, expanded and are then transmitted into the atmosphere, by two computer-controlled steerable 45° mirrors. The range of zenith angles is ± 105° and that of azimuth angles is ± 360°.

The backscattered lidar signals are collected by a 300 mm Newtonian telescope (1050 mm focal length) coaxially and sequentially in time by the same photomultiplier (PMT EMI 9783 R). The PMT output signal after amplification is digitized by a transient digitizer at 20 MHz and with a 10 bit resolution. A geometrical signal compression is performed by a suitable diaphragm placed in the focal plane of the receiver optics. The main characteristics of the mobile ARGOS system are presented in Table 3.14.

Table 3.14: The GKSS DIAL system characteristics.

Lasers	Two Nd:YAG-pumped dye lasers
Wavelengths	280.91 nm (ON), 282.72 nm (OFF)
Output Energy	15-25 mJ
Pulse repetition frequency	10 Hz
Beam-expansion	× 20
Output beam divergence	0.1 mrad
Telescope diameter	300 mm
Telescope focal length	1050 mm
Compression of signals dynamics	Geometrical
Receiver field of view	0.95 mrad
Digitizer	10 bit, 20 MHz
Detector	PMT EMI 9783 R

3.6.5 IAO ozone DIAL system

V.D. Burlakov and V.V. Zuev

The ozone DIAL transmitter developed at the Institute of Atmospheric Optics in Siberia is based on a Cu-vapour laser. The CVL second harmonic (SHG) and frequency mixing (SFG) generation experimental set-up is displayed in Fig. 3.17. The pump laser comprises a master oscillator (MO) and a power amplifier (PA) unit.

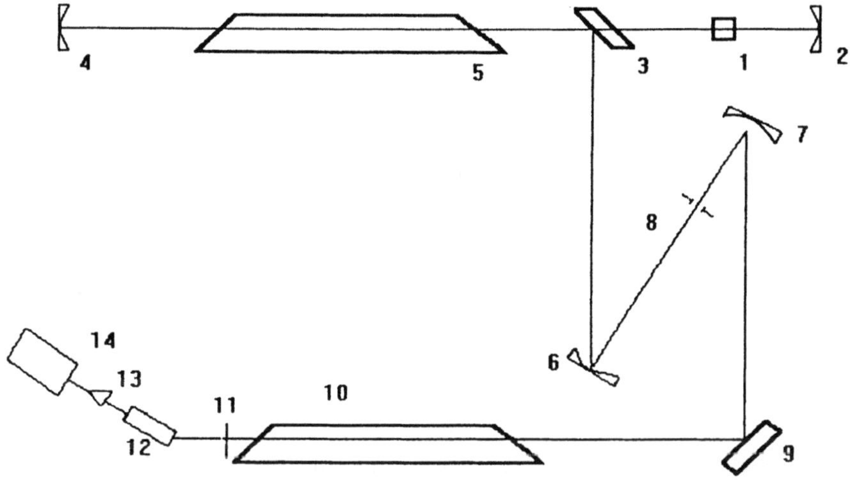

Fig. 3.17: Schematic diagram of the experimental layout used for the sum-frequency and second harmonic generation of the IAO DIAL system.

The MO is made of a gas discharge tube (5) and of an unstable telescopic resonator, which is defined by two total reflecting spherical mirrors (2, 4) of 6 cm and 100 cm focal length, respectively. A Glan prism (1) polarises the output beam. A holed plane output mirror (3), inside the unstable resonator, sends the output laser MO beam to a reflecting spatial-filter collimator (RSFC), comprising two confocally arranged mirors (6,7) of 60 cm and 80 cm, focal length, respectively and a 5 cm-diameter diaphragme located in the focus of the mirror (8). The gas-discharge tubes used in the MO and the PA (5 and 10, respectively) are of the "Kulon" (active length: 30 cm, diameter: 1.2 cm) and "GL-201" type (length: 75 cm, diameter: 2 cm).

The PA outcoming laser radiation is then focused to a BBO crystal by a biconvex lens of 1.5 m focal length. The BBO crystal ($\Theta = 48°$, $\phi = 90°$) has a size of 4 × 4 × 6 mm^3. A quartz prism (11) separates the output laser lines the energy of which is measured by a laser power meter. The conversion efficiencies of the SHG and SFG are 25 % and 14 %, respectively. The output UV laser lines have an average power of 0.5 W and a divergence of 0.2 mrad. The laser pulse repetition frequency is 7 kHz.

In the case of tropospheric ozone measurements the wavelength pair 271.2-289.1 nm has been used, while for stratospheric ozone measurements the wavelength pair 301-313.153 nm has been taken. Simultaneously emitted laser wavelengths in the visible channel permit the multifrequency sounding of the aerosols from the lower troposphere up to the higher stratosphere at wavelengths 510.6, 578.2 and 627.8 nm.

Additionally, atmospheric measurements of the Raman scattering on the atmospheric nitrogen (excitation wavelength 277.1 nm) and oxygen (excitation wavelength 268 nm) have been made up to altitude of 2.5 km, with a range resolution of 120 m and a time resolution of 15 min.

3.6.6 IFU ozone DIAL system

W. Carnuth, U. Kempfer and T. Trickl

The IFU ozone DIAL system is based on a narrow-band KrF Raman-shifted pulsed laser. The KrF laser yields energies higher than 250 mJ with a repetition rate of up to 80 Hz. The divergence of the KrF laser is less than 0.2 mrad. The KrF laser radiation is Raman-shifted to yield laser radiation at 277.1 and 313 nm simultaneously. The output energies are about 70 mJ for each wavelength under optimum conditions. Additionally, an operation at 291.9 is planned.

The backscattered laser light is collected by two receiving telescopes, which are used to divide the enormous dynamic range of the lidar signals. In this configuration the first telescope (0.13 m in diameter), located at a distance of 0.2 m from the emitted UV beam, is used to collect the lidar signals in the near-range (80 m up to 2.5 km height). The second telescope (0.5 m in diameter) located at a distance of 1.6 m from the emitted UV beam, is used to collect the lidar signals in the far-range (1.4 km up to 10 km height) [55, 10].

In the present configuration ozone measurements can be performed, either at 248.5 nm/277.1 nm or at 277.1 nm/313.2 nm. The former wavelength pair has rarely ben used because of range limitations. The wavelength selection is performed by the use of interference filters having a bandwith of 7-12 nm (maximum transmittance: 15 %). The detection bandwidth has recently been improved by the use of two 1.1 m grating spectrometer (about 0.3 nm bandwidth FW), replacing the interference filters.

Two types of PMTs were used; EMI 9813 QB for the two detection channels of the small telescope and EMI 9893 QB for those of the large one. The performance of the 9813 PMT is inferior. Thus, these tubes are currently replaced by 9893 ones. The gain of the PMTs was range-gated to prevent excessive anode currents next to the peak signals.

The resulting analog signals are digitized by four 8-bit, 30 MHz transient digitizers with amplifiers and 24-bit signal averagers (all DSP Technology). The amplifier gain was set at 10. Due to minor problems with averaged single bit steps a compatible 12-bit, 40 MHz system was recently purchased. In addition, a dithering technique will be used to obtain optimum linearity.

For the 277.1 nm channels a 50 MHz photon counting multichannel scaler (1 μs/channel) is applied in addition and it is connected to the PMT parallel to the analog output. Faster photon counters (500 MHz, 100 ns dwell time) which became recently available, would reduce the measurements starting altitude down to 4 km and would also improve the vertical resolution. The data of the four transient digitizers and the photon counter are transferred (every 15 000 laser shots) to a personal computer (PC) and stored on optical and mini disks. The entire data acquisition is controlled by a PC program. A schematic lay-out of the experimental set-up of the IFU DIAL system is presented in Fig. 3.18, while the main characteristics of the system are presented in Table 3.15.

Table 3.15: The IFU DIAL system characteristics

Lasers	KrF Raman-shifted laser
Wavelengths	277.1 nm, 313.153 nm
Output energy	72 mJ, 72 mJ
Pulse repetition frequency	80 Hz
Beam-expansion	5:1
Output beam divergence	< 0.2 mrad
Telescope diameter	130 mm (*f*/5.5, Newton), 500 mm (*f*/4, Newton)
Receiver field of view	4 mrad, 2.3-4 mrad
Receiver bandwidth	7-12 nm
Digitizer (four channels)	30 MHz, 8bits
Photon counting	50 MHz, 1 μs
Detectors	PMT EMI 9813 & 9893

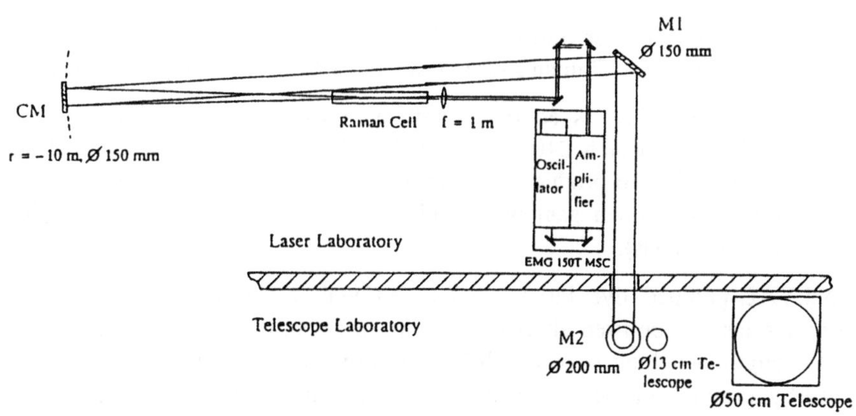

Fig. 3.18: Experimental layout of the IFU DIAL system.

3.6.7 LIT ozone DIAL system

H. Edner

The LIT mobile ozone DIAL system is a flexible system constructed for measurement of various trace gases [4, 56]. The whole system is mounted in a truck with a specially designed cargo compartment of size 6.0 × 2.3 × 2.1 m (Fig. 3.19).

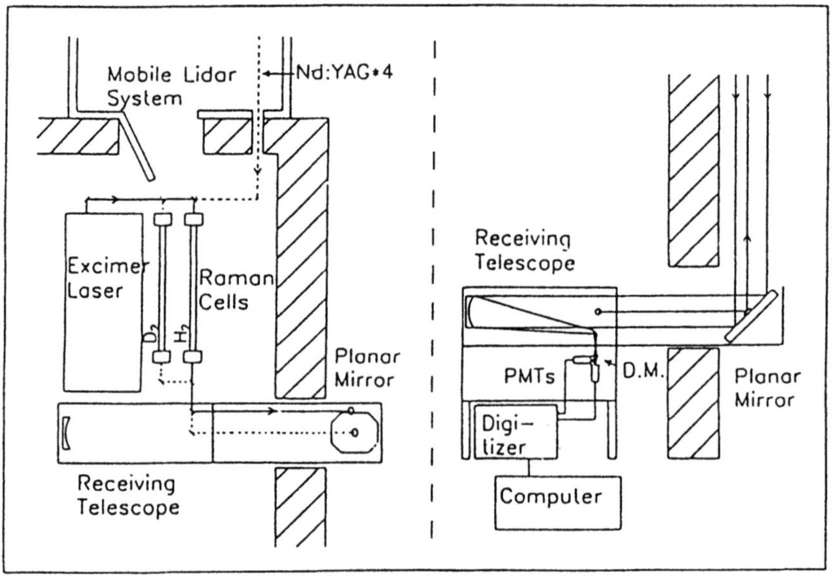

Fig. 3.19: Experimental layout of the LIT DIAL system.

Two different types of laser transmitters have been used to generate the various DIAL wavelengths: a KrF-Raman shifted laser and a Nd:YAG-pumped dye laser.

The KrF laser transmitter is equipped with an unstable resonator. The output energy is 420 mJ, of which 240 mJ are within 0.4 mrad divergence. The laser beam is Raman shifted to produce wavelengths at 248.5 nm, 277.1 and 313.2 nm (using hydrogen as Raman gas) or at 248.5 nm, 268 nm and 292 nm (using deuterium as Raman gas).

The Nd:YAG-pumped dye laser is equipped with a dual-wavelength option, enabling the laser to be fired alternately at two different pre-set wavelengths. The fast wavelength switching is accomplished by directing the beam alternately onto two nearby gratings inside the oscillator cavity. One of these gratings can be individually adjusted to change the wavelength difference, whereas both gratings are moved together when the wavelength pair is scanned. The repetition rate of the Nd:YAG laser is 20 Hz, thus giving 10 Hz repetition frequency at each wavelength. Different wavelength pairs can be used regarding tropospheric ozone measurements. Since the wavelengths are continuously tunable it is possible to minimise the SO_2 interference when measurements in polluted areas are considered (*i.e.*, using wavelengths at 278.7 nm and 286.4 nm). The output energy at these wavelengths is about 10 mJ when using two KD*P doubling crystals in tandem, angle tuned for each wavelength. The dye laser wavelength scale is calibrated against known Ne spectral lines, observed with the optogalvanic signal from a Ne-filled hollow cathode lamp, illuminated by a small part of the laser beam. The absorption in a small gas cell can be used for fine tuning of the wavelengths.

The laser beams emitted from both systems are directed coaxially to a vertically mounted telescope and are transmitted into the atmosphere by a flat mirror in a retractable transmitting/receiving dome on the roof. A quartz window seals off the dome. Stepping motors are used to turn the dome and to tilt the mirror. Thus, it is possible to perform measurememts in 360° direction (horizontally) and 45° direction (vertically). An extra folding mirror was installed on the roof to facilitate vertical measurements. Two video cameras are used to control and supervise the measurement direction.

The lidar signals are received by a Newtonian telescope of 40 cm in diameter and of 2-5 mrad adjustable field of view. After passing a set of intererference filters the signals are detected by PMTs. To prevent detector overload and to reduce the dynamic range of the lidar signals the PMT gains are modulated. The PMT output signals are then amplified and digitized by a 8-bit, 100 MHz transient digitizer. Normally 2000 channels, each 10 ns wide, are recorded, corresponding to lidar ranges of 3 km. After digitization, the data are transfered *via* a GPIB bus to a PC, where they are added into a 32-bit data array. The main caracteristics of the LIT DIAL system are summarized in Table 3.16.

Table 3.16: The LIT DIAL system characteristics.

Lasers	KrF Raman-shifted laser	Nd:YAG-pumped dye laser
Wavelengths	277.1, 313.2 or 268, 292 nm	278.7, 286.4 nm
Output energy	72, 35 mJ or 70, 60 mJ	10 mJ
Pulse repetition frequency	80 Hz	10 Hz
Output beam divergence	< 0.1 mrad	< 0.1 mrad
Telescope diameter	400 mm	400 mm
Receiver field of view	2-5 mrad	2-5 mrad
Digitizer	100 MHz, 2048 channels, 8 bits	100 MHz, 2048 channels, 8 bits
Detector	PMT	PMT

3.6.8 MPI ozone DIAL system

J. Bösenberg and T. Schaberl

The MPI ozone DIAL system is based on a narrow-band KrF Raman-shifted pulsed laser. The KrF laser yields energies higher than 250 mJ with a repetition rate of up to 80 Hz. The divergence of the KrF laser is less than 0.2 mrad. The KrF laser radiation is Raman-shifted to yield laser radiation at 277.1, 313.2, 268 and 291.9 nm. The respective output energies are: 27 mJ, 62 mJ, 33 mJ and 44 mJ. During the experiment either the 248.5, 277.1, 313.2 nm set or the 248.5, 268 and 291.9 nm set has been used [42].

The system has been set-up in a bistatic configuration with 40 cm separation between transmitter and receiver axes in order to supress the close-range returns. The backscattered laser radiation is collected by a 40 cm-diameter telescope directly coupled to a 1.2-mm core diameter quartz fiber fixed in the focal plane. The output of the telescope is transmitted to the spectrograph by the quartz fiber to separate the various wavelengths. The spectrograph is a modified Czerny-Turner design with the fiber-optic input directly on the axis of a parabolic collimating mirror. A ruled plane grating blazed at 250 nm provides a linear dispersion of 0.36 mm/nm at the output. Six quartz fibers with a core diameter of 1.5 mm are mounted in the exit plane to collect the different transmitted wavelengths and divert them to the PMTs.

A second telescope with a primary mirror of 150 mm diameter is installed to receive the close-range returns separately, thus avoiding an overload of the far-range detectors. The output is fiber-coupled, too, and the same spectrometer as for the 40 cm telescope is used. To accomplish this double use of a single spectrometer, just a second fiber input and another row of output fibers had to be installed, demonstrating the high flexibility of this particular design.

The PMTs output signals are amplified and digitized simultaneously by 12-bit, 20 MHz transient recorders. Averages of typically 100 shots (software selectable) are computed by hardware averagers, transferred to the computer and stored on hard disk for off-line evaluation and displayed for almost real-time.

The system is mounted in a standard 20 ft container to make it easily transportable to different experimental sites. All optical elements are set up on two optical tables, which are mounted shock-proof during transport and fixed to the supporting structure of the container during operation. A schematic lay-out of the experimental set-up of the MPI DIAL system is presented in Fig. 3.20 while the main characteristics of the system are presented in Table 3.17.

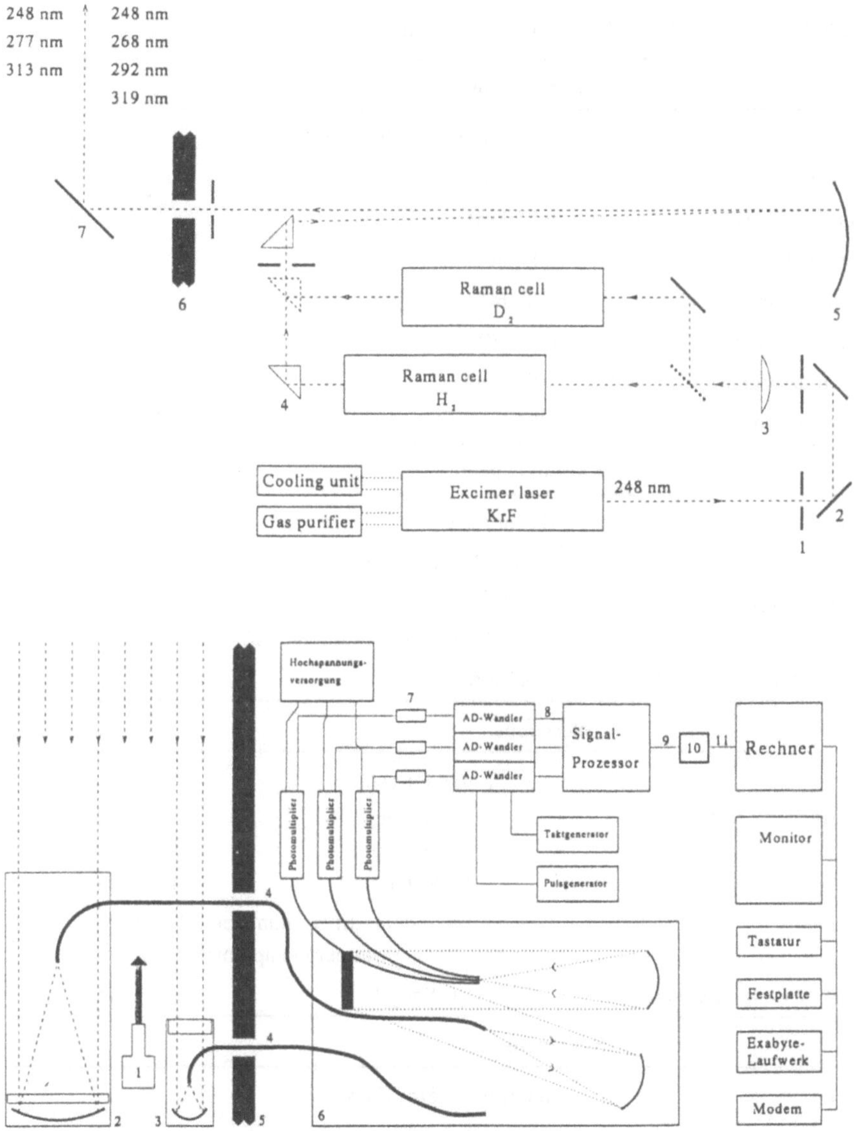

Fig. 3.20: a,b: Experimental layout of the MPI DIAL system.

Table 3.17: The MPI DIAL system characteristics.

Excimer laser		
Type	Lambda Physik, EMG 150 MSC-T	
Wavelength	248 nm	
Output energy	≈ 250 mJ	
Bandwidth (FWHM)	0.8 nm	
Tunable range	0.8 nm	
Repetition rate	max. 80 Hz, typ. 10 Hz	
Beam divergence (full angle)	< 0.2 mrad	
Pulse length (FWHM)	22 ns	
Raman cells		
Gases	H_2 and/or D_2	
Typical filling	40 bar H_2; 40 bar D_2	
Output energy	62/19/19 mJ for 248/277/313 nm (H_2), 62/23/13/4 mJ for 248/268/292/319 nm (D_2)	
Beam divergence (full angle)	< 0.3 mrad	
Telescopes		
	Large telescope	Small telescope
Mirror	parabolic, f = 1000 mm, ∅ 400 mm	parabolic, f = 450 mm, ∅ 150 mm,
Output	quartzfiber, diameter 1.25 mm numerical aperture 0.2	
Field of view (full angle)	1.0 mrad	2.7 mrad
Spectrometer		
Type	modified Czerny-Turner	
Grating	plane, ruled reflection grating, 1200 grooves/mm, blazed at 250 nm, 120 × 140 mm^2	
Linear dispersion	2.8 nm/mm	
Collimating mirror	parabolic, f = 280 mm, ∅ 130 mm	
Imaging mirror	spherical, f = 280 mm, ∅ 150 mm	

(Table 3.17 continued)

Monochromator	
Type	Seya-Namioka
Grating	concave, holographic reflection grating, 1200 grooves/mm, blazed at 250 nm, 42.4 x 42.4 mm^2
Linear dispersion	7 nm/mm
Detectors and data acquisition	
Photomultiplier tubes	EMI 9883 QB, 14 dynodes BeCu,solar blind filter
Preamplifier	Analog Modules 310-3, 10 Hz - 10 MHz
ADC	PENTEK 12 bit, 10 MHz
Computer	SUN workstation
Photoncounter	Stanford Research Systems SR 430

3.6.9 MPI DIAL-Raman system

V. Matthias

The measurement of atmospheric Raman scattering of nitrogen or oxygen allows direct determination of aerosol extinction coefficient when no absorbing species is present [57]. Thus, the MPI Ozone DIAL system is equipped with an additional lidar-Raman channel to get independent information about the aerosol particle extinction. The Raman backscattered light should be in the solar blind region to make daytime measurements possible, but in the UV spectral region the light absorption by ozone is predominant. Therefore, a new evaluation scheme, using two elastically scattered signals and one Raman-scattered signal has been developed and is presently being tested.

As excitation wavelengths for the Raman scattering, 277.1 nm or 268 are used. The vibrational band of nitrogen for the excitation at 277.1 nm is at 296 nm, which is at the border of the solar blind region; the vibrational band of oxygen for the excitation at 268 nm is at 280 mm. These wavelengths are chosen in a way that there are no interferences with elastically scattered light in the spectrometer. As the Raman cross section at 277.1 nm is 6.5×10^{-4} times smaller than the Rayleigh cross section, one needs a stray light rejection of at least 5×10^{-6} to get less than 1 % elastically scattered light in the Raman channel. This is achieved by the combination of a spectrometer and a grating monochromator for the Raman wavelength.

The spectrometer used is a Czerny Turner type, consisting of two mirrors and a ruled plane grating (Fig. 3.21) and having a linear dispersion of 2.8 nm/mm (Table 3.17).

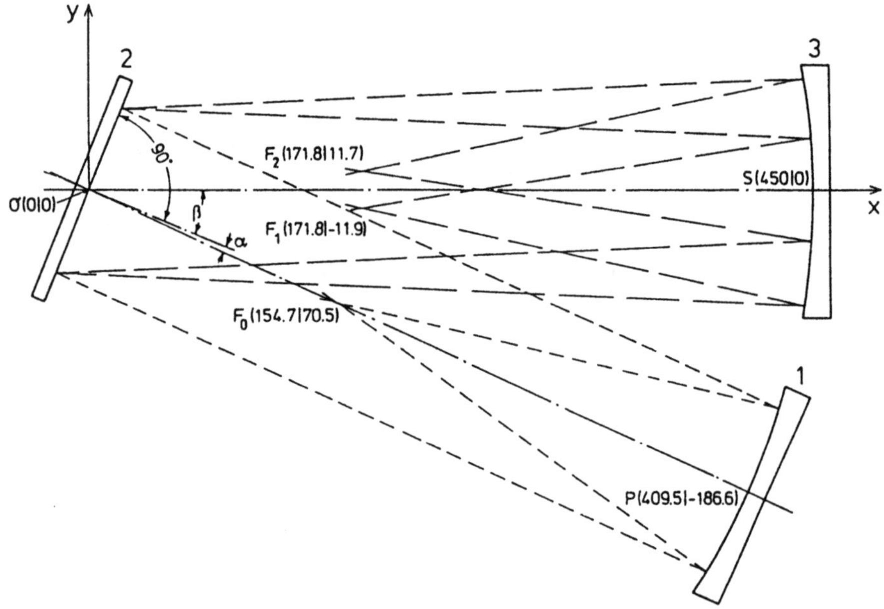

Fig. 3.21: Optical layout of the MPI-spectrometer.

The spectrometer is also used for the separation of the elastically scattered wavelengths. There are two fibers (diameter 1.2 mm) at the entrance (which arrive from the two telescopes) and two rows of fibers (diameter 1.5 mm) at the exit. Using deuterium as the Raman shifter, one obtains the wavelengths 248.5 nm, 268 nm, 280 nm (Raman), 292 nm; using hydrogen as Raman shifter, one obtains the wavelengths 248.5 nm, 277.1 nm, 296 nm (Raman) and 313.2 nm. The output of the Raman channel of the spectrometer is coupled to the input of a monochromator (Seya Namioka type) through a quartz fiber (diameter 1.2 mm). The concave holographic grating has a linear dispersion of 7 nm/mm and can be rotated to select the desired wavelength.

The spectrum of the laser after Raman shifting in hydrogen has been measured with this spectrometer and monochromator combination (Fig. 3.22). The straylight rejection is 3×10^{-8} at 230 nm. It has been found that the laser emits a broad and background light between 260 and 330 nm. It is of the order of 10^{-6} of the intensity of the laser wavelength at 248.5 nm. This result has been proved by the repetition of the experiment with a wavelength-selective filter, which was placed directly behind the laser. The filter had high transmission at 248.5 nm, and only low transmission above 280 nm. Less background light was measured above 280 nm, which showed that the broadband background light is emitted by the laser and not caused by fluorescence on optical elements. However, the intensity of this light is small enough that the measurement of the Raman-scattered light should not be disturbed.

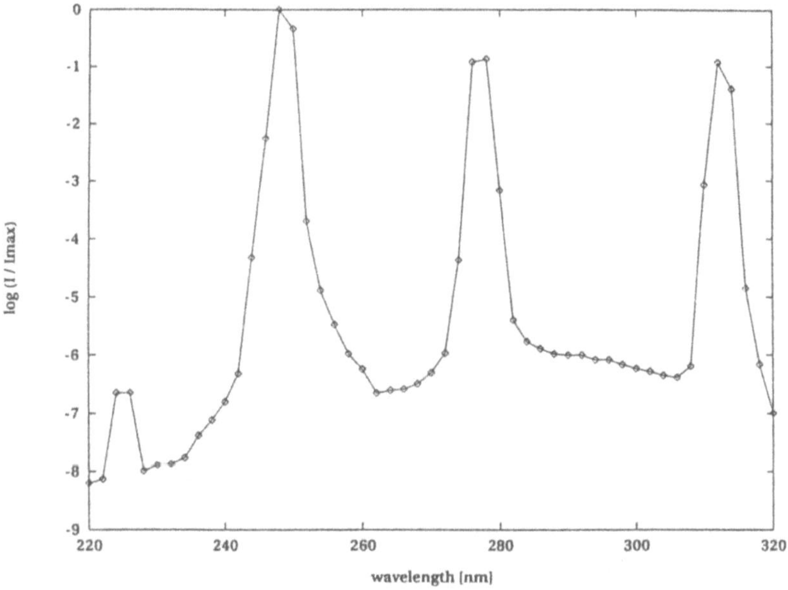

Fig. 3.22: Spectrum for hydrogen-wavelengths using the spectrometer-monochromator combination.

The Raman scattered light is detected by a photomultiplier (THORN EMI 9883QB), which is placed directly behind the exit slit of the monochromator. The photomultiplier is protected by a solar blind filter and it is operated in the photon counting mode.

A photon counter with internal discriminator and no dead time between time bins (Stanford Research Systems SR 430) is used for the storage of the data. The system photomultiplier-photon counter has a dead time which must be taken into account. The measured dead time of 8.4 ns means, that counting rates of more than 1 MHz have to be corrected for dead time. For evaluation, counting rates should less than 10 MHz, since the error in the correction gets too large for these counting rates.

3.6.10 The RIVM ozone DIAL system

A. Apituley

The RIVM ozone DIAL system uses two pulsed Nd:YAG lasers, one pumping a Raman cell filled with D_2, the other pumping a cell filled with H_2. Under this configuration the first laser generates the wavelengths 266 nm and 289 nm, while the second laser generates the wavelengths 266 nm and 299 nm. The beams from both lasers are directed vertically into the atmosphere via two steering mirrors, mounted to the side of the receiving telescope in a double-biaxial configuration. The steering mirrors are positioned symmetrically with respect to the telescope (Fig. 3.23).

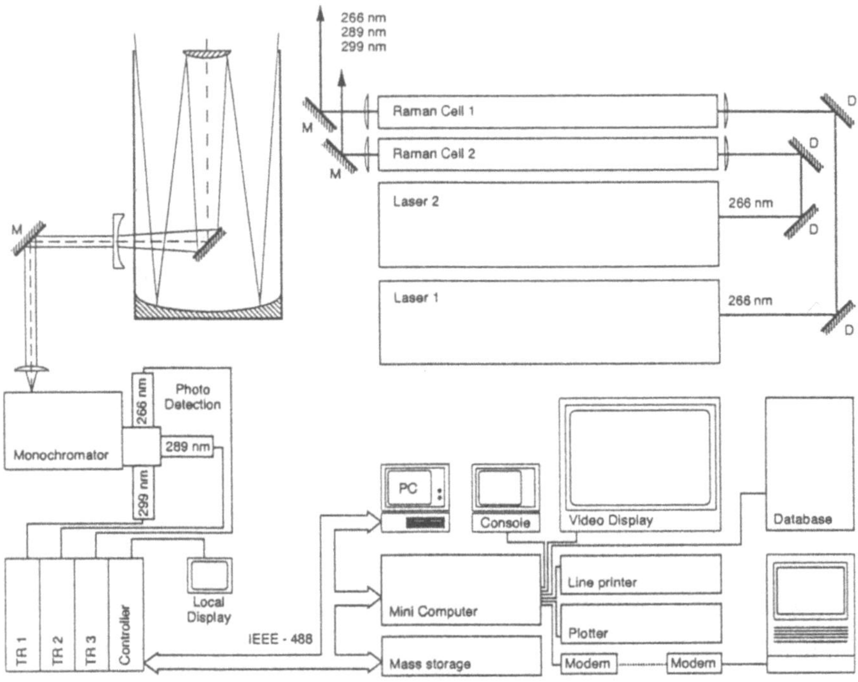

Fig. 3.23: Experimental layout of the RIVM DIAL system.

The backscatterd signals are collected by a Dall-Kirkham telescope (60 cm in diameter, *f*/12, 2 mrad field of view). A grating monochromator is directly coupled to the receiving telescope, through a negative lens ($f = -300$ mm) placed at the exit of the receiving telescope to produce a collimated beam of about 25 mm in diameter. The beam is folded 90 degrees by a UV mirror towards a $f = 125$ mm lens to focus it on the entrance pupil of the monochromator. The aperture ratio of the obtained beam is *f*/5 which equals that of the monochromator. The lenses are AR coated to maximize throughput. The monochromator is used to suppress non-resonant background skylight and to separate the various wavelengths. At the exit focal plane of the monochromator a mask is placed with three pupils where the wavelengths are focussed. The beams are directed towards the detectors using small aluminum, MgF_2 coated mirrors for the 266 nm and 299 nm signals, while the 289 nm signal is left unobstructed. Daylight background could only be detected at 299 nm and barely exceeded the photomultiplier dark current.

Each of the three wavelengths is detected by a separate photomultiplier tube. The entire system is supported by a 1.5×4.5 m^2 optical table. The digitization of the lidar signals is performed a CAMAC based transient recorder system. A mini computer system is used for overall system control and data can be viewed on a graphics display and copied to a plotter. The system parameters are summarized in Table 3.18. The monochromator was later replaced by a system consisting of

Table 3.18: The RIVM DIAL system characteristics.

(a) **Emitter**

Subsystem	Specifications
Laser	Nd:YAG, 10 Hz rep. rate 600-700 mJ/pulse at 1064 nm, 8-10 ns pulse 80-90 mJ/pulse at 266 nm, 4-5 ns pulse
Raman cells	2 m length $f = 1100$ mm lenses uncoated suprasil windows, 20 mm thick
Gas filling	289 nm (S1 D_2): 10 bar D_2, 25 bar Ar 299 nm (S1 H_2): 5 bar H_2, 23 bar Ar
Gas purity	D_2: 99.7 %; H_2: 99.999 %; Ar: 99.999 %
Emitted energy	266 nm, 30 mJ/pulse, < 0.7 mrad divergence 289 nm, 25 mJ/pulse, < 0.6 mrad divergence 299 nm, 40 mJ/pulse, < 0.6 mrad divergence

(b) **Receiver**

Subsystem	Specifications
Telescope	Dall-Kirkham configuration ∅ 60 cm, *f*/12.2 mrad FOV (max) MgF_2 coated
Wavelength separation	Czerny-Turner configuration 300 mm focal length, *f*/4.2 1800 grooves/mm, passband < 6.5 mm at 1.2 mrad FOV 20 % throughput adapted for simultaneous detection of 3 wavelengths
Detectors	Thorn-EMI 9817QA, linearly focussed S20 cathode (15-20 % quantum efficiency) 12 dynodes (BeCu), 3×10^6 gain
Digitizers	LeCroy TR6810, 5 MHz, 12 bits 409.6 mV maximum input sensitivity
Computers	HP1000/A600+ LeCroy 6010 CAMAC controller

dichroicmirrors and interference filters. Fig. 3.24 displays the experimental set-up, as well as the spectral transmission of the entire optical system.

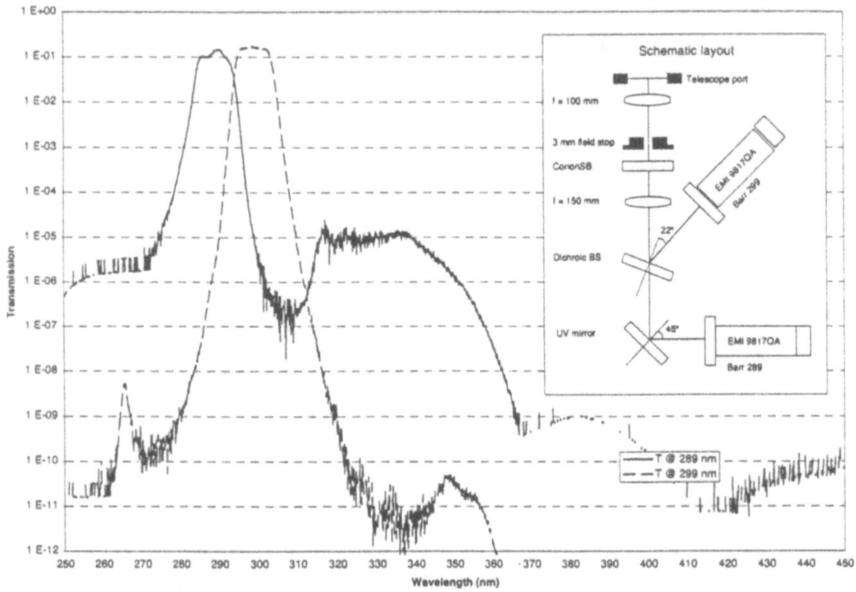

Fig. 3.24: System parameters of the RIVM DIAL system.

To alleviate problems with signal induced noise, a fast mechanical shutter (7200 rpm, 28 cm diameter) was installed after the *f*/12 cone of light from the Dall-Kirkham telescope was converted into a *f*/3 cone. In this configuration the chopper acts as master oscillator and the delay between the chopper and the lasers is adjusted to select the desired measurement range. The lower limit of the ozone profiles starts at about 1 km. The upper limit in analog mode is about 8 - 12 km). Photon counting channels will be added in the very near future to extend the upper range. It is also foreseen that an additional channel will be added for detection of Raman-scattered light at 283 nm, as well as a small (*ca.* 20 cm diameter, *f*/3 telescope) for measurements within the planetary boundary layer.

3.6.11 CNRS airborne ozone DIAL system

G. Ancellet

During the TESLAS Project the Service d'Aeronomie du CNRS developed, tested and operated the first European airborne LIDAR system ALTO (Airborne Lidar for Tropospheric Ozone) for tropospheric ozone monitoring. This new system is used to study the ozone budget on a regional scale and is mounted in the French Atmospheric Research Aircraft (ARAT) Fokker 27.

In the case of the airborne LIDAR system only one Nd:YAG laser in the FHG configuration was used, in contrast to its ground-based version (*cf.* section 3.6.1). The Raman shifting unit comprised only one Raman cell filled with deuterium in order to generate simultaneously three laser wavelengths: 266 nm, 289 nm and 316 nm. The emitted laser wavelengths are then transmitted into the atmosphere, through an afocal telescope and a series of total reflectors (mirrors, prisms). The role of the afocal telescope is to reduce the laser beam divergence. The lidar configuration is of the coaxial type, thus permitting the first ozone data to be collected at ranges around 300 m.

Compared to its ground-based version, the airborne system presented a reduced instrument volume/weight as well as an improved alignment in the emitted laser beams, since all wavelengths are emitted collinearly by the same laser source. On the other hand the main drawback of the airborne lidar version is that using the wavelengths available (266 nm, 289 nm, 316 nm), the DIAL ozone measurements become twice more sensitive to the aerosols interference for the wavelength pair 289-316 nm. Another important drawback is that the background skylight becomes larger at 316 nm than is at 299 nm (by a factor of 5-10). However, during the testing period of the airborne lidar system it was shown that the airborne version provided similar results as the ground-based one, with less sensitivity to alignment problems in the short ranges and a better determination of the aerosol layers as the reference wavelength at 316 nm is completely free of ozone absorption in the troposphere.

The receiving telescope (40 cm diameter) is coupled to a grating spectrometer through a quartz optical fiber of 1.5 mm diameter. The optical fiber transmits the lidar signal from the telescope focal plane to the entrance slit of the grating spectrometer (transmitance greater than 80 % at wavelengths larger than 280 nm). Although, this somewhat increases the sensitivity of the lidar to laser beam alignment (especially at ranges shorter than 700 m) it allows a rugged design of the optical receiving unit. The receiving telescope can be rotated in a vertical plane in order to accomodate both a nadir and a zenith pointing direction. The whole lidar system is fixed to a carbon fiber optical Table. The total weight of the lidar system is 440 kg.

The wavelength separation is performed by a Czerny-Turner spectrometer equipped with a holographic grating (3600 grooves/mm). The spectrometer bandpass is 2 nm, thus the daytime background skylight is efficiently reduced. At the exit slits of the spectrometer six photomultiplier tubes are mounted to detect the lidar signals at the three wavelengths simultaneously. Two different types of photomultiplier tubes are used for the detection of the lidar signals from the near (analog detection mode/current measurement) and the far ranges (photon counting mode). At the output of the analog mode a low frequency amplifier is used (5 MHz, 30 dB) followed by a waveform recorder (12 bits, 10 Mhz). In the photon counting mode a high frequency amplifier is used (500 MHz, 40 dB) followed by a photon counter (300 MHz, 1 μs dwell time)

This improves the design of the amplifier module between the PMT and the processing unit (waveform recorders or pulse counters) as the requirements are completely different for each recording mode. As a consequence, there is also an improvement in the overlap between the ranges where ozone is calculated using the waveform recorder and the ranges where ozone is calculated using the pulse counters output data. The electronic system operates following the principle developed for the ground-based lidar system (except for the optical separation between analog and photon countig detection modes), but the unit was redesigned to reduce its volume by a factor two. A schematics of the lidar implementation in the Fokker 27 aircraft is presented in Fig. 3.25.

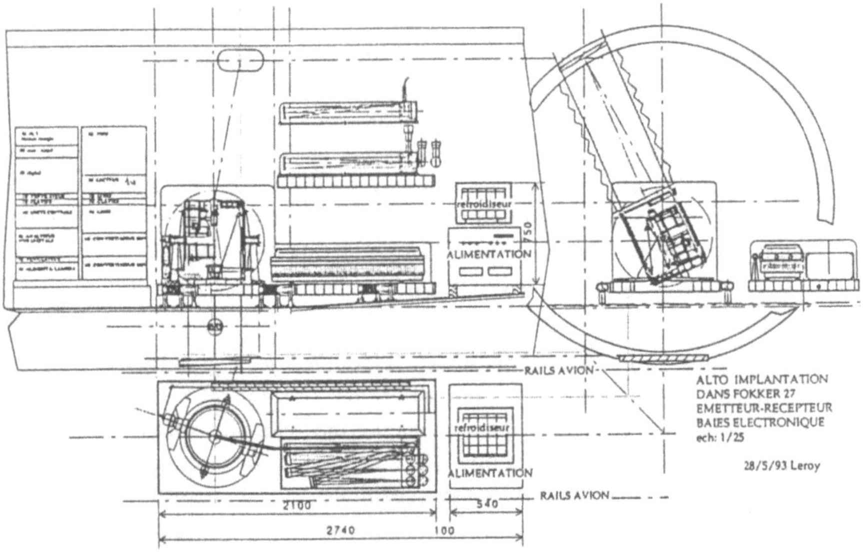

Fig. 3.25: The ALTO DIAL system implanation into the Fokker 27 airplane by the SA/CNRS.

The expected performances for the ALTO lidar system are summarised in Table 3.19, where two different cruising altitudes are considered and thus both tropospheric and stratospheric ozone measurements can be assessed. In the case of lower stratospheric ozone measurements the gas in the Raman cell has to be changed to hydrogen (instead of deuterium), in order to use the wavelength pair 299-341 nm, since the 289 nm wavelentgh is too strongly absorbed by the large stratospheric ozone concentrations.

This system was flown successfully in February 1995 onboard the ARAT research aircraft (see Chapter 6).

Table 3.19: ALTO lidar expected performances.

Tropospheric Ozone DIAL system Wavelengths: 266/289/316 nm)	
Altitude range	0-5 km (nadir)
(flying altitude = 5 km)	5-12 km (zenith)
Altitude range	
(flying altitude = 12 km)	5-12 km (nadir)
Vertical resolution	250 - 1000 m
Horizontal resolution	10 km
Measurements precision	5 - 10 %
Stratospheric Ozone DIAL system (Wavelengths: 299/341 nm)	
Altitude range	
(flying altitude = 12 km)	12-25 km (zenith)
Vertical resolution	500 - 1000 m
Horizontal resolution	10-20 km
Measurements precision	10 %

3.7 References

1. J. Pelon, G. Mégie; Ozone monitoring in the troposphere and the lower stratosphere: evaluation and operation of a ground-based lidar station, *J. Geophys. Res.* **22** (1982) 522-534.
2. E.V. Browell, A.F. Carter, S.T. Shipley, R.J. Allen, C.F. Butler, M.N. Mayo, J.H. Siviter, Jr., W.M. Hall; NASA multipurpose airborne DIAL system and measurements of ozone and aerosol profiles, *Appl. Opt.* **22** (1983) 522-534.
3. G. Mégie, G. Ancellet, J. Pelon; Lidar measurements of ozone vertical profiles, *Appl. Opt.* **24** (1985) 3454-3463.
4. H. Edner, K. Fredriksson, A. Sunesson, S. Svanberg, L. Uneus, W. Wendt; Mobile remote sensing system for atmospheric monitoring, *Appl. Opt.* **26** (1987) 4330-4338.
5. G. Ancellet, G. Mégie, J. Pelon, R. Capitini, D. Renaut; Lidar measurements of SO_2 and O_3 in the boundary layer during the FOS-Berre campaign, *Atmos. Environ.* **21** (1987) 2215-2226.
6. U. Goers, J. Glauer, W. Lahmann, C. Weitkamp, W. Michaelis; Tropospheric Ozone measurements with the mobile Differential Absorption Lidar ARGOS, in: *EUROTRAC Annual Report '91*, Part 7, TESLAS, EUROTRAC ISS, Garmisch-Partenkirchen 1992, pp. 17-23.

7. E. Durieux, B. Calpini, A. Clappier, L. Fiorani, M. Flamm, L. Jaquet, H. van der Bergh; An innovative high repetition rate shot-per-shot acquisition system and its application to a new kind of tropospheric lidar measurements, in: L.D. Laude (ed.), *Excimer Lasers*, Kluwer Acad. Publ., Dordrecht 1994, pp. 245-267.

8. O. Uchino, M. Totunaga, M. Maeda, Y. Miyazoe; Differential Absorption Lidar measurements fo tropospheric ozone with excimer Raman-hybrid lasers, *Opt. Lett.* **8** (1983) 347-349.

9. W. Grant, E. Browell, N. Higdon, S. Ismail; Raman shifting of KrF laser radiation for tropospheric ozone measurements, *Appl. Opt.* **30** (1991) 2628-2633.

10. U. Kempfer, W. Carnuth, R. Lotz, T. Trickl; A wide-range system for tropospheric ozone measurements: development and application, *Rev. Sci. Instrum.* **65** (1994) 3145-3164.

11. T. Schaberl; Messung des Ozonflusses in der unteren Troposphäre mit einem neuen Ozon-DIAL-System und einem Radar-RASS, *Ph.D. Thesis*, Universität Hamburg, 1995.

12. G. Ancellet, A. Papayannis, G. Mégie, J. Pelon; Tropospheric ozone measurements using a Nd:YAG laser and the Raman shifting technique, *J. Ocean. Atmos. Technol.* **6** (1989) 832-839.

13. A. Papayannis, G. Ancellet, J. Pelon, G. Mégie; Multiwavelength lidar for ozone measurements in the troposphere and the lower stratosphere, *Appl. Opt.* **29** (1990) 467-476.

14. D. Haner, S. McDermid; Stimulated Raman shifting of the Nd:YAG fourth harmonic (266 nm) in H_2, HD and D_2, *IEEE J. Quant. Electr. QE*-**26** (1990) 1292-1299.

15. I.S. McDermid, D. Haner, M. Kleiman, D.T. Walsh, M. White; Differential Absorption Lidar system for tropospheric and stratospheric ozone measurements, *Opt. Eng.* **30** (1991) 22-30.

16. L. Stefanutti, Castagnoli, M. Del Guasta, M. Morandi, V.M. Sacco, L. Zuccagnoli, S. Godin, G. Mégie, J. Porteneuve; The Antarctic ozone LIDAR system, *Appl. Phys.* **B55** (1992) 3-12.

17. J.A. Sunesson, A. Apituley, D.P.J. Swart; Differential absorption lidar system for routine monitoring of tropospheric ozone, *Appl. Opt.* **33** (1994) 7045-7058.

18. K. Asai, T. Igarashi; Detection of O_3 by differential absorption using CO_2 laser, *Opt. Quant. Electr.* **7** (1975) 211-215.

19. K. Asai, T. Itabe, T. Igarashi; Range-resolved measurements of atmospheric ozone using a Differential Absorption CO_2 laser radar, *Appl. Phys. Lett.* **35** (1979) 60-62.

20. R. Stewart, J. Bufton; Development of a pulsed 9.5 m lidar for regional scale ozone measurements, *Opt. Eng.* **19** (1980) 503-507.

21. T. Itabe, K. Asai, M. Ishizu, T. Agura, T. Igarashi; Measurements of the urban ozone vertical profiles with an airborne CO_2 DIAL, *Appl. Opt.* **28** (1989) 931-934.

22. R. Barbini, F. Colao, A. Palucci, S. Ribezzo; Tropospherеic ozone measurements from the ENEA lidar/DIAL station, in: *EUROTRAC Annual Report '92*, Part 7, TESLAS, EUROTRAC ISS, Garmisch-Partenkirchen 1993.

23. P.A. Bokhan, V. Burlakov, V. Gerasimov, V. Solomonov; Mechanism of the stimulated emission and energy characteristics of the manganese vapour laser, *Sov. J. Quantum Electron.* **3** (1976) 1239-1243.

24. G. Evtusgenko; Gold-vapour laser. Present state-of-the-art. Design and application, *Proc. SPIE* **2110** (1993) 204-214.

25. A. Papayannis, G. Ancellet, J.L. Conrad, C. Laqui, J. Pelon; Active optoelectronic system for stabilizing pulsed lasers output energy at the 4th harmonic frequency, *Rev. Sci. Instrum.* **64** (1993) 612-617.

26. T. Loree, R. Sze, D. Barker, P. Scott; New lines in the UV: SRS of excimer laser wavelengths, *IEEE J. Quant. Electr. QE*-**15** (1979) 337-342.

27. G. Kunz, F. Swart; Light source for dynamic testing of photo detectors, *TNO-report* FEL 1989-69, TNO, Den Haag 1989.

28. A. Luches, V. Nassissi, M.R. Perrone; Improved conversion efficiency of XeCl radiation to the first Stokes at high pump energy, *Appl. Phys.* **B 47** (1988) 101-105.

29. B. Scott, N. Djeu; Efficient Raman energy extraction in HD, *Appl. Opt.* **29** (1990) 2217-2218.

30. D. Diebel, M. Bristow, R. Zimmermann; Stokes shifted laser lines in KrF-pumped hydrogen: reduction of beam divergence by addition of helium, *Appl. Opt.* **30** (1991) 626-628.

31. W. Fenner, H. Hyatt, J. Kellam, S. Porto; Raman cross-section of some simple gases, *J. Opt. Soc. Amer.* **63** (1973) 73-77.

32. J.H. Newton, G.M. Schindler; Numerical model of multiple-Raman-shifting excimer lasers to the blue-green, *Opt. Lett.* **6** (1981) 125-127.

33. A. DeMartino, R. Frey, F. Pradere; Near-to-infrared tunable Raman laser, *IEEE J. Quant. Electr. QE*-**16** (1980) 1184-1191.

34. J. Ottusch, D.A. Rockwell; Measurement of Raman gain coefficients of hydrogen, deuterium ant methane, *IEEE J. Quant. Electr. QE*-**24** (1988) 2076-2080.

35. B. Bobbs, C. Warner; Absence of second Stokes in a Raman generator with no four-wave mixing, *Opt. Lett.* **11** (1986) 88-89.

36. K. Leung, M. Oron, D. Klimek, R. Holmes, A. Flusberg; Observation of parametric gain suppression in rotational Raman transitions of N_2 and H_2, *Opt. Lett.* **13** (1988) 33-35.

37. H. Komine, E. Steppaerts; Higher Stokes order Raman conversion of XeCl laser in hydrogen, *Opt. Lett.* **7** (1982) 157-159.

38. K. Kuroda, T. Shimura, T. Omatsu; *Tech. Digest, Conf. Lasers Electro-Opt. Soc. Amer.* Washington, DC. 1988, paper WV4.

39. C.D. Marshall, *et al.*; Ultraviolet laser emission properties of Ce:LiSAF, *Technical Digest 'Advanced solid state lasers'*, Optical Society America, Washington, DC.1994, p. 113.

40. R. Measures; *Laser Remote Sensing: Fundamentals and Applications*, Wiley-Interscience Publ., New York 1984.

41. D.P.J. Swart, J. Spakman, H.B. Bergwerff; RIVM's stratospheric ozone lidar for NDSC station Lauder: System description and first results, in: *Abstracts of Papers ILRC* 17, Sendai 1994, pp. 405-408.

42. J. Bösenberg, G. Ancellet, A. Apituley, H. Bergwerff, G.V. Cossart, H. Edner, J. Fiedler, B. Galle, C.N. de Jonge, J. Mellquist, V. Mitev, T. Schaberl, G. Sonnemann, J. Spaakman, D.J.P. Swart, E. Wallinder; Tropospheric ozone lidar intercomparison experiment, Trolix '91: Field phase report, *Report* No. 101, Max-Planck-Institut für Meteorologie, Hamburg 1993.

43. J.A. Sunesson, A. Apituley; RIVM tropospheric ozone lidar: system description and first results, *RIVM Report* 222201006, National Institute of Public Health and Environmental Protection, Bilthoven 1990.

44. R.E.W. Pettifer; Signal induced noise in lidar experiments, *J. Atmos. Terr. Phys.* **37** (1975) 669-673.

45. Y. Iikura, N. Sugimoto, Y. Sasano, H. Shimizu; Improvement on lidar data processing for stratospheric aerosols measurements, *Appl. Opt.* **26** (1987) 5299-5306.

46. H. Sang Lee, G.K. Schwemmer, C.L. Korb, M. Dombrowski, C. Prasad; Gated photomultiplier response characterization for DIAL measurements, *Appl. Opt.* **29** (1990) 3303-3315.

47. I.S. McDermid, S.M. Godin, D.T. Walsh; Lidar measurements of stratospheric ozone and intercomparisons and validation, *Appl. Opt.* **33** (1990) 4914-4923.

48. R. Barbini, A. Ghigo, M. Giorgi, K.N. Iyer, A. Palucci, S. Ribezzo; Injection locked single mode high power low divergence TEA CO_2 laser using SFUR configuration, *Opt. Comm.* **60** (1986) 239-243.

49. R. Barbini, A. Ghigo, A. Palucci, S. Ribezzo; Line tunable TEA CO_2 laser using SFUR configuration, *Opt. Comm.* **68** (1988) 41-46.

50. R. Barbini, F. Colao, F. D'Amato, M. Giorgi, S. Marchetti; A long pulse TE CO_2 laser with a negative branch unstable resonator, *Nuovo Cimento* **12 D** (1990) 967-974.

51. R. Barbini, P. Belli, G. Bitelli, F. D'Amato, E. Galletti, E. Stucchi, A. Ferrario; A long pulse, high energy, narrow bandwidth CO_2 laser for Doppler wind Lidar applications, *Nuovo Cimento* **12 D** (1990) 1633-1639.

52. E. Galletti, E. Stucchi, A. Ferrario, R. Barbini, P. Belli, G. Bitelli, F. D'Amato, M. Giorgi; Development of a CO_2 pulsed laser for spaceborne coherent Doppler Lidar, *SPIE Proc.* **1181** (1989) 113.

53. C. Weitkamp, U.-B. Goers, J. Glauer, W. Lahmann, P. Bisling, S. Koehler, W. Michaelis; Das mobile Lidarsystem ARGOS zum ortsaufgelösten Fernmessung gasfömiger Luftschadstoffe, in: Verein Deutscher Ingenieure (VDI) (ed.), *Umwelt-Messtechnik, VDI Tagungsbericht*, Düsseldorf 1992, pp. 105-114.

54. U.B. Goers; Laserfernmessung von Schwefeldioxid und Ozon in der unteren Troposphäre mit Hilfe der differentiellen Absorption und Streuung unter den Bedingungen des mobilen Einsatzes und der besonderen Berücksichtigung des Einflusses von Grenzschicht-Aerosolen, *Ph.D. Thesis*, University of Hamburg, 1994, GKSS-Forschungszentrum Geesthacht GmbH, GKSS-94/E/52.

55. U. Kempfer; Entwicklung und Anwendung einew differentiellen Absorptions-LIDAR-Systems zur Messung der troposphärischen Ozonkonzentration, *Ph.D. Thesis*, University of Munich 1992.

56. H. Edner, P. Pagnarson, S. Svanberg, E. Wallinder; Vertical Lidar probing of ozone and related trace species, *EUROTRAC Annual Report '90*, Part 7, TESLAS, EUROTRAC ISS, Garmisch-Partenkirchen 1991, pp. 37-40.

57. J. Cooney, J. Orr, C. Tomasetti; Measurements separating the gaseous and aerosol components of laser atmospheric backscatter, *Nature* **224** (1969) 1098-1099.

Chapter 4

Data Processing

Eric Durieux and Luca Fiorani

EPFL-LIDAR Group, Bellerive 34, CH-1007 Lausanne, Switzerland

4.1 Introduction

The methodology of lidar measurements for tropospheric ozone concentrations has been discussed in Chapter 2 of this report, and a presentation of the instrumental setups developed within the frame of the TESLAS subproject has followed. Here the data processing procedures will be discussed.

The treatment of the lidar signals is divided into two successive steps. Following the general outline in Chapter 2, the ozone concentration is retrieved using the DIAL equation (2.2). A first estimate of the ozone concentration is obtained by taking the spatial derivative of the logarithm of the ratio of the signals obtained for the two used wavelengths. Corrections to that value must then be applied to remove from these results the contributions of different effects which are possibly wavelength dependent. These effects are responsible for a differential intensity of the backscattered signals of the experiment at the two wavelengths, which is not related to the presence of ozone.

There are variations between the algorithms proposed by the members of the collaboration. Different paths are followed to increase the signal-to-noise ratio of the data sets and to take the derivative of the DIAL equation for instance. Although the correction scheme employed for the final determination of the ozone concentration is basically the same for the different groups, they are based on free parameters which are usually not fixed at the same values. Such differences between the data analysis procedures have been suspected to lead to important biases on the final results. This suspicion has materialized at the occasion of the 1991 TROLIX intercomparison field campaign. Whereas profiles measured by four instruments probing the same atmospheric column have been shown to be compatible within about 10 % in homogeneous meteorological conditions, important discrepancies were observed for heterogeneous conditions, when gradients in the aerosol density are present. They have been mainly attributed to differences in the data processing. The two situations are respectively illustrated in Fig. 4.1a.

It has been decided to study the contribution of the data processing algorithms to the observed differences on the measurements, by applying four of them on the same data sets. This intercomparison of the data processing pathway was not meant to provide an absolute determination of the effect of each step and choice applied over the analysis procedures on the obtained final results. This would require a systematic study of the sensitivity of the final results on each option of the method followed to extract ozone concentrations from raw lidar returns. It could be done from a simulated data set, for instance, in which lidar returns would be generated from a model of the interaction of light with backscattering matter in the atmosphere, and of the different components of a lidar experimental system.

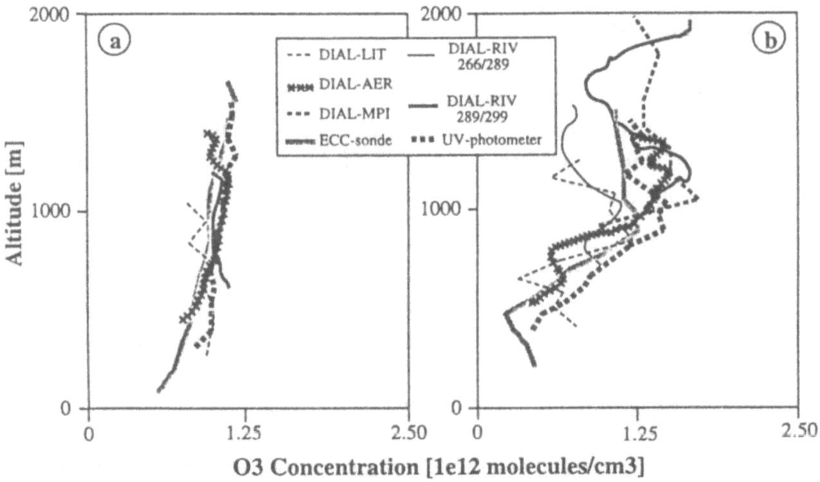

Fig. 4.1: a) Intercomparison of tropospheric ozone profiles for different instruments probing the same atmospheric column in homogeneous meteorological atmospheric conditions. b) Same as a) for heterogeneous atmospheric conditions.

The approach chosen within the collaboration has been different. It has been assumed that the processings applied by the different groups are representative of the options which can reasonably be chosen, once an agreement has been found to follow the general methodology described in Chapter 2 of this report. The intercomparison of the results obtained by the application of these algorithms on a same data sample is thus presented below as a good tool to estimate the quality of their homogeneity. It was found necessary to define such a qualitative appreciation of the variations observed between the different paths in complement to their raw description which would have brought no information on their effects on the final results.

The algorithms will be described for some groups in the following, before the data sets on which they have been applied are presented. The results obtained by the application of one analysis to four data sets will then be detailed to illustrate the different steps of the processing. This will preceed the presentation of the results obtained for the intercomparison and their discussion.

4.2 Data processing algorithms

The choices defining a processing path are based on the appreciation by each group of the steps to follow for an optimization of the data treatment with respect to the instrumentation and to the conditions of measurements. Whereas the details of each algorithm and their justification has been described in the next subsections by their developers themselves, the main steps of the processing are reminded below.

First of all, the signal to noise ratio is commonly increased by an averaging of the raw signals. This averaging is usually operated both over a number of successively recorded signals, and over a number of successive sampling channels of the converters within a bin of measurement. Linearisations and polynomial or exponentional fits are occasionally used to smooth the data. A background level is subtracted from the signals. It depends on the experimental configuration and is made of a sum of various contributions. The pedestal level on the analog to digital converter, the contamination of the recorded signals with photons coming from the sun, and the memory effect of the detecting PMT, coming from an exposure to a too large number of photons and leading to an artificial enhancement of the signal especially in its tail are the three main ones.

Then the total range of the measurement is determined by identifying the distance at which the cones for the light emission and the light collection overlap, and that at which the signal-to-noise ratio becomes too low.

A raw concentration is finally calculated from the spatial derivative of the natural logarithm of the ratio of the two signals obtained for the two wavelengths. Statistical errors can be evaluated at this stage from a propagation to the results of the fluctuations either of the raw collected signals or of the concentrations themselves.

In the next stage, corrections have to be applied to these first results with respect to differential aerosol backscatter and extinction, and possible absorption by other gases, as explained in Chapter 2. For the latter, different molecules are considered. Their absorption cross sections are not always univoquely defined in the literature for the considered wavelengths, and atmospheric models are used to estimate their concentrations. They depend on phenomenological parameters which values have to be chosen following a qualitative appreciation of the atmospheric conditions at the time of the measurements. The estimation of the contribution of the aerosols to the differential signals is usually based on an inversion of the lidar equation following a formulation originated by Klett in 1981, which was later improved and adapted to a broader range of experimental conditions by several authors (see Chapter 2). Some more parameters are given a relevant value for this operation.

4.2.1 The SA/CNRS analysis scheme

Summary

The background is determined by a least square fit of a 4-th order polynomial function. It is applied to the portion of the lidar backscattered returns $P(t)$ for $t > t_0$ where t_0 is defined by $P(t_0 - 20\ \mu s) - P_0 > 0.03$ counts/μs, and P_0 is the average of the signal between 13.5 and 15.0 km. A smoothing of the slope of the Napierian logarithm of the range corrected signals is then applied by using a binomial filter, whereas the slope itself is obtained with a polynomial derivative filter.

Block diagram for the lidar signals 289/299 nm

- Filters:
  - Analogue:
    * Polynomial derivative filter with a parabolic variation of N between the two following values:
      · $N = 10$ pts at 2.4 km
      · $N = 25$ pts at 8.4 km
    * Binomial filter with $N = 10$ pts = constant for all altitudes
    * Rayleigh correction
  - Photon counting:
    * Polynomial derivative filter:
      · $N = 1$ pt at 4.5 km
      · $N = 8$ pts at 13.5 km
    * Binomial filter with $N = 1$ pt
  - Note : N corresponds to half the points needed in the window centered at the point where the calculation is performed: $[P_i - N, P_i + N]$
- Background estimate:
  - Analog:
    * Averaging of the signal between 13.5 and 15 km
  - Photon counting:
    * Averaging of the signal between 13.5 and 15 km

* Leastsquares fit of a 4th-order polynomial to the signal for $t > t_0$ and $P(t_0 - 20\ \mu s) - P_0 > 0.03$ counts/µs

- Next steps:
  - Correction of the delay between the beginning of the record and the laser firing
  - Calculation of $\ln(P \cdot R^2)$
  - Signal slope calculation with a polynomial derivative filter
  - Alignment correction using an overlap function model of the telescope (optional)
  - Rayleigh correction using a density profile which is dependent on the ozone cross section calculated from Molina at the temperature corresponding to the density value
  - Ozone profile calculation

4.2.2 The EPFL analysis scheme

Summary

The data sample is first analysed on a shot per shot basis in order to identify possible ambiguities in the collected records. Signals are then averaged and a standard deviation for each sampling channel is calculated. The background level is estimated by averaging the last channels of sampling. This has been shown to be in agreement with its determination from a pretrigger sample. It is subtracted from the mean signal. Bins are then defined by an averaging over a few channels of sampling. The raw concentration of ozone is obtained by the discretisation of the DIAL equation, and corrections are applied to subtract the contribution of particles and other molecules to the differential absorption of the light.

Block diagram

Sample check and temporal averaging. The lidar returns are recorded and stored after each laser shot. The data sample is examined to check the absence of corrupted events or of temporal drifts in the returned signals. The successive signals are averaged after the distributions of the number of triggers over the reachable ADC values for a given sampling channel have been reconstructed and the corresponding standard deviations calculated.

Background subtraction. After calculating a constant background level

$$BG = \frac{1}{100} \sum_{i=1901}^{2000} P_i^{raw}$$

where i is an index over the sampling channels, it is subtracted from the temporally averaged mean lidar signal: $P_i = P_i^{\mathrm{raw}} - BG$.

BG is calculated on the signal-less zone of the ADC sampling (2048 channels). In the EPFL sample file the channel width is 40 ns, corresponding to a 6 m resolution.

Distance averaging. The distance averaging is operated over N_{av} sampling channels. Taking the j-th ensemble of N_{av} channels, its altitude and average signal are respectively defined as R_j and P_j with:

$$R_j = \tfrac{1}{2}(r_1 + r_{N_{\mathrm{av}}}) \text{ and } P_j = \tfrac{1}{N_{\mathrm{av}}} \textstyle\sum_1^{N_{\mathrm{av}}} P_i$$

For the sample file below, $N_{\mathrm{av}} = 20$, corresponding to a resolution of 120 m.

Range determination: choice of the lower and upper limits R_A and R_B. R_A is the first, and R_B is the last distance considered for the O_3 profile. The choice is made after analysis of the data plots: R_A is chosen after the overlap of the emission and detection cones (600 m in the EPFL data sample), and R_B before a bad signal to noise ratio (1800 m in the EPFL data sample).

Calculation of uncorrected O_3 concentration. The uncorrected ozone concentration is calculated as follows [2]

$$C_0\left(\frac{R_{j+1}+R_j}{2}\right) = \frac{1}{2[\sigma_{O3}(\lambda_{\mathrm{on}}) - \sigma_{O3}(\lambda_{\mathrm{off}})]\cdot[R_{j+1}-R_j]} \cdot\left(\ln\left[\frac{P_{j+1}(\lambda_{\mathrm{off}})}{P_{j+1}(\lambda_{\mathrm{on}})}\right] - \ln\left[\frac{P_j(\lambda_{\mathrm{off}})}{P_j(\lambda_{\mathrm{on}})}\right]\right) \quad (4.1)$$

Calculation of corrections. An atmospheric density profile following the USSA '76 [3] model is assumed:

$$N(R) = N_0\left(\frac{288.15 - 0.006\,500\cdot(R+H)}{288.15}\right)^{4.255\,877} \quad -1524 \le R+H < 11000 \quad (4.2)$$

where, $N_0 = 2.5470 \times 10^{19}$ molecules/cm^3, and H is the altitude of the measurement station (R and H are in m).

Molecular extinction. The molecular extinction is [4]

$$\alpha_{\mathrm{mol}}(\lambda) = N\frac{8\pi^3}{3}\left\{\frac{n_0^{\,2}(\lambda)-1}{N_0}\right\}^2 \frac{1}{\lambda^4}\left(\frac{6+3\gamma}{6-7\gamma}\right) \quad (4.3)$$

where $\gamma = 0.0279$ is the depolarization factor of air (the depolarization factor is related to the anisotropy in the polarizability of the molecules), and n_0 is the refractive index of air for $R = 0$ [5], given by:

$$(n_0 - 1)\cdot 10^8 = A + \frac{B}{D-\lambda^{-2}} + \frac{C}{E-\lambda^{-2}} \tag{4.4}$$

where: A = 8342.13; B = 2 406 030; D = 130; C = 15 997; E = 38.9; and λ is in μm.

O_2 *extinction*. The oxygen extinction is given by [6]

$$\alpha_{O2}(\lambda) = N_{O2}\,\sigma_{O2}(\lambda) \tag{4.5}$$

where $N_{O2} = 0.209\,476\, N(R)$

NO_2 *extinction*. The nitogen dioxide extinction is given by [7]

$$\alpha_{NO2}(\lambda) = N_{NO2}\,\sigma_{NO2}(\lambda) \tag{4.6}$$

with the assumption $N_{NO2} = 10^{-8}\, N(R)$

SO_2 *extinction*. The oxygen extinction is given by [8-10]

$$\alpha_{O2}(\lambda) = N_{O2}\,\sigma_{O2}(\lambda) \tag{4.7}$$

with the assumption $N_{SO2} = 10^{-8}\, N(R)$

***Atmospheric backscattering and extinction*.** The aerosols backscattering and extinction coefficients are evaluated for λ_{off} according to the Klett algorithm [1] in a formulation due to Browell, Ismail and Shipley [11]. The backscattering coefficient is decomposed in the aerosols and molecular contributions as $\beta = \beta_{aer} + \beta_{mol}$. With $\alpha_{mol} = (8\pi/3)\,\beta_{mol}$ [12], assuming $\alpha_{aer} = (1/P_\pi)\,\beta_{aer}$ where $P_\pi = 0.028\ \text{sr}^{-1}$, and defining

$$\mu(R) = \ln\left[p(R)\,R^2\right] - 2\int_R^{R_M}\left[\sigma_{O3}C(R') + \sum\alpha_i(R') + \left(1-\tfrac{3}{8\pi P_\pi}\right)\alpha_{mol}(R')\right]\mathrm{d}R'$$

where $i = O_2$, NO_2, and SO_2 yields to the following differential equation:

$$\frac{\mathrm{d}\mu(R)}{\mathrm{d}R} = \frac{1}{\beta(R)}\frac{\mathrm{d}\beta(R)}{\mathrm{d}R} - \frac{2}{P_\pi}\beta(R) \tag{4.8}$$

which has the solution

$$\beta(R) = \frac{\exp(\mu(R)-\mu_M)}{\frac{1}{\beta_M} + \frac{2}{P_\pi}\int_R^{R_M}\exp(\mu(R')-\mu_M)\,\mathrm{d}R'} \tag{4.9}$$

where $\mu_M = \mu(R_M)$ and $\beta_M = \beta(R_M)$.

To calculate this solution R_M must be chosen and β_M evaluated. Neglecting the logarithmic term in (1) it can be assumed that:

$$\beta_M = \frac{P_\pi}{2} \frac{\mu(R_M - \delta R) - \mu(R_M + \delta R)}{2\delta R} \tag{4.10}$$

From (2) and (3) R_M must be taken after R_B and in a zone where $\ln[P(\lambda_{off})R^2]$ is linear. So the proposed choice of $R_M = 4020$ in the EPFL data sample.

The wavelength dependence of the backscattering coefficient is assumed as:

$$\beta_{aer}(R, \lambda_{on}) = \beta_{aer}(R, \lambda_{off}) \left(\frac{\lambda_{on}}{\lambda_{off}} \right)^{k_\beta}$$

where $k_\beta = -1$ for the EPFL data sample.

***Calculation of corrected* O_3 *concentration*.** The corrected ozone concentration is calculated as:

$$C_1 = C_0 - \frac{1}{[\sigma_{O3}(\lambda_{on}) - \sigma_{O3}(\lambda_{off})]} \left(\frac{1}{2} \frac{d}{dR} \ln \left[\frac{\beta(R, \lambda_{off})}{\beta(R, \lambda_{on})} \right] + \sum_i \Delta\alpha_i \right)$$

where $\Delta\alpha_i = \alpha_i(\lambda_{on}) - \alpha_i(\lambda_{off})$; i = mol, aer, O_2, NO_2, SO_2.

***Iteration*.** The O_3 concentration just calculated is now used in the definition of μ obtaining a new aerosols correction and a new concentration. The iteration is stopped when:

$$C_i - C_{i-1} < 0.01\, C_0 \quad i = 1, 2, 3, \ldots$$

No such a variation in C_i has ever been observed even after the first iteration. To observe an effect of the iterations, the weight of the backscattering correction has had to be artificially enlarged assuming $P_\pi = 0.28\ \text{sr}^{-1}$.

***Parameter dependence*.** The values of the next parameters have been chosen up to this point as: $P_\pi = 0.028\ \text{sr}^{-1}$, $k_\beta = -1$, and $R_M = 4020$ m.

They have been varied in the following ranges: $0.010\ \text{sr}^{-1} < P_\pi < 0.050\ \text{sr}^{-1}$; $0 > k_\beta > -2$; and $2820 < R_M < 5220$ m.

The observed differences in the corrected ozone concentration have always remained lower than 10 %.

***Error determination*.** The statistical errors on the ozone concentrations are obtained by a propagation of the standard deviations from the distributions of events reconstructed for each sampling channel. The systematic errors on the measurements must be estimated from the sensitivity of the results on the parameters of the analysis for the specific experimental conditions of the studied profile.

4.2.3 The MPI analysis scheme

Summary

The raw signals are corrected for a DC-offset as determined in the time intervall (typically 5 μs) before the laser is fired. A second correction is applied for some small electromagnetic interference caused by the laser power supply; this is determined in separate records with the laser operating but the beam blocked. The amplitude of this interference signal is less than about one digitizer count and can only be determined by averaging over many shots, typically 2000 (≈ 3 min) per data file.

In the data acquisition, each shot is recorded by the ADC and transferred to a signal processor, where it is checked for overload in each range bin. Valid signals are averaged and their variance calculated, again for each range bin. This is done for a selectable number of shots, typically 100, and the results are stored on disk for later detailed analysis. This process runs continuously until it is stopped on purpose.

The trigger point of the laser is determined from the signals, it is associated with a small spike. It was found to be constant over several hours of operation. The algorithm used for calculating ozone takes into account, that signals may not be uniformly distributed over the range cell used for taking the derivative. This is performed by calculating the center of gravity of the signal in each range cell, and using that distance rather than the middle of the cell for calculating ΔR.

The aerosol backscatter profile is calculated using a backward/forward integration scheme as described by Sasano [13]. Forward integration is only used in special cases, and for a short distance only, when a sufficiently distant calibration range cannot be found for an extended time series of measurements. Else of course the stable backward integration is preferred.

If ozone absorption is not negligible at the offline wavelength, *e.g.* for $\lambda_{off} = 292$ nm, first raw ozone is calculated without aerosol correction. The offline signal used for the backscatter retrieval is then corrected for ozone absorption, and the backscatter correction calculated. Further corrections applied are for extinction by known other gases (in particular O_2 at 268 nm), and by differential extinction due molecular and aerosol scattering. The whole process can be iterated, but this is generally not necessary more than once, the convergence is very fast.

Block diagram

Input signals:	2 channels
Signal corrections:	Select pretrigger sample for offset determination Average pretrigger samples for offset determination Subtract offset from signal Determine electromagnetic interference signal (EMI): normal data acquisition but laser beam blocked Subtract EMI
Signal normalisation:	Determine time of laser transmission t_0 Set R_0 Calculate $P \cdot R^2$ (corrected and normalised data)
Range restriction:	Determine point of full overlap R_A Detemine lower limit for signal R_B
average in time:	Sum n records
average in range:	n sample gliding average
Calculate β_{aer}:	Calculate air density profile Select calibration height R_M Select β_{aer}/β_{mol} at R_M Select $S = \alpha_{aer}/\beta_{aer}$ Use backward/forward calibration scheme (*e.g.* Sasano [13])
Calculate raw ozone:	Set range cells Average signal in range cells Calculate center of gravity for signals in range cells R_i Calculate ozone concentration from the discrete DIAL equation
Apply corrections:	Subtract α_{mol} Subtract α_{O2} Subtract α_{aer} Correct for $\Delta\beta_{aer}$

4.2.4 The RIVM analysis scheme

Summary

Data acquisition from the lidar system is done in bursts. The number of laser shots averaged locally in the transient recorders in one burst is set by the operator. Bursts contain typically 300 to 600 shots (*i.e.* half a minute or one minute averaging at 10 Hz). These individual bursts are stored. The data processing software reads the bursts from file. The operator can decide to average several bursts from one or more files to increase the signal to noise ratio while rejecting bursts containing clouds.

Because a fast mechanical shutter is used to block backscattered light from close ranges, the pretrigger signal can not be used as the background is also blocked. Therefore, the background level of each burst is determined in the tail of the signal. The standard (default) samples where this is done are at 27 km altitude, averaged over a distance of 900 m. The data acquisition is triggered by optical triggering units attatched to the lasers. No electromagnetic interference is present on the baseline of the signals, which is regularly checked by recording signals with a closed receiver (while the rest of the system is running).

Block diagram

1. Input signals

   2 channels (289 and 299 nm)
2. Background subtraction
   (a) Average 30 to 50 datasamples in the tail of the signal to determine signal background.
   (b) Subtract baseline
3. Signal summation
   (a) Determine cloud base height
   (b) Select data records free of low clouds (*N* records)
   (c) Sum *N* datarecords
   (d) Calculate average
4. Raw ozone calculation
   (a) Set altitude resolution
   (b) Calculate DIAL curve: $\ln(P_{\lambda_{on}}/P_{\lambda_{off}})$
   (c) Calculate ozone profile by sweeping a linear least squares fit at the given range resolution over the DIAL curve
5. Corrections
   (a) Calcuate number density profile of air
      i. Set pressure at ground level (optional)
      ii. Set temperature at ground level (optional)
   (b) Calculate Rayleigh extinction correction, α_{mol}
   (c) Calculate oxygen exctinction correction, α_{O2}
   (d) Calculate aerosol interference (optional)
      i. Set calibration height R_M
      ii. Set β_{aer}/β_{mol} at R_M
      iii. Set $S = \alpha_{aer}/\beta_{aer}$
      iv. Set wavelength dependence k_β
      v. Calculate $S_K = \beta_{mol}/(\beta_{mol} + \beta_{aer})$ following Browell [11].
      vi. Calculate inhomogeneous aerosol bacscatter correction by sweeping linear least squares fit at the given range resolution over S_K
      vii. Calculate aerosol extiction correction α_{aer}
   (e) Apply corrections

The aerosol correction is optional. The scheme that was implemented follows the procedure proposed by Browell *et al.* [11], using the lidar signal at 299 nm. No iteration is applied.

4.2.5 The IFU analysis scheme

Data structure

A total of four analogue channels (two wavelengths (277 and 313 nm), two receiving telescopes) are used, consisting of 4000 points each. 512 of these points are assigned to the pretrigger region. The rest covers a range of about 17.8 km. In each channel, detector range gating is applied. As soon as the signal reaches a low level the PMT high voltage is enhanced to improve the signal-to-noise ratio for the higher altitudes. Typical range-gate switching points are 0.3 to 0.4 km for the near-range telescope and 3 km for the far-range telescope. In addition to the four analogue channels a single-photon channel is used at the large telescope for the 'on' wavelength (277 nm).

General description of the computer program

The PC program (Turbo Pascal) consists of three blocks which can be run alternatively:

a) Graphical representation of the original backscatter profiles with different options

b) Logarithmic representation of the data; this routine contains all algorithms for the background subtraction, slope adjustment at range-gate switching points by refinement of signal offsets and signal-induced nonlinearity in channel 4 (277 nm, large telescope)

c) Calculation of ozone density profiles; this routine contains again all signal-adjustment procedures listed under b) (which are, however, mostly carried out in routine b) because of the logarithmic plots available), the computation of Rayleigh and total backscatter coefficients, and the ozone density calculation. An iterative solution is chosen. A graphical representation of either the ozone density distribution or the distribution of the backscatter coefficients can be selected.

The global variables are available to all blocks. Thus, adjustments made in one block remain valid if one switches into another one.

Calculation of the different quantities

a) The ozone density is calculated from the signals by numeric derivative formation. The signals may be Rayleigh and/or aerosol corrected if requested. A third-order polynomial fit is applied to the logarithm of the 'off'/'on' signal ratio in an interval around the actual ΔR window. The interval size is chosen to be three times ΔR. From the polynomial the slopes in the ΔR window are computed which finally yield the ozone number density. In order to avoid any bias caused by the asymmetry of the noise in a logarithmic representation we prefer near the end of the data field to fit the polynomial to the signal ratio prior to forming the logarithm. The data points within the rising edge of the range-gate are excluded from the evaluation, but the least-squares fits are continued across the switching zones in order to minimize discontinuities.

b) Rayleigh backscatter coefficients may be obtained either from a simple exponential model for the atmospheric pressure up to 10 km, or from LOWTRAN midlatitude models or radiosonde data. The value for the cross section is based on NBS expressions for the refractive index of humid air. A depolarization correction factor of 1.0608 is applied.

c) The calculation of the total backscatter coefficients starts with a Klett inversion for 313 nm which includes a treatment of the ozone and Rayleigh contributions in the signal. If not specified differently, the reference backscatter coefficient is automatically set to the Rayleigh value at the upper end of the evaluation range. The 277 nm backscatter coefficients are derived from those for 313 nm by assumung a λ^{k_β} wavelength dependence of the aerosol scattering coefficients.

Sequence of the evaluation of ozone density profiles

1. Initialize the global variables
2. Check and optimize background signal above range-gate switching points
3. Check and optimize background signal below range-gate switching points; criterion: d $\ln(R^2 \cdot \text{signal})/\mathrm{d}R$ has to be equal on both sides of the switching points (minor deviations exist in the early data files); a least-squares algorithm for the slope optimization is provided, but has to be handled with care.
4. Remove signal-induced contribution in channel 4 (presumably caused at the beam splitter in front of the PMTs of the large telescope); this can be done manually or by a least-squares fit. Compare corrected analogue and photon-counting traces and modify correction, if necessary. Check and correct for equal slopes of $\ln(P(R) \cdot R^2)$ on both sides of the range-gate switching points after the correction.
5. Define switching point between small and large telescope (about 1.5 km).
6. Define switching point between analogue and photon counting data (about 5.5 km).

7. Select vertical resolution at point of highest signal; ΔR is dynamically changed according to the local signal-to-noise ratio which gives an approximately constant statistical error in the entire operating range.

8. Select aerosol parameters.

9. Calculate Rayleigh backscatter coefficients; atmospheric models as well as experimental data (pressure, temperature, humidity) can be included.

10. Start iterative calculation of the ozone density and the backscatter coefficients; in the first round a preliminary ozone distribution is calculated, which is just Rayleigh corrected; this is followed by the calculation of the aerosol backscatter coefficients. Afterwards, a second-round ozone distribution is derived which now is also aerosol corrected. A third round is usually not necessary due to the use of an almost true 'off' wavelength (313 nm).

List of the most important specifications and parameters

Operating wavelengths:	277 and 313 nm
Range:	Maximum: 0.1 to 12 km above ground; typical: 0.2 to 8 km (summer), 0.2 to 12 km (winter)
Vertical resolution:	Dynamically varied from 0.03 km to 0.8 km
Typical ozone density error:	7.5×10^{16} m^{-3}
Altitude of lidar station:	740 m a.s.l.
Backscatter-to-extinction ratio for aerosols:	0.030 sr^{-1}
Wavelength dependence of aerosol backscattering (λ^{k_β}):	$k_\beta = 0.7$ (winter), $k_\beta = 1.5$ (summer)
Klett reference value:	Chosen at upper end of evaluation range (typ. 8 to 12 km), outside cirrus clouds; value set to that Rayleigh backscatter coefficient at the same wavelength (313 nm).

4.2.6 The LIT analysis scheme

Raw signals. The signals are averaged during typically 1-2 minutes, with the laser firing alternately on the two wavelengths, 278.7 / 286.4 nm chosen to minimize interference with SO_2. The background light is measured during the measurement cycle with the outgoing laser beam blocked each tenth shot. The background is subtracted before storing the averaged signals. This procedure is necessary since the gain of the PMT is modulated.

Smoothing. The raw signals are normally smoothed with a gaussian function of a few channels width (1 channel = 10 ns) before division.

Calculation of the concentration. The concentration is evaluated by a linear or exponential fit to different parts of the ratio curve. It can be done in some different ways. A quick fit consists of measuring the slope in a small interval by taking the ratio of the mean of a few channels in the beginning and end of this interval. The evaluation interval is swept across all distances. A more accurate and time consuming approach is to make an exponential fit with a Gaussian function of all the channels within the interval. Also here the evaluation interval is swept over all the channels. The last approach also gives the standard deviation of the evaluated concentration as a function of the distance.

Corrections. The concentrations are finally corrected for differential Rayleigh attenuation. Any corrections for aerosols have not yet been implemented.

Scanning. Sometimes the concentration profiles from a few subsequent measurements are averaged. The mobile Lidar system also has the possibility to change the measurement direction and present the concentration in a 2D chart. A vertical profile can be inferred from a vertical scan with a projection procedure.

4.2.7 The IAO analysis scheme

The algorithm for lidar return signal processing based on smoothing cubic splines are detailed described in the papers [14-16]. Here we present the brief description of the algorithm only.

The determination of the vertical profile of ozone content from lidar returns obtained with a bifrequency lidar is reduced to the problem of differentiation of the function $f(z)$:

$$f(z) = \frac{1}{2}\ln\frac{U_{\text{off}}(z)}{U_{\text{on}}(z)} + \psi(z) \tag{4.11}$$

$$\psi(z) = \frac{1}{2}\ln\frac{\beta_{\text{on}}(z)}{\beta_{\text{of}}(z)} - \left[\tau_{\text{on}}(z) - \tau_{\text{off}}(z)\right]$$

where $U_{\text{off}}(z)$ and $U_{\text{on}}(z)$ are the lidar returns at the distance z at the wavelengths λ_{off} and λ_{on}, $\beta_{\text{off}}(z)$ and $\beta_{\text{on}}(z)$ are the backscattering coefficients at the wavelengths λ_{off} and λ_{on}, and $\tau_{\text{on}}(z)$ and $\tau_{\text{off}}(z)$ are the optical thickness of molecular scattering and aerosol extinction.

It is assumed in eq. (4.1) that $U_{\text{off}}(z)$ and $U_{\text{on}}(z)$ are free of the signals of the atmospheric background radiation. The function $\psi(z)$ is assigned starting from model representations or is determined from an independent experiment.

The ozone concentration is defined as

$$\rho(z) = \frac{1}{2\Delta K(z)} \phi(z)$$

where $\phi(z)$ is the regularized analog of the derivative $f'(z)$ of the function $f(z)$ and ΔK is the differential cross section of the O_3 absorption.

Let the function $f(z)$ be present by its measured values $f(z_i)= f_i$ at the nodes of the grid z_i: $a = z_1 < z_2 < \ldots < z_{n-1} < z_n = b$. A smoothing cubic spline $S_{n\Delta}$ is the solution of the variational problem

$$F_\Delta = \inf\left\{\alpha \int_a^b [S''(z)]^2 dz + \sum_{i=1}^{n} p_i [f_i S_i]^2\right\} \tag{4.12}$$

$$S(z) \in C^2[a,b]$$

and is represented on each segment $h_i = z_{i+i} - z_i$ by the third-degree polynomial

$$S_{\text{na}} = a_i + b_i\,(z - z_i) + c_i\,(z - z_i)^2 + d_i\,(z - z_i)^3$$

with the continuous derivatives $S'_{\text{na}}(z)$ and $S''_{\text{na}}(z)$.

The spline coefficients a, b, c, and d are expressed in terms of the elements of the vector of the second-order derivatives $M_{i+1} = \{m\}_i$ which are the solution of the system of linear equations

$$(\alpha\, \boldsymbol{H}\, \boldsymbol{P}^{-1} \boldsymbol{H}^{\mathrm{T}} + \boldsymbol{A})\, \boldsymbol{m} = \boldsymbol{H} \boldsymbol{f}$$

with the boundary conditions $M_1 = M_n = 0$.

Here α is the scalar ($\alpha > 0$) which has the meaning of the smoothing parameter. The components of the matrices $\boldsymbol{H}$ and $\boldsymbol{A}$ are expressed in terms of the elements of the sequence $h_1 = z_{i+1} - z_i$:

The accuracy of construction of the spline $S_{\text{na}}(z)$ depends strongly on the value of the smoothing parameter. We use a criterion for selecting α in the case of the unknown matrix of the measurement noise [14].

4.3 Intercomparison

Four of the algorithms described above have been used to process four data sets recorded with different instruments in various experimental conditions. These data samples have been selected for their representativity of either standard or difficult measurement conditions caused for instance by the presence of aerosol inhomogeneities or optical distortions. Their detailed analysis with the algorithm of the EPFL group is presented to illustrate the main steps of the data processing.

4.3.1 Data samples

The main features of the four data sets are briefly presented in Table 4.1.

Table 4.1: Main features of the data sets used for the intercomparison.

Institution date	T_{start}(UT)	T_{stop}(UT)	range / m	sampling (points)	mode	λ_{on} / nm	λ_{off} / nm
CNRS 20.02.92	18:29:41	19:26:41	4376-15626	76	counting	289	299
EPFL 03.08.93	20:55:00	20:58:20	6-6000	1000	analog	272.5	299
MPI 20.06.91	08:25:00	08:39:34	5-3605	481	analog	277	313
RIVM 03.08.90	11:17:38	11:19:18	5-14975	500	analog	289	299

Data have been collected with the telescopes pointing vertically. The full data sets are presented in tabular form in the appendix A to provide the interested user with some sets of real data which can be used to check any newly developed algorithm against the results obtained within the collaboration. For the presentation, the original data have been processed as follows:

- the Lidar signals have been averaged over the time of the measurements defined as $t_{stop} - t_{start}$, and over the distance intervals recommended for their analysis as precised in Table 4.2.
- the signals have been background subtracted.

Figs. 4.2 to 4.5 present graphically the natural logarithm of the range corrected signals $\ln(P(R) \cdot R^2)$ for all four data sets.

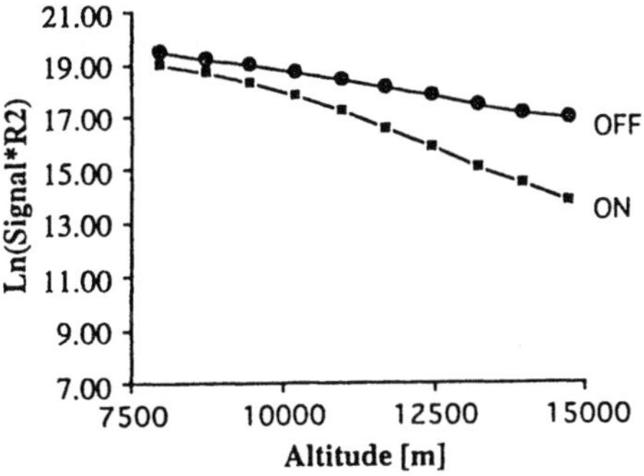

Fig. 4.2: Natural logarithm of the range corrected lidar signals for the CNRS system.

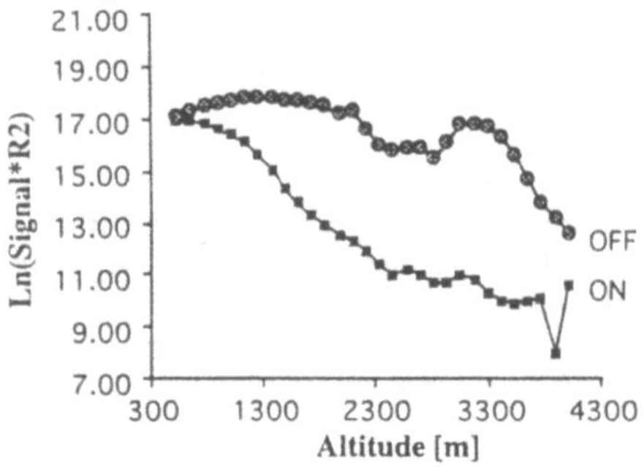

Fig. 4.3: Natural logarithm of the range corrected lidar signals for the EPFL system.

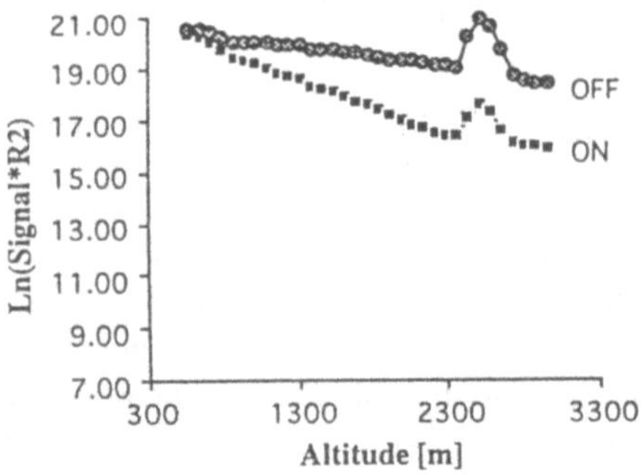

Fig. 4.4: Natural logarithm of the range corrected lidar signals for the MPI system.

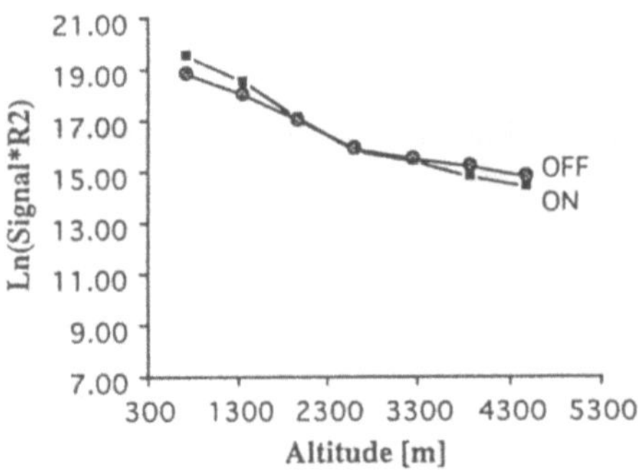

Fig. 4.5: Natural logarithm of the range corrected lidar signals for the RIVM system.

4.3.2 Data analysis

The results of the analysis of the four data sets with the EPFL algorithm are presented below. Values quoted in the Table 4.2 for the parameters of the data processing have been used to obtain them.

Table 4.2: Recommended values for the analysis parameters.

	range / m	interval / m	β_M / m^{-1} sr^{-1}	R_M / m	k_β	P_π	O_2, NO_2, SO_2 correction	α_{mol}
CNRS	7000-15626	not specified					no	CNRS
EPFL	540-1860	120	6.46×10^{-5}	4020	–1	0.028	yes	USSA
MPI	600-3000	75	1.1 β_{mol}	2105-2300	0	0.04	no	USSA
RIVM	720-4500	630	1.2 β_{mol}	4500	–1	0.028	yes	USSA

The variables have been discussed in the methodology section of this document, and the notations are those from the description of the EPFL data processing. These values have been recommended in each case by the group providing the data set and they are optimized for one algorithm. The values of k_β and P_π come from the literature. They depend on the measurement conditions and the choices $k_\beta = -1$ and P_π =0.028 sr^{-1} correspond to a rural atmosphere, probed at 300 nm (but see also the discussion in Section 2.5.2). The parameter β_M is the backscattering coefficient estimated at a distance R_M further than the last point at which the ozone concentration is reconstructed. As R_M is supposed to be in the free troposphere, β_M can be either estimated from the molecular contribution to the backscattering coefficient β_{mol}, or from the data themselves.

The backscattering coefficients and the corrections ΔC to the ozone concentration are presented before the raw and corrected results C_0 and C_1. The ratio $\Delta C/C_1$ quantifies the contribution of the corrections to the presented results. Their precision depends on the reliability of the correction scheme and on the estimate of its parameters: the larger the corrections, the larger their contribution to the uncertainty on the final results.

Analysis of the CNRS data set

Lidar signals are provided in that data set beyond 3855 m. The recommended starting point for the analysis has been placed at 7000 m. The contribution of the aerosols to the backscattering coefficient is supposed to be negligible due to their low concentration at this altitude. The effect of the Rayleigh extinction has been estimated from a table provided by the CNRS, and the corresponding backscattering coefficient and corrections are shown in Fig. 4.6a and 4.6b.

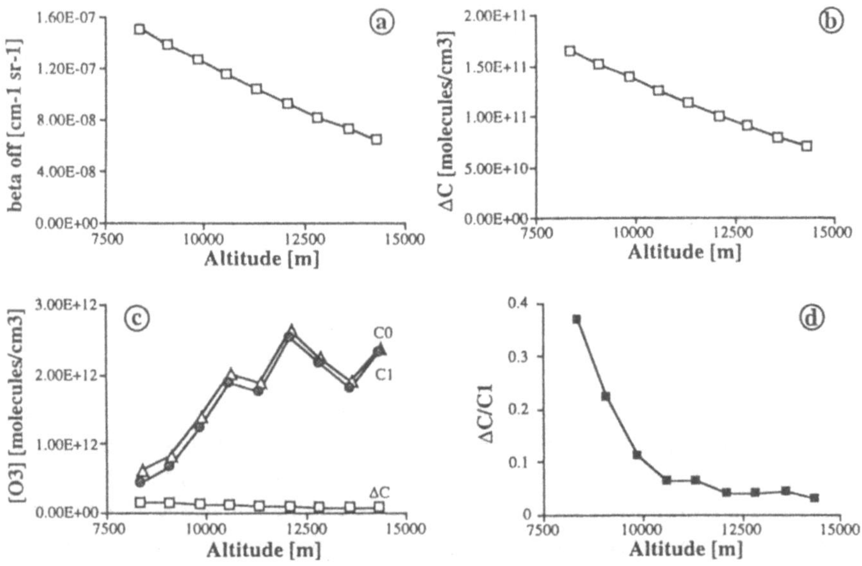

Fig. 4.6: Analysis of the CNRS data set with the EPFL analysis procedure. a) Retrieved backscattering coefficient for the OFF wavelength; b) Corrections (ΔC) to the ozone concentrations; c) Uncorrected (C_0) and corrected (C_1) ozone concentrations; d) Relative level of the corrections on the ozone concentrations.

The concentrations are shown before and after the application of the correction scheme in Fig. 4.6c and the relative level of the correction is presented in Fig. 4.6d.

Analysis of the EPFL data set

The lidar returns for 200 successive laser shots are shown in Fig. 4.7a, and the corresponding distributions are presented in Fig. 4.7b.

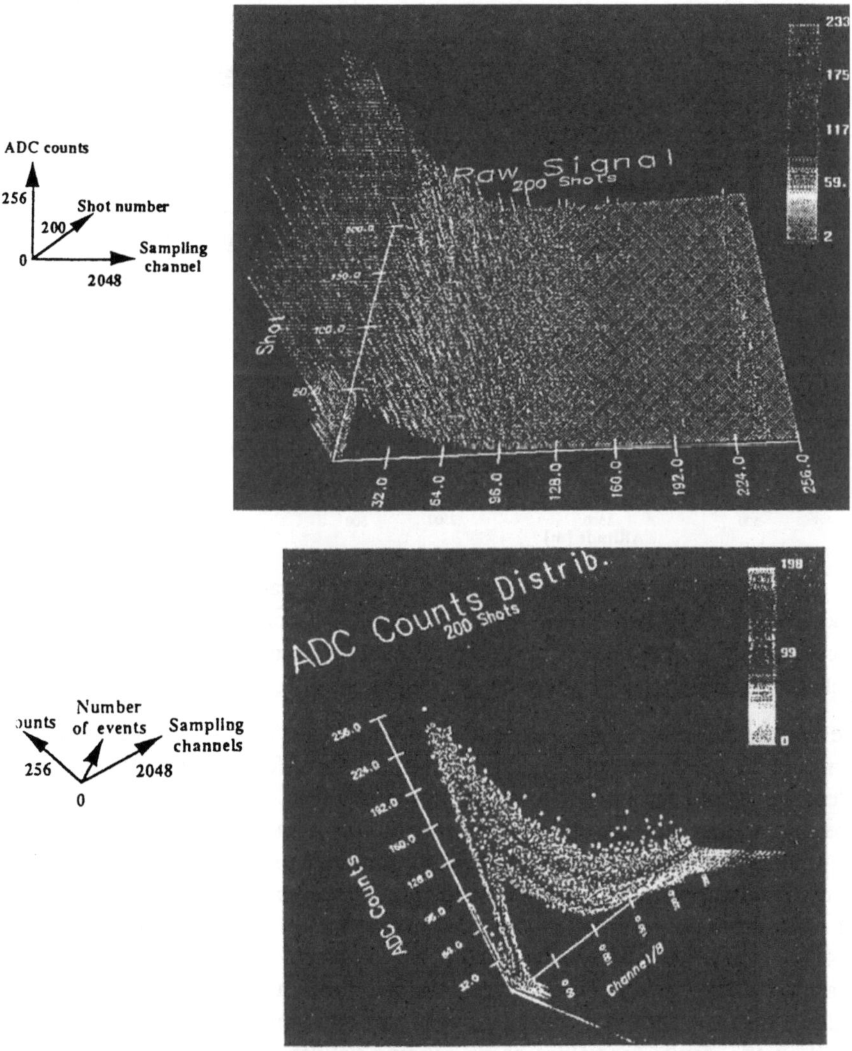

Fig. 4.7: a) A sample of 200 successive lidar returns: Raw signals b) A sample of 200 successive lidar returns: distribution of the number of laser shots over the reachable ADC values.

The contributions of both the molecules (β_{mol}) and aerosols (β_{aer}) to the backscattering coefficient reconstructed for the off wavelength are shown in Fig. 4.8a.

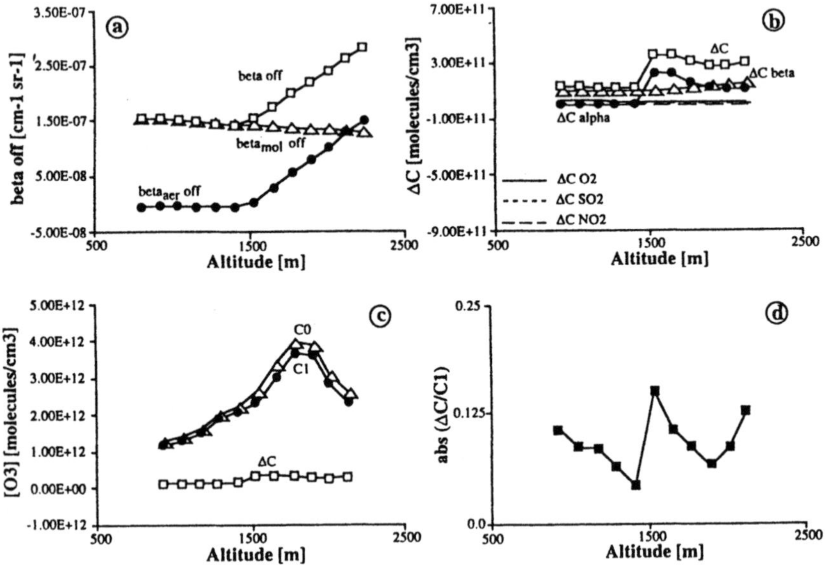

Fig. 4.8: Analysis of the EPFL data set with the EPFL analysis procedure. a) Retrieved backscattering coefficient for the OFF wavelength; b) Corrections (ΔC) to the ozone concentrations; c) Uncorrected (C_0) and corrected (C_1) ozone concentrations; d) Relative level of the corrections on the ozone concentrations.

The unexpected raise of the β_{aer} coefficient after 1500 m is interpreted as coming from a misalignement of the optical system: the effect of the aerosols should indeed be more important below than above 1500 m. The data sample analysed here was acquired during the first runs of tuning of the instrumentation, and it is preliminary. Optical distorsions can fake atmospheric effects. They are thus treated as such by the processing algorithms which validation by an intercomparaison does not suffer from the presence of that kind of artefacts in the data sample. The value of 4020 m has been recommended for the parameter R_M because the Neperien logarithm of the range corrected signals is linear at that point. A sensitivity analysis on that value has been pursued, and no significant difference could be found on the results for values of R_M ranging between 2800 m and 5200 m. The different contributions to the final correction factor ΔC are shown in Fig. 4.8b: ΔC_β represents the correction term due to the spatial derivation of the Neperien logarithm of the ratio of the backscattering coefficients for the off to the on wavelengths; ΔC_α is the correction due to the differential absorption of the aerosols and molecules for the on- and off-wavelengths; ΔC_{O2}, ΔC_{SO2} and ΔC_{NO2} represent the correction factors for the differential absorptions of the on- and off-wavelengths by these molecules respectively. The reconstructed concentrations of ozone, and the relative weight of the corrections on the final results are respectively presented in Fig. 4.8c and 4.8d.

Analysis of the MPI data set

The contribution of the aerosols to the backscattering coefficient in the MPI data set is shown in Fig. 4.9a to be more important at low altitude for which their concentration is higher. No correction was recommended on that data set to take into account the effect of the O_2, NO_2, and SO_2 molecules. There is an oscillating structure which can be observed over the distance for the correction factor ΔC_β in Fig. 4.9b.

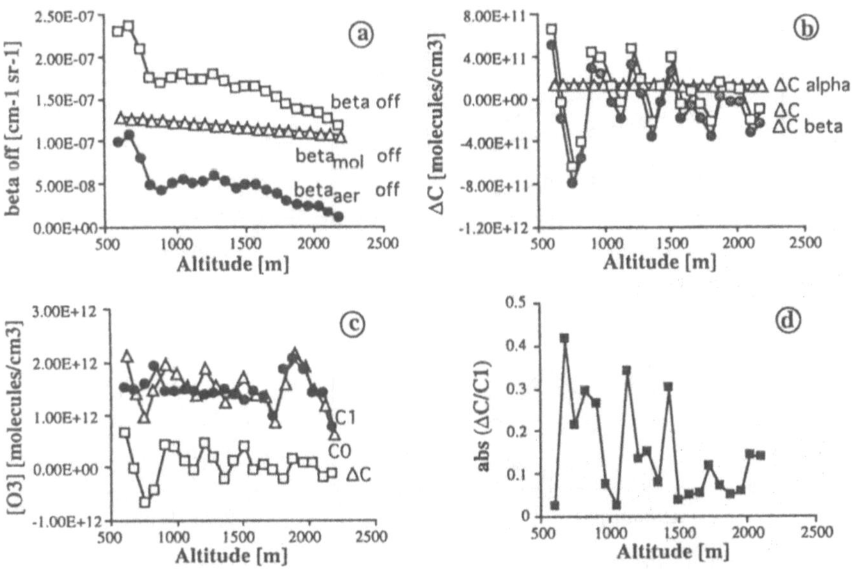

Fig. 4.9: Analysis of the MPI data set with the EPFL analysis procedure. a) Retrieved backscattering coefficient for the OFF wavelength; b) Corrections (ΔC) to the ozone concentrations; c) Uncorrected (C_0) and corrected (C_1) ozone concentrations; d) Relative level of the corrections on the ozone concentrations.

When applied to the raw concentration C_0, this correction leads to a more constant value of the final concentration C_1 presented in Fig. 4.9c. This shows the capability of the correction sheme to correct the results not only for a level, but also for a structure. Although the recommended range was reaching 3000 m, results are not presented up to that altitude, as the correction scheme is not capable to cope with the cloud which is visible in the raw data from about 2300 m. The importance of these corrections is outlined in addition in Fig. 4.9d where they appear to reach a mean level of about 20 %.

Analysis of the RIVM data set

The recommended interval for the analysis of the RIVM data set is 630 m. This value is explained by the necessity to increase the signal to noise ratio of the data by an averaging over many sampling channels, the averaging over repeated samples having not been sufficient in that case to cope with the atmospheric and experimental fluctuations. The presentation of the different steps discussed above for the previous data sets is repeated in Fig. 4.10. The contribution of the corrections to the presented final results is here again reaching a mean value of about 20 %.

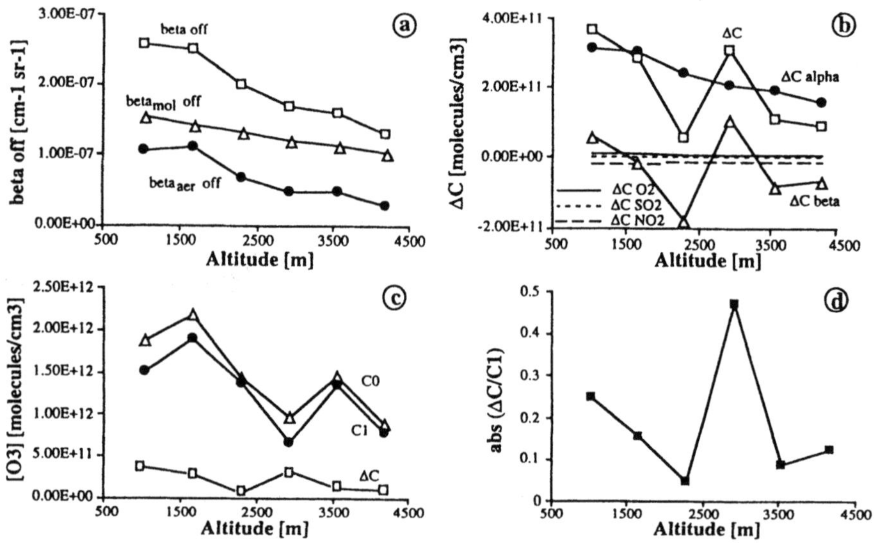

Fig. 4.10: Analysis of the RIVM data set with the EPFL analysis procedure. a) Retrieved backscattering coefficient for the OFF wavelength; b) Corrections (ΔC) to the ozone concentrations; c) Uncorrected (C_0) and corrected (C_1) ozone concentrations; d) Relative level of the corrections on the ozone concentrations.

Conclusion

The same data processing algorithm has been applied to four data sets without any need for specific adaptations. The recommendations of the groups who produced the data have been followed for the choice of the analysis parameters. The corrections on the final results have been shown to reach values as high as about 50 % with a mean around 20 %. Although these high levels depend on the meteorological conditions at the time of the data taking, they confirm the necessity to carefully assess the precision of the measurements when working in the troposphere.

4.3.3 Intercomparison

The four data sets have been processed in a next step with up to four algorithms. The intercomparison of the results is shown in Fig. 4.11.

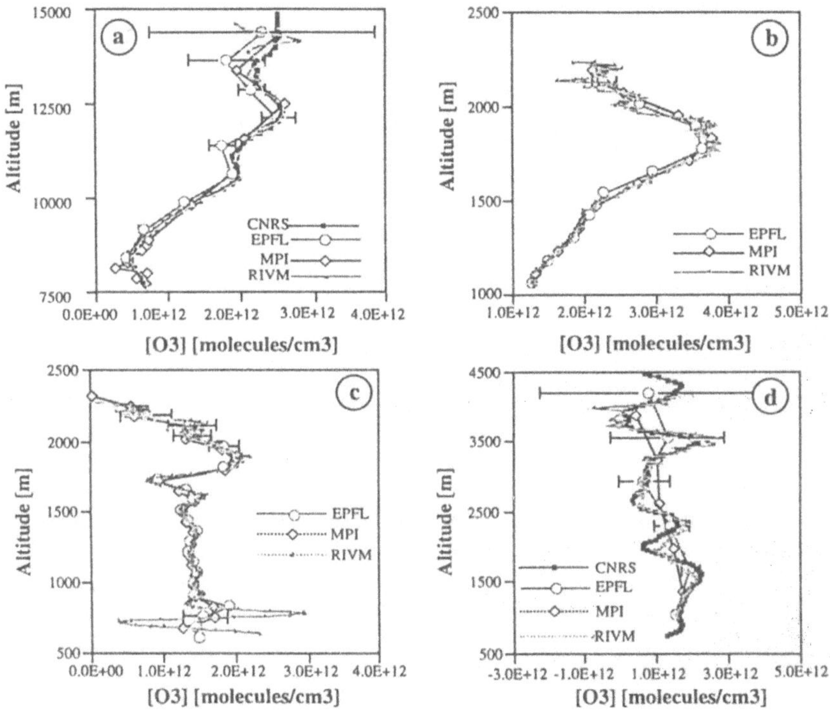

Fig. 4.11: Application of four data analysis procedures on four data sets: a) CNRS data set, b) EPFL data set, c) MPI data set, d) RIVM data set.

An attempt has been made to quantify the overall agreement reached for these analyses. Mean values have been calculated for each profile from the different retrieved concentrations. The absolute differences between the means and each individual profiles have been calculated. Concentrations have been either taken or linearly interpolated at the altitudes corresponding to the recommended intervals, as used for the EPFL analysis. The distribution of the number of occurences over the reached differences is presented in Fig. 4.12 for the 142 measurement from the four profiles.

The corresponding cumulative distribution function is presented in Fig. 4.13.

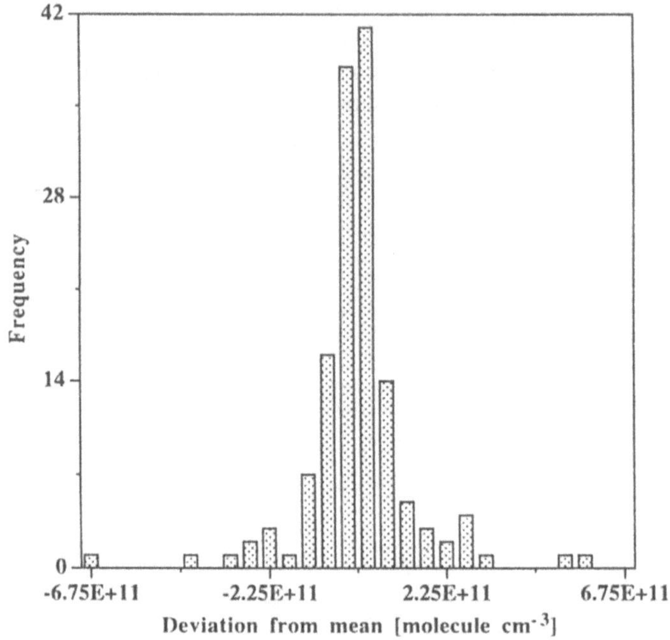

Fig. 4.12: Distribution of the differences from the mean for the measurement data points from Fig. 4.11.

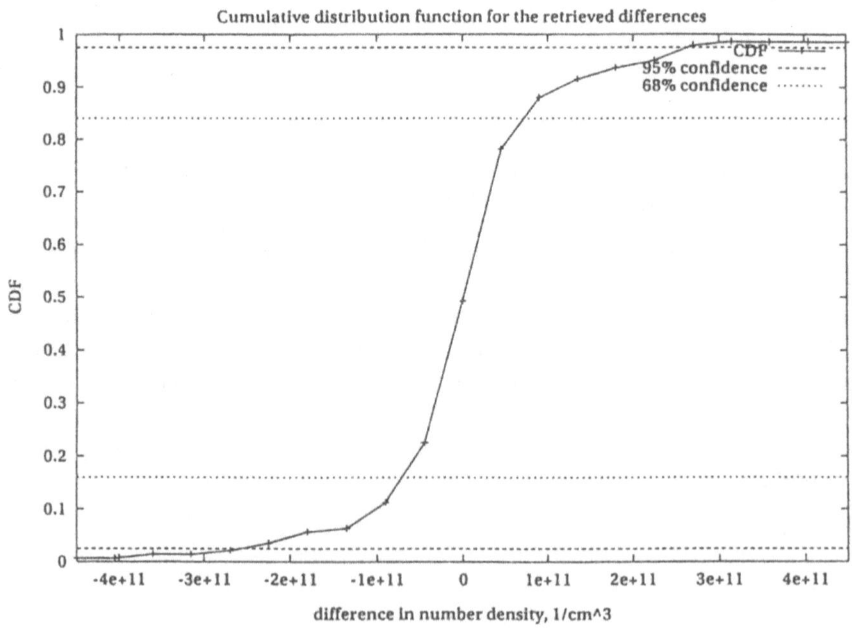

Fig. 4.13: Cumulative distribution function for the observed differences.

The distribution clearly is not Gaussian. From the 68 % confidence limits a standard deviation of 6.5×10^{10} molecules/cm^3 is inferred, while the 95 % confidence levels would suggest a value of 2.6×10^{11} molecules/cm^3. Referencing this to the mean concentration of 1.68×10^{12} molecules/cm^3, the relative errors are about 4 % and 16 %, respectively.

From Fig. 4.11 it is visible that the differences between the results are most pronounced for the RIVM dataset, and the smallest deviations occur for the MPI dataset. Presumably the rather large amount of noise in the RIVM dataset is the main reason for the larger discrepancies in the evaluation of this case. This high noise level, as reflected in the large estimated errors, requires averaging or smoothing over many samples. Differences in the processing schemes show up in this case. The algorithms which use a gliding window for smoothing, show sharp structures in the retrieved ozone concentration, which are handled differently by the different algorithms. It should however be noted, that the differences between the results from different algorithms are usually smaller than the random errors as estimated from the signal-to-noise ratio above 2500 m, so that the rather large deviations observed from that altitude are not significant.

For the case showing the smallest discrepancies, the MPI dataset, the noise level is much lower, allowing a much higher vertical resolution. As a result, there appears to be only little structure in the ozone distribution which is not resolved by all retrievals, with the exception of the lowest levels.

To be complete, this intercomparison should include an analysis of the statistical uncertainties of the lidar signals. Unfortunately this information is not available for most of the data set. So this has been done only for the results obtained with the EPFL analysis of the EPFL data set. For this case, error bars are for the statistical errors only. Because the EPFL data sample was recorded shot per shot, it was possible to calculate the statistical distributions for each channel of sampling. A standard deviation can thus be reconstructed from these distributions, and then propagated up to the ozone concentrations. The Fisher formula has been employed for the other data sets, after the range corrected signals have been linearised over each bin. The standard deviation thus obtained is then propagated also up to the ozone concentrations. Although this method was shown to overestimate the errors when compared to the one reconstructed from a shot-per-shot data sample, it is applied here on the data sets for which this information is not available, to give an upper limit on the statistical error affecting the results.

4.4 Conclusions

The data processing algorithms used within the frame of TESLAS have been shortly presented. They have been applied to four different data sets in order to evaluate how sensitive the calculated ozone concentrations are on the chosen options for the data analysis. The results showed good agreement in a major part of the range of conditions covered by the actual data, but rather large discrepancies remained especially in cases of low signal-to-noise ratio and/or large variation of the ozone density within the resolved height interval. This is not completely satisfying, since the data processing should not introduce major errors into the results. Further work is required to identify exactly the source of the observed differences and to come to an agreement on the correct solution.

4.5 References

1. J.D. Klett; Stable analytical inversion solution for processing lidar returns, *Appl. Opt.* **20** (1981) 211.
2. L.T. Molina, *et al.*; Absolute absorption cross sections of ozone in the 185- to 350-nm wavelength range, *J. Geophys. Res.* **91** (1986) 14501.
3. D.R. Lide (ed.), *Handbook of Chemistry and Physics*, CRC Press, 1992.
4. L. Elterman; UV, Visible, and IR attenuation for altitudes to 50 km, *AFCRL*-68-0153 (1968).
5. P.M. Teillet; Rayleigh optical depth comparisons from various sources, *Appl. Opt.* **29** (1990) 1897.
6. M.W.P. Cann, *et al.*; Oxygen absorption in the spectral range 180 - 300 nm for temperatures to 3000 K and pressures to 50 atm, *Can. J. Phys.* **62** (1984) 1738.
7. A.M. Bass, *et al.*; Extinction coefficients of NO_2 and N_2O_4, *J. Res. Nat. Bur. Stand.* **80A** (1976) 143.
8. D.J. Brassington; Sulfur dioxide absorption cross-section measurements from 290 nm to 317 nm, *Appl. Opt.* **20** (1981) 3774.
9. C. Weitkamp, *et al.*; Sulfur dioxide absorption cross-section between 265 and 298 nanometers, GKSS 91/E/16 (1991).
10. S.L. Manatt, A.L. Lane; A compilation of the absorption cross sections of SO_2 from 106 to 403 nm, *J. Quant. Spectrosc. Radiat. Transfer* **50** (1993) 267.
11. E.V. Browell, S. Ismail, S.T. Shipley; Ultraviolet DIAL measurements of O_3 profiles in regions of spatially inhomogeneous aerosols, *Appl. Opt.* **24** (1985) 2827.
12. R.M. Measures; *Laser Remote Sensing*, Krieger Publishing Company, 1992.
13. Y. Sasano, E.V. Browell, S. Ismail; Error caused by using a constant extinction / backscattering ratioin the lidar solution, *Appl. Opt.* **24** (1985) 3929-3932.
14. Yu.E. Voskoboinikov, A.A. Mitsel'; The use of smoothing splines for restoration of the H_2O absorption molecular coefficient profile, *Izv. Akad. Nauk, Fiz. Atmos. Okeana* **17** (1981) 175.
15. V.E. Zuev, Yu.S. Makushkin, V.N. Marichev, A.A. Mitsel', V.V. Zuev; Lidar differential absorption and scattering technique: Theory, *Appl. Opt.* **22** (1983) 3733.
16. A.V. El'nikov, V.V. Zuev, M.Yu. Kataev, V.N. Marichev, A.A. Mitsel'; Sounding of stratospheric ozone with an UV bifrequency DIAL: Methods for solving the inverse problem and results of the field experiment, *Atmos. Oceanic Opt.* **5** (1992) 362.

Chapter 5

Assessment of Accuracy

Arnoud Apituley

Rijksinstituut voor Volksgezondheid en Milieu, Laboratory for Air Research, P.O. Box 1, NL- 3720 BA Bilthoven, The Netherlands

5.1 Introduction

It has often been claimed that the DIAL method is self calibrated. For tropospheric ozone this is true only if the following effects are properly accounted for: 1) differential extinction and backscatter due to molecules, 2) differential extinction and backscatter due to aerosols, and 3) differential absorption by other gases (see Chapter 2). In addition to the methodological limitations, technical limitations of the lidar instruments have to be taken into account. The most important instrumental errors are linked with: detector non-linearity, digitization errors and alignment errors (see Chapter 3). For a particular lidar system, these errors will largely determine what the minimum and maximum achievable altitude ranges, and what the achievable spatial and temporal resolutions actually are.

The accuracy of the retrieval of the tropospheric ozone distribution by lidar has to be assessed by comparing the measurements from lidar instruments to those from conventional (*e.g. in situ*) instruments. Comparative measurements of ozone in the lower troposphere are particularly difficult due to high temporal and spatial variability. Matters are further complicated by the different spatial and temporal sampling properties of different types of instruments. It is therefore important that the difference between measurement volumes of the instruments is as small as possible. Because a situation with exactly equal measurement volumes cannot be achieved, as many parameters of the ozone distribution (*e.g.* average, variance and trend as well as vertical distribution) as possible should be determined. For altitudes above the mixed layer height, the vertical ozone distribution is generally more homogeneous. Consequently, the difference between measurement volumes becomes less critical. However, in this case the temporal variability of the ozone concentration becomes more important because longer data integration times are required at higher altitudes.

Several instruments can be used in comparative analyses. Among them are:

- ***Electrochemical sonde***

  The measurement principle of both Brewer-Mast sondes [1] and ECC sondes [2] is based on the electrochemical detection of ozone in an iodine/iodide solution. The observed current between both electrodes minus the background current is proportional to the partial pressure of ozone.

  In the troposphere, Brewer-Mast sondes generally show lower ozone values than ECC sondes. For example, during the second International Ozone Sonde Intercomparison campaign 1978 at Hohenpeissenberg, Brewer-Mast sondes showed ozone values 12 % smaller than ECC sondes in the first five km of the troposphere [3] and during BOIC 1984 [4] 25 % smaller values. Differences of this magnitude cannot be neglected when using tropospheric ozone profiles from Brewer-Mast and ECC sondes in a common data base. However, an other experiment has shown that sondes of the Brewer-Mast type can also give ozone values larger than ECC sondes [5].

  The precision and systematic errors of ECC and Brewer-Mast sondes in the troposphere due to experimentally known error sources (calibration at ground, solution evaporation, pump efficiency, response time and altitude restitution) are of the order of a few percent [6]. However, two error sources, which could largely determine the systematic error and precision of electrochemical sondes in the troposphere, are not well established: the evolution of the background current and the contamination of the ozone sonde by reducing agents. In the stratosphere, these two error sources become less important compared to the uncertainty in the pump-efficiency correction which limits systematic error and precision of Brewer-Mast sondes to about 5 %. Overall accuracies for ECC sondes are +3.5 ± 4.5 % in the troposphere. For Brewer-Mast sondes, the corresponding values are –9.5 ± 9.5 % [7].

- ***UV-absorption ozone monitor***

  The UV-absorption ozone monitor, or UV-photometric analyzer, is a point sampling device. The measurement principle is based on the optical absorption of ozone around 254 nm (the ozone absorption maximum in the UV). Several types of this instrument are commercially available and the operational properties are well established. This type of instrument is mostly used in ground-level monitoring. However, reliable in situ measurements of the vertical ozone distribution are possible when the instrument is mounted on an aircraft [8, 9]. The typical response time is about 10 s. The precision of this type of instrument is about 4 %. The instrument is quite insensitive to interferences from other gases such as SO_2. However, contamination of the air suction system can cause systematic differences (*i.e.*, ozone depletion in tubes).

- ***Differential optical absorption spectroscopy (DOAS)***

  The DOAS technique is based on the broadband absorption features of specific gases and the fine structure superimposed on them [10]. The mean gas concentration over a measurement path (typically in the order of 1-2 km) is obtained by comparing recorded spectra of the light transmitted along the measurement path to the spectral signature of the (broadband) light source and reference spectra of atmospheric gases. The use of differential absorption features and polynomial short pass filtering suppresses instrument factors, aerosols and slowly varying gaseous absorption. Usually, the measurement path is horizontal, which makes the DOAS technique very suitable for intercomparison with horizontally operated lidar systems. Typical response times are in the order of 10 - 20 minutes.

Interference of hydrocarbons (HC) to DOAS or other conventional monitors of ozone may be in the order of a few percent in case the ozone measurements are taken in highly polluted areas (*i.e.*, high HC concentrations). However, specially designed instruments normally take this into account so that the HC interference becomes nearly negligible.

The comparison experiments carried out by members of the TESLAS community can be divided into two categories: 1) horizontal intercomparisons and 2) vertical intercomparisons. The systems involved in the experiments were quite different in design with respect to operating wavelengths, range resolution, minimum and maximum range, etcetera. Also, the meteorological conditions under which the systems have been tested were quite different, as were the types of comparative instruments used. It is therefore difficult to make general statements about the DIAL method for tropospheric ozone, or which system design is more successful than another. In Sectin 5.3 a table shows the results from the individual comparison campaigns presented here, which are based on the various lidar instruments described in Chapter 3.

5.2 Experiments

5.2.1 Intercomparison between lidar, ECC and Brewer-Mast sondes, autumn 1989

M. Beekmann, G. Ancellet, G. Mégie, H.G.J. Smit and D. Kley

An intercomparison campaign was conducted at the Observatoire de Haute Provence (OHP) in Southern France in September 1989 in order to compare the three instruments used for vertical tropospheric ozone profiling in the European TOR (Tropospheric Ozone Research Project) network: balloon borne ECC and Brewer-Mast sondes and a ground based tropospheric ozone UV-DIAL [6]. Additionally, a stratospheric lidar system and the Dobson spectrophotometer of the

OHP were operated. The participating instruments and institutes are listed in Table 5.1.

The tropospheric ozone lidar at OHP for measurements in the free troposphere [11] uses the two wavelengths (289 nm, 299 nm) obtained from the fourth harmonic of a Nd:YAG laser and subsequent stimulated Raman shifting in hydrogen and deuterium. The data acquisition consisted of a photon counter with a 1 MHz sampling frequency (150 m height resolution); the bandpass of 300 MHz allowed measurements down to 5 - 6 km. Analog detection for recording of the signals from lower altitudes was not possible during this campaign.

Table 5.1: Instruments and institutes involved in the intercomparison at Observatiore de Haute Provence, France, September 1989.

Instrument	Institute
ECC sondes	Research Centre Jülich, Germany
Brewer-Mast sondes	Service d'Aéronomie du CNRS, France
Dobson	Service d'Aéronomie du CNRS, France
Stratospheric ozone lidar	Service d'Aéronomie du CNRS, France
Tropospheric ozone lidar	Service d'Aéronomie du CNRS, France

Measurement conditions

Measurements of vertical ozone profiles were performed on 8 days from September 5 to September 13, 1989. The meteorological situation at the beginning of the campaign was characterized by a high-pressure system centered over Western Europe and a cut-off low located over Spain. The winds measured at the nearby radio sounding station of Nimes were weak throughout the troposphere (< 10 m/s), ensuring a good comparability of the ozone profiles. On the 6th however, the air masses were advected by south-westerly wind southward of the OHP and by north-easterly wind northward of the OHP. This made the comparison between lidar and sondes doubtful. Between the 8th and the 10th, several perturbations, related to a Mediterranean low, passed over the OHP. On these days, lidar measurements were inhibited (8th and 9th) or perturbed below 7 km (10th) by the presence of alto-stratus clouds. Between the 11th and the 13th, the winds were stronger, especially at higher altitudes, but unidirectional (westerly) with respect to height. Performing the tropospheric lidar measurements 15 minutes after the sonde launch, ensured sampling of identical air masses by lidar and sondes.

Results

Fig. 5.1 shows the superposition of the vertical ozone profiles obtained by electrochemical sondes and the lidar system. Ozone profiles from electrochemical sondes are presented without applying the Dobson normalization. Nevertheless, the Dobson normalization factor was calculated for the different flights since it gives a first indication of the data quality. The mean Dobson normalization factors for all flights were well within the range of commonly recommended correction factors [12].

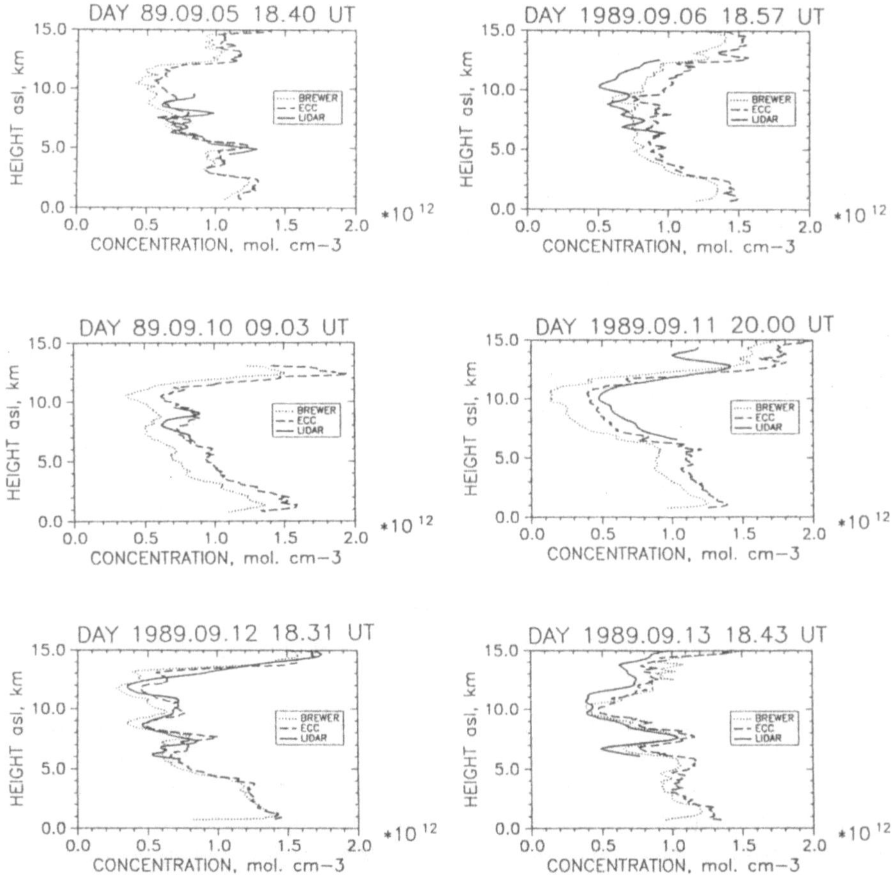

Fig. 5.1: a-f) Ozone profiles measured by lidar, ECC-sonde and Brewer-Mast sonde at the Observatoire de Haute Provence, France, September 1989.

The altitude range of the tropospheric lidar profiles was between 5 and 10 - 12 km at the beginning of the campaign and shifted to higher altitudes, 6 - 15 km, for the last three days when the laser output energy was higher. For the daytime measurement performed on the 10th, the profile ended at 10 km as the background skylight reduced the signal to noise ratio at higher altitudes.

Discussion

Structures of the ozone profile with a vertical extension greater than 1 km were quite similar for lidar and electrochemical sonde profiles. However, the fine structure was different as the height resolution of the lidar was lower and as lidar and sondes probably sampled slightly different air masses.

The differences between the lidar and electrochemical sondes in the troposphere were analyzed in the height range between 6.5 and 9.5 km, where six simultaneous lidar and sonde profiles were available (Fig. 5.2).

The mean absolute difference (*i.e.*, without considering the sign) between the lidar and ECC sondes was 15 %. Considering the sign of the differences, the ECC-sonde values were 6 ± 9 % higher than lidar values. Therefore, the mean difference between both instruments was not significant at the 95 % confidence level. Compared to Brewer-Mast sondes, lidar ozone concentrations were 18 ± 10 % higher, with a significance close to the 95 % confidence level. Hence, lidar ozone concentrations were on the average inbetween the ozone values of both sonde types, but closer to those of ECC sondes. At the lower limit of the altitude range (6.5 km), the mean differences were close to those averaged between 6.5 and 9.5 km. Potential error sources of the lidar system which were most important at lower altitudes (alignment of the optical system, saturation of the photoncounting system) apparently did not cause a particular bias of the lidar measurements, while they may have played a role in the variability of the differences, which were slightly higher at 6.5 km. Above the ozonopause, generally at 12 km, the lidar showed larger differences compared to ECC sondes. This can partly be explained by the larger atmospheric variability of ozone due to its strong vertical gradient at these altitudes. Also, the reduced height resolution of the lidar (1.5 km at 15 km) resulted in differences up to 15 $\mu g/m^3$ between the two instruments at heights with large gradients in the vertical ozone distribution.

At its upper measurement limit, the lower accuracy of the lidar (21 %), due to the lower signal to noise ratio, also contributed to larger differences. In the stratosphere, absolute differences between lidar and sondes were typically between 5 and 10 %. Considering the sign of the differences, the lidar showed on the average slightly higher ozone values than the ozone sondes (about 5 %), but the differences were not significant.

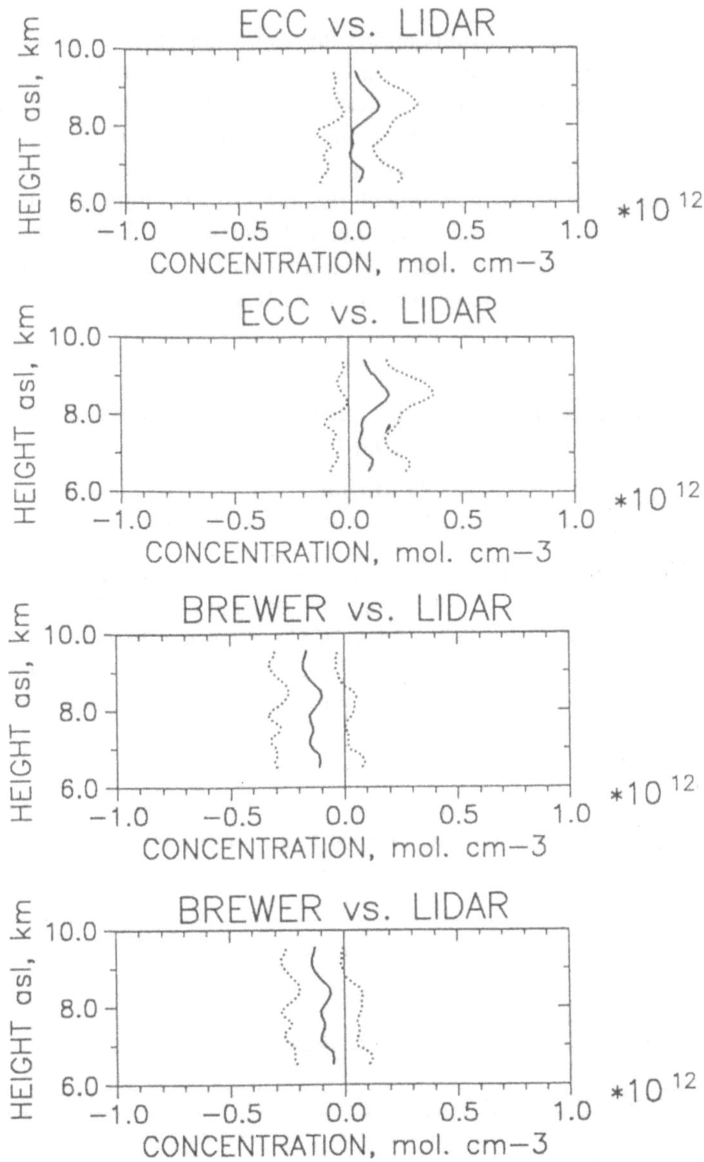

Fig. 5.2: a-d) Comparison of results from ECC-sondes and lidar (upper half) and Brewer-Mast sondes and lidar (lower half). Figures on the right hand side are without Dobson normalization. Figures on the left hand side are with Dobson normalization.

Conclusions

Lidar ozone values were above those measured by Brewer-Mast sondes (which probably underestimated the ozone concentrations) and below those measured by ECC sondes (which showed a potential for ozone overestimation). It is concluded that the lidar gave unbiased results during this campaign, (6 ± 9 %), although the absolute difference remained of the order of 15 %. This result agreed well with the estimated systematic error for lidar measurements of less than 7 % in the height range 6.5-9.5 km. The limited precision was probably due to the important uncertainty of the lidar data below 8~km because analog recording of the signals was not possible. The normalization with the total ozone column measured by the Dobson spectrophotometer is not well adapted for an accurate correction of the bias between the lidar and the sondes in the troposphere, although it may partially reduce the large bias between the Brewer-Mast sondes and the two other instruments.

5.2.2 Intercomparison between lidar, electrochemical sondes and an airborne UV photometer, spring 1991

M. Beekmann, G. Ancellet, D. Martin, C. Abonnel, G. Duverneuil, F. Eideliman, P. Bessemoulin, N. Fritz and E. Gizard

The intercomparison campaign held at the Observatoire de Haute Provence (OHP) from March 23 to April 3, 1991, had the objective to further elucidate the differences between Brewer-Mast sondes, ECC sondes and the tropospheric ozone lidar system, and to validate the ozone measurements by UV photometry on the French research aircraft Merlin [7]. The participating groups and instruments are listed in Table 5.2.

Table 5.2: Instruments and institutes involved in the intercomparison at Observatoire de Haute Provence, France, March - April, 1991.

Instrument	Institute
ECC sondes	Météo France, CNRM/GMEI, France
Brewer-Mast sondes	Service d'Aéronomie du CNRS, France
UV photometric analyzer	Météo France, CNRM, France
Tropospheric ozone lidar	Service d'Aéronomie du CNRS, France

Measurement conditions

The meteorological situation was characterized by two different periods. From March 22 to 26, a baroclinic wave developed with a trough extending from Scandinavia to northern Africa, and later formed a cut-off low centered over the Mediterranean Sea. From March 31 to April 3, the conditions over southern France were anti-cyclonic with a cloud free sky.

An ozone profile was constructed from each Merlin flight regardless of the flight pattern. Data points were selected in order to coincide most closely in time to the lidar and sonde measurements. The maximum distance of the Merlin aircraft from the OHP was about 6 km for flights with extended constant vertical levels and about 3 km for flights with spiral ascends and descents.

ECC sondes and Brewer-Mast sondes were released simultaneously. After 20 min of flight the balloons reached 7 km altitude at typically 5 - 12 km away from the OHP. Sonde and lidar measurements were taken typically 30 min apart, due to the 60 min integration time for lidar measurements. Given a typical wind speed of 60 km/h at 7 km altitude, this corresponds to a horizontal displacement of 30 km.

UV-photometer measurements on a horizontal flight track of 10 km length show a scatter of typically 1 to 2 ppb (standard deviation) in the middle troposphere, which is in the range of the precision of the instrument. This small horizontal ozone variability should ensure a good comparability of the ozone profiles from the different instruments. Near the tropopause the ozone variability on the 10 km horizontal flight track was up to 5 ppb. Indeed, during the first part of the campaign, the OHP was located near a cut-off low and stratosphere-troposphere exchange processes could have contributed to an enhanced ozone variability in the upper troposphere.

Results

Tropospheric ozone profiles obtained for the four instruments during this campaign are presented in Fig. 5.3 for the four instruments. The profiles show the same general structures, but there are differences in the vertical distribution and in the mean level of the measured tropospheric ozone concentration. These differences were expected as the instruments did not measure exactly at the same place and time. In addition, vertical structures in the lidar ozone profiles were smoothed due to the chosen lower height resolution of the lidar, in particular in the upper measurement range (400 m at 6 km, 800 m at 10 km). Furthermore, for several days (March 28, March 30, April 4), residues of cirrus clouds were detected in the 8 - 10 km altitude region using the lidar signal at the least absorbing wavelength. They caused strong vertical variations of the atmospheric scattering, inducing artificial vertical structures in the ozone profile. The small ozone peak at 9.5 km height on March 30 is an example of this effect.

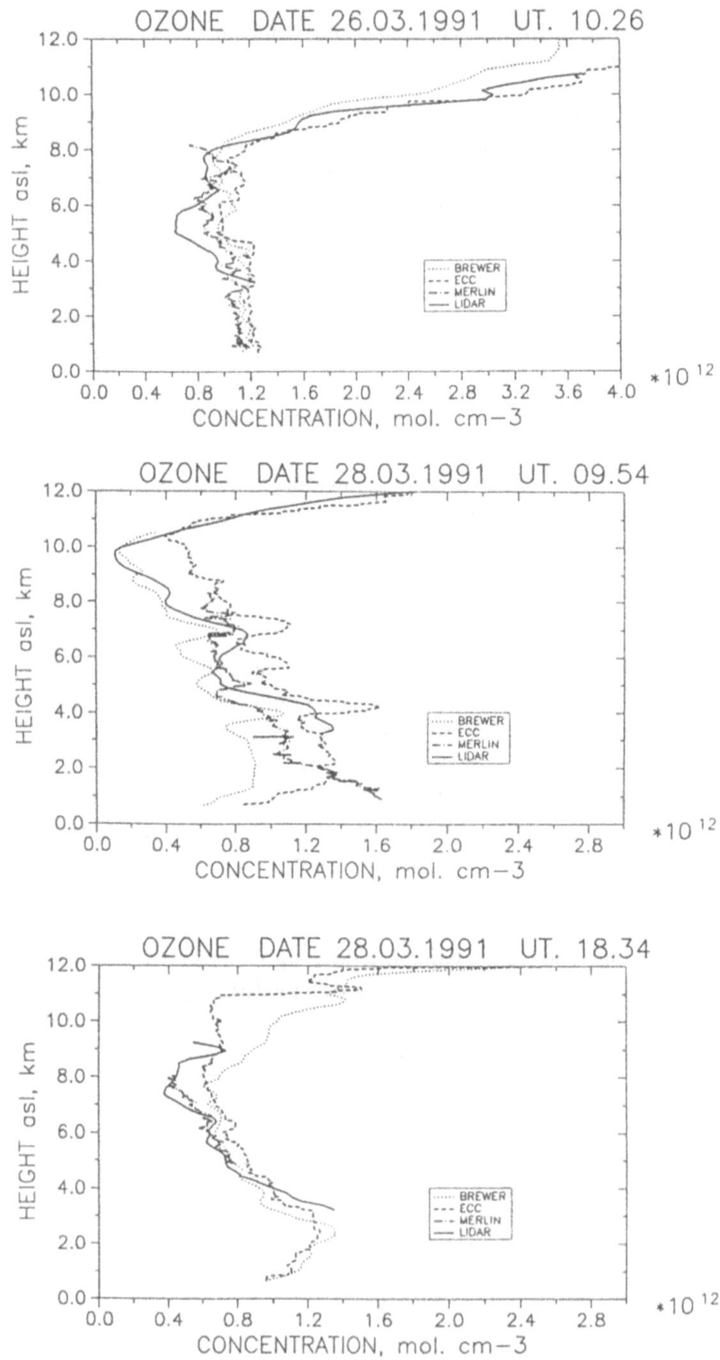
OZONE DATE 26.03.1991 UT. 10.26
HEIGHT asl, km
12.0
10.0
8.0
6.0
4.0
2.0
0.0
0.0 0.4 0.8 1.2 1.6 2.0 2.4 2.8 3.2 3.6 4.0
*10 12
CONCENTRATION, mol. cm−3
BREWER
ECC
MERLIN
LIDAR
OZONE DATE 28.03.1991 UT. 09.54
HEIGHT asl, km
12.0
10.0
8.0
6.0
4.0
2.0
0.0
0.0 0.4 0.8 1.2 1.6 2.0 2.4 2.8
*10 12
CONCENTRATION, mol. cm−3
BREWER
ECC
MERLIN
LIDAR
OZONE DATE 28.03.1991 UT. 18.34
HEIGHT asl, km
12.0
10.0
8.0
6.0
4.0
2.0
0.0
0.0 0.4 0.8 1.2 1.6 2.0 2.4 2.8
*10 12
CONCENTRATION, mol. cm−3
BREWER
ECC
MERLIN
LIDAR

Fig. 5.3: a-c

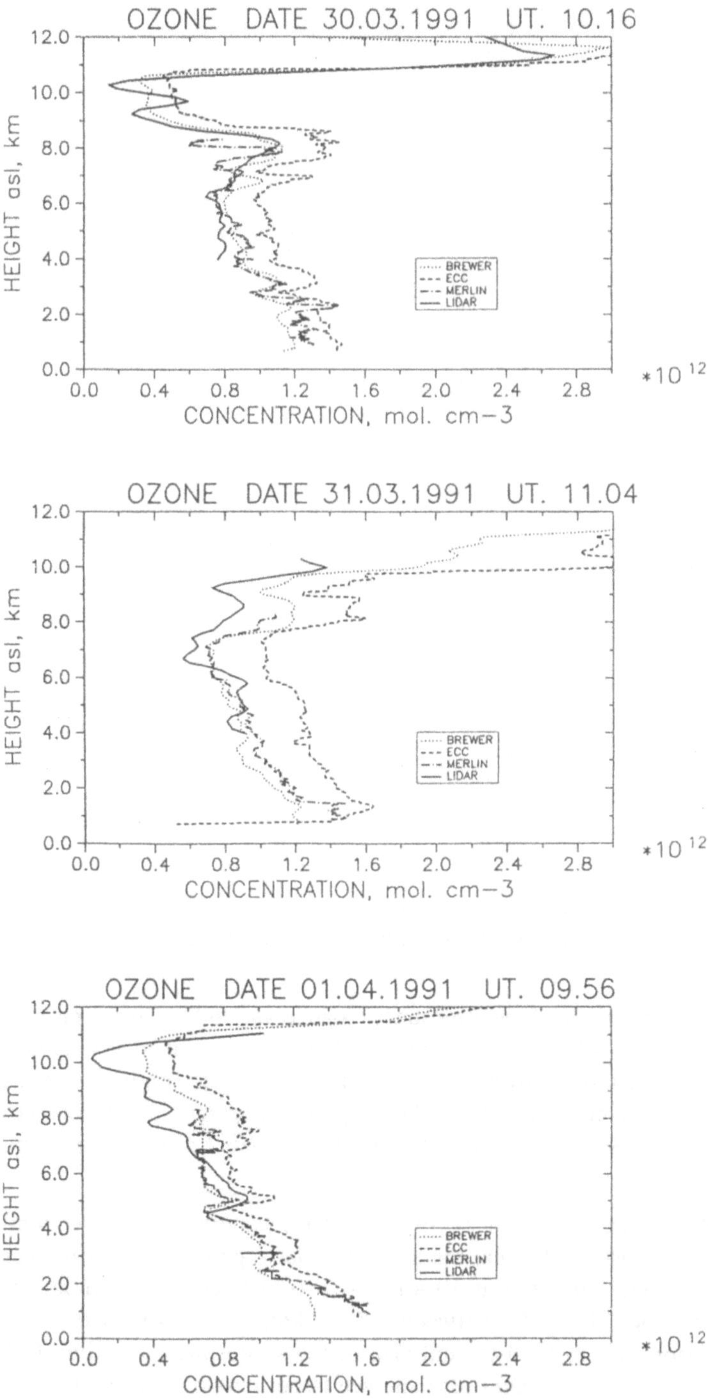
OZONE DATE 30.03.1991 UT. 10.16
HEIGHT asl, km
CONCENTRATION, mol. cm−3
*10 12
BREWER
ECC
MERLIN
LIDAR
OZONE DATE 31.03.1991 UT. 11.04
HEIGHT asl, km
CONCENTRATION, mol. cm−3
*10 12
BREWER
ECC
MERLIN
LIDAR
OZONE DATE 01.04.1991 UT. 09.56
HEIGHT asl, km
CONCENTRATION, mol. cm−3
*10 12
BREWER
ECC
MERLIN
LIDAR

Fig. 5.3: d-f

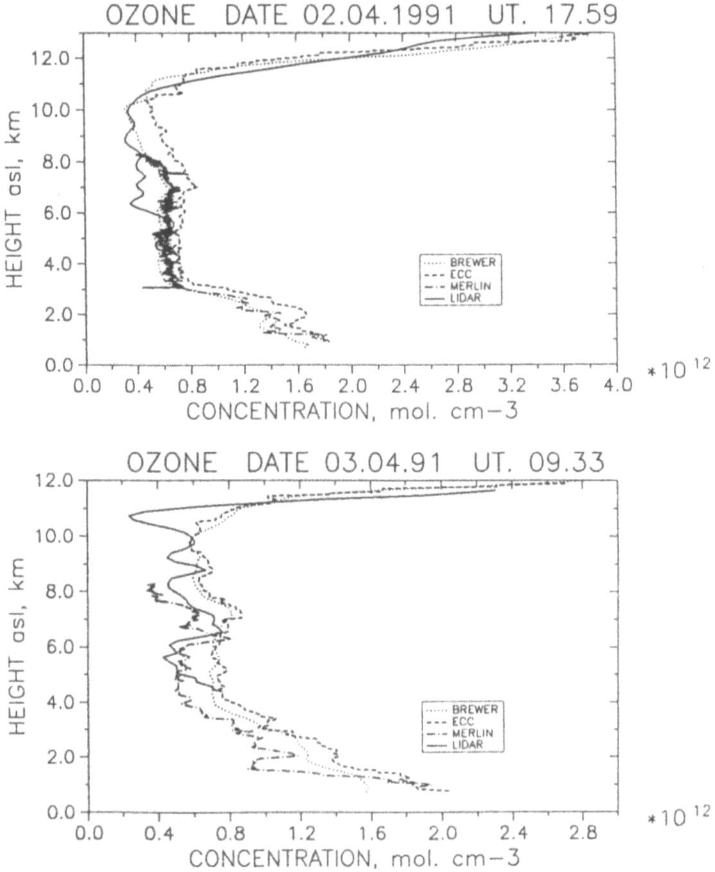

Fig. 5.3: a-h) Ozone profiles measured by lidar, ECC sonde, Brewer-Mast sonde and UV-photometric analyzer (indicated by 'Merlin') at Observatoire de Haute Provence, France, March-April, 1991.

In order to minimize the effect of the atmospheric ozone variability, ozone profiles were integrated between 4.5 and 8 km altitude before comparison. Systematic differences between pairs of instruments were calculated by averaging over the eight simultaneous measurements performed during this campaign. The differences were considered statistically significant if they are larger than 2 times the standard deviation of the mean.

Larger ozone values were measured by ECC sondes compared to Brewer-Mast sondes. The difference was 15 % near the ground, a maximum relative difference of 30 % near the ozone minimum at 10 km, 5-10 % difference throughout the lower stratosphere and 10-15 % between 25-28 km. The differences were significant in the troposphere (> 3 σ) and also in the stratosphere for almost all height ranges (> 2-3 σ). ECC sondes also showed larger ozone concentrations in the troposphere than the airborne UV photometer and the lidar (Fig. 5.4a). The differences were about 15 μg/m³ with the Merlin instrument up to 8 km, and with the lidar up to 6.5 km. Above 6.5 km the difference was about 20 μg/m³ with the lidar.

Brewer-Mast sondes, airborne UV photometer and lidar showed good agreement in the troposphere. Average differences between the UV photometer and Brewer-Mast sondes were statistically insignificant (Fig. 5.4b and Fig. 5.4c). The differences between lidar and Brewer-Mast sondes were also insignificant, except between 7-8 km, where the lidar ozone values were lower. There was a larger scatter of the differences above 9.5 km in the tropopause region. This fact, already observed during the intercomparison campaign at OHP in September 1989 [4], was attributed to the larger atmospheric variability in this altitude region with large vertical ozone gradients. The low height resolution of the lidar at this

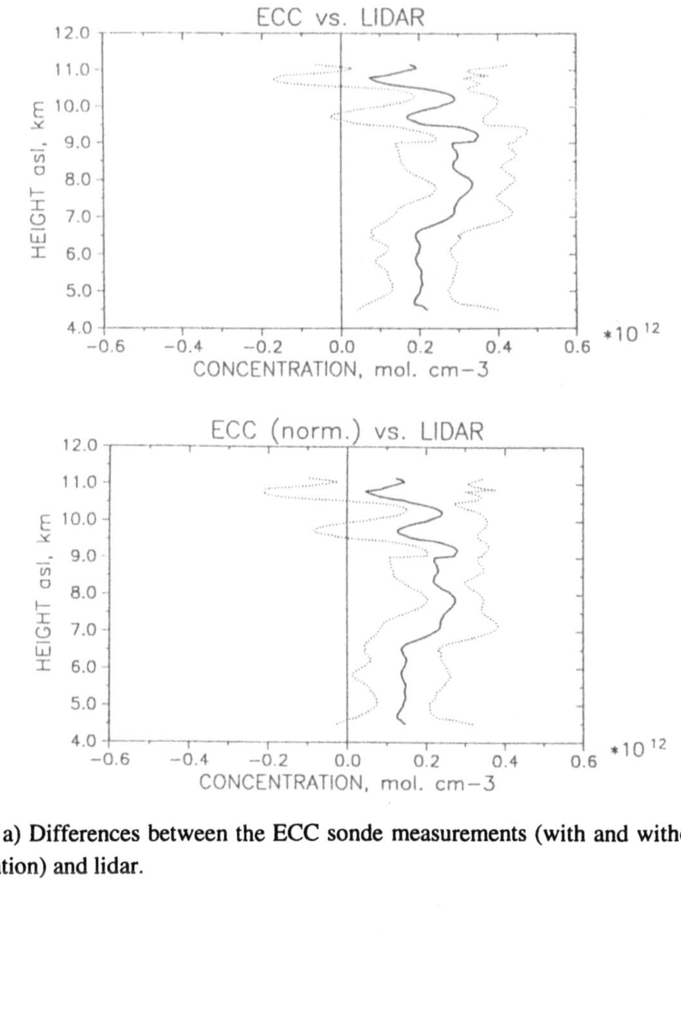

Fig. 5.4: a) Differences between the ECC sonde measurements (with and without Dobson normalization) and lidar.

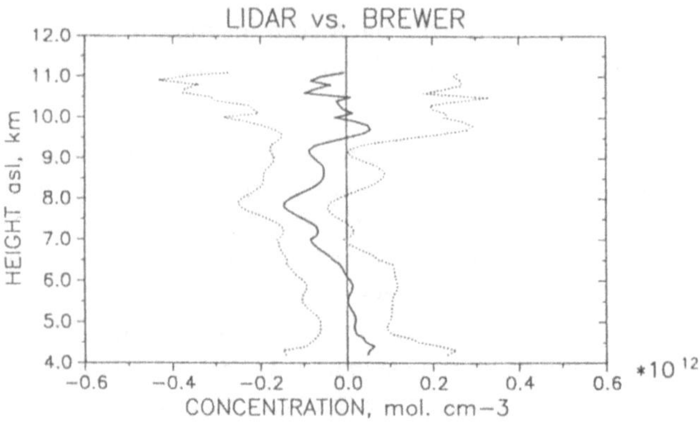

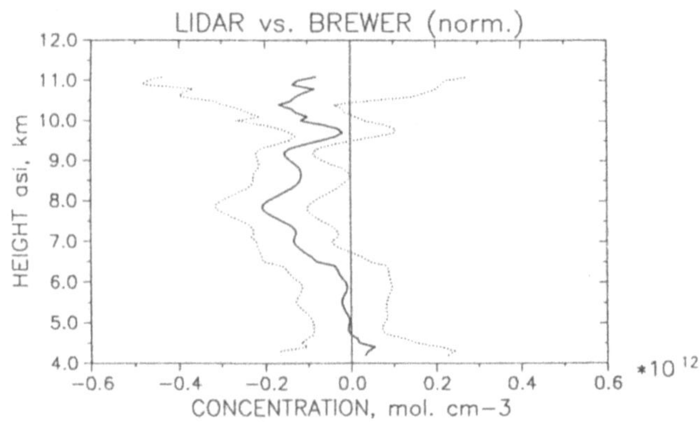

Fig. 5.4: b) Differences between the Brewer-Mast sonde measurements (with and without Dobson normalization) and lidar.

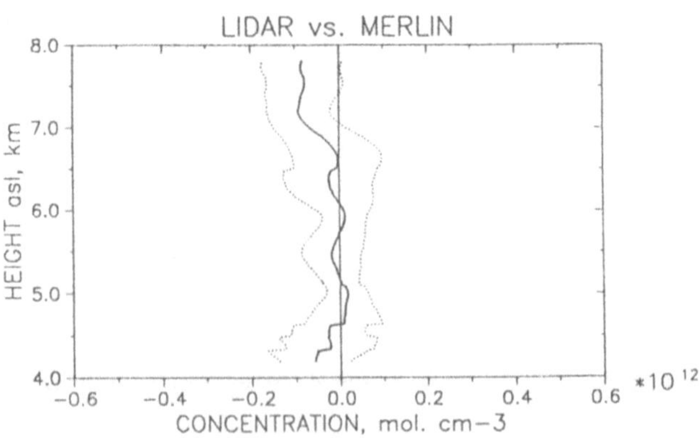

Fig. 5.4: c) Differences between the UV-photometric analyzer (indicated by 'Merlin') and lidar.

altitude (*e.g.*, 800 m at 10 km) and the lower precision of the lidar near its upper measurement limit have contributed to this variability. For the pair UV photometer/lidar, again, no significant differences were observed, except between 7-8 km altitude, where the lidar ozone values were smaller (6.3 $\mu g/m^3$ or 10 %).

Discussion

When averaging the lidar ozone concentrations over the 4.5 - 8 km height range, the impact of the statistical error was only minor (1 %) and the alignment correction (6 %) and the desaturation of the photon counting system (3.5 %) became the most important contributions to the overall accuracy of 7 %. Signal-induced noise was not important at altitudes below 8 km. The observed lidar precision for the ozone concentration averaged between 4.5 and 8 km altitude was 5 %. Hence, it can be concluded that a substantial part (roughly a half) of the error related to the misalignment correction and to the desaturation of the photoncounting system is of random nature.

The structure of ozone profiles obtained by the lidar was sometimes different from that of electrochemical sondes and the UV-photometer, especially around 6 km height. At this altitude range, artificial structures can be introduced into the ozone profiles by discrepancies between the analog and photon counting modes of detection. A striking example is the profile recorded on April 2 when the analog mode indicated about 20 $\mu g/m^3$ higher ozone concentrations than the photon counting mode. This was related to a large difference in the photon counting signal for λ_{on}, compared to the λ_{off}, caused by a higher laser output energy at λ_{on} on this particular day. The optimum altitude range for the photon counting mode was shifted upward while the gain of the analog mode was not adjusted to shift the measurement range to higher altitudes.

Conclusions

ECC sondes have measured significantly higher ozone concentrations than the other three instruments during this campaign. These differences were independent of the actual ozone concentration. The reason could not be determined.

The negative bias of –7 ± 3 % (absolute difference of 7 %) deduced for the lidar between 4.5 and 8 km is in good agreement with its estimated systematic error of better than 10 %. The negative deviation between 7 - 8 km (about –10 % compared to Brewer-Mast sondes and the UV-photometer) was unexpected, as the systematic error was estimated to be minimum in this height range (1 - 6 % at 8 km).

The moderate bias (–7 %) and the good precision (5 %) of the lidar observed during this campaign make this instrument a valuable tool for climatological studies, which need in general measurements of the ozone content averaged over a given altitude range. The main limitation related to a systematic distortion of the vertical ozone distribution for some profiles can easily be traced back and, therefore, removed during the data analysis.

5.2.3 Tropospheric ozone lidar intercomparison experiment, summer 1991

A. Apituley, J. Bösenberg, G. Ancellet, H. Edner, B. Galle and G. Sonnemann

TROLIX '91 was organized to assess the accuracy of four tropospheric ozone lidar systems developed within TESLAS when operating under realistic conditions. The campaign was held between June 10 and June 28, 1991. The main objectives defined for the campaign were:

- to determine how many and which wavelengths are needed for a sufficiently accurate retrieval of ozone concentrations in the lower troposphere;
- to compare the performance of different, co-located, simultaneously operating ozone lidar systems;
- to compare these systems to *in-situ* instruments; and
- to assess the accuracy of these systems that can be achieved under realistic conditions.

It was decided to compare different lidar systems to *in-situ* instruments carried by a helicopter circling around the lidar site in discrete height steps. Further, horizontally pointing lidars to could be compared to DOAS using a nearby path, and to ground based point sampling monitors.

The instruments that participated in the campaign are listed in Table 5.3 The location of the experimental site, at the Dutch National Institute of Public Health and Environmental Protection (RIVM) in Bilthoven, The Netherlands was chosen because there was excellent support for the operation of the lidar systems, and all necessary amending measurements could be provided. Additional meteorological data (weather forecasts, weather analysis maps, PTU-sonde data) were kindly provided by the Royal Meteorological Institute (KNMI) in de Bilt). Furthermore, the surrounding terrain was quite flat and homogeneous, minimizing the problems of non-exact co-location of instruments. Also, extensive support was kindly provided for the helicopter measurements by the Royal Dutch Air Force from a nearby airport.

Table 5.3: Instruments and institutes involved in TROLIX '91, June 10 - June 28, 1991, Bilthoven, The Netherlands.

Instrument	Institute
GROUND BASED INSTRUMENTS	
UV photometric analyzer	RIVM, the Netherlands MPI, Germany
Ground-level chemistry	RIVM, the Netherlands
Tropospheric ozone lidar	RIVM, the Netherlands Service d'Aéronomie du CNRS, France MPI, Germany LIT, Sweden
DOAS	IVL, Sweden
AIRBORNE INSTRUMENTS	
UV photometric analyzer	RIVM, the Netherlands
ECC sonde	OAK, Germany
NO_x photoluminescence	RIVM, the Netherlands

Weather conditions were extraordinarily bad, with a large amount of rain and low clouds. In spite of these unfavorable conditions a large data set was collected. The results of almost all instruments were laid down in technical reports separately [13-19] and an overview of the data collected during TROLIX '91 was presented in a field-phase report [20].

Horizontal path measurements

The following systems were involved in the horizontal intercomparison: LIT-lidar, MPI-lidar, IVL-DOAS, RIVM ground level monitor, MPI ground level monitor. Both lidar systems were pointing in a direction very close to the DOAS path. The height of the lidar paths was about 5 m above sea level. The DOAS instrument was set up 20 m above sea level in a laboratory. Both lidars and the DOAS were pointing slightly upward. The DOAS pointing to retro reflectors on top of two buildings, the lidars to get free of obstacles in the beam path about one hundred meters away. The working range and spatial resolution of the instruments are listed in Table 5.4. The measurement volumes were not the same. However, this was alleviated by averaging all measurements over 15 minutes, the time resolution of the DOAS.

Table 5.4: Working range and spatial resolution of instruments in horizontal intercomparison.

Instrument	Range resolution / m
IVL-DOAS	0 - 884
MPI-lidar	625 - 775
LIT-lidar	300 - 1050

The results from two cases are shown in Figs. 5.5 and 5.6. The LIT lidar values were slightly larger than the monitor and DOAS values by about 10 μg/m^3 for the data presented in Fig. 5.5 The MPI lidar values were 25 to 17 μg/m^3 higher depending on the wavelength pair used. This was as expected, since the lidar results were not corrected for aerosol extinction. For the measurement period presented in Fig. 5.6 all instruments agreed very well.

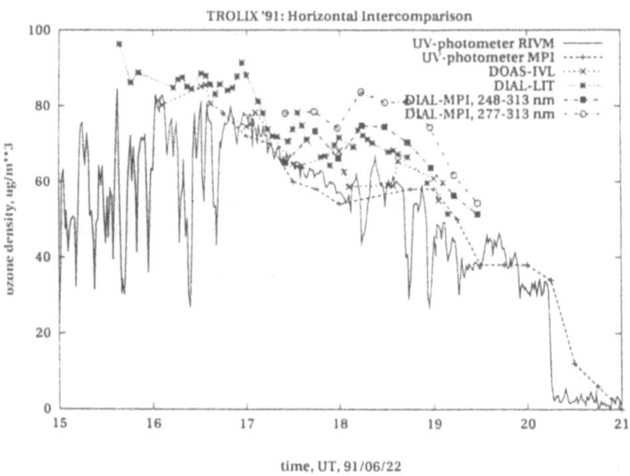

Fig. 5.5: Results from the horizontal intercomparison on June 22, 1991.

The aerosol extinction error was quite small, < 4 μg/m^3 and was estimated by the slope method. This method could be used, since the aerosol interference problems are much reduced in horizontal measurements when averaged over several minutes. It was not likely that horizontal gradients in the aerosol backscatter coefficient had persisted at the same location. In this respect the atmosphere could be assumed almost homogeneous. Therefore, the differential backscatter gradient errors were small, while the differential aerosol extinction error could be estimated by the slope method. The interference due to SO_2 was negligible due to low abundance.

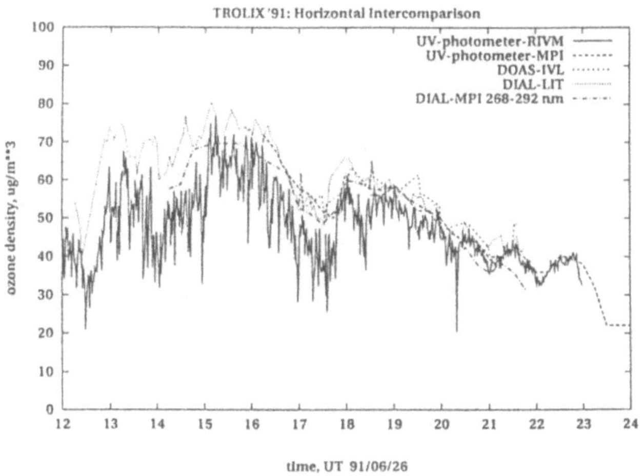

Fig. 5.6: Results from the horizontal intercomparison on June 26, 1991.

These results show that, in a homogeneous airmass, excellent agreement was achieved between lidar, DOAS and standard *in-situ* instruments.

Flight intercomparison

During the campaign three helicopter flights were carried out, during one of which technical problems occurred. Both successful flights provided in situ measurements of the vertical ozone profile up to about 1500 m. The meteorological conditions were not in favor of lidar measurements because of low clouds. The low clouds made proper alignment of the lidars very difficult.

On June 20, low clouds were broken and highly variable. The aerosol load was also high and variable. The results for this flight had been shown already in Chapter 4, but will be repeated here in Fig. 5.8 for convenience. They show considerable differences between the instruments, but the main feature – a steep increase of ozone towards the top of the mixed layer – can be seen in all traces. It is unclear what caused the large difference (12 - 22 $\mu g/m^3$) between the two in-situ instruments. Both systems were calibrated prior to and after the flight (including the air sampling system), showing no significant deviations [20]. Low clouds were also present on June 24, but the situation was more homogeneous. The results are shown in Fig. 5.7, again repeated from Chapter 4.

Much better agreement was found between the instruments: about 8 $\mu g/m^3$ over most of the measurement range. These results show that the agreement between the instruments is quite good if the atmosphere is rather homogeneous. Inhomogeneous situations are more difficult to interpret.

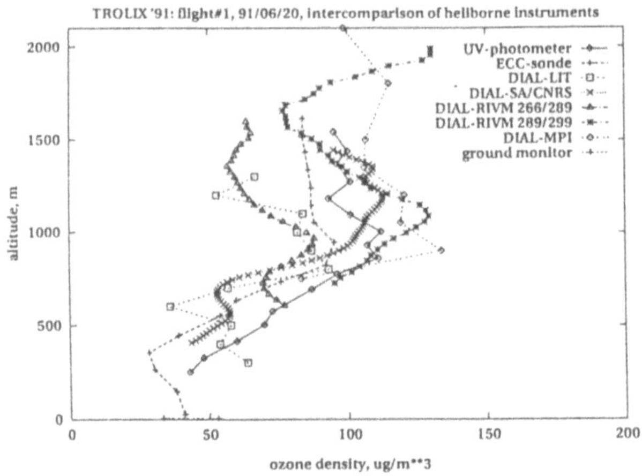

Fig. 5.7: Results from the helicopter flight on June 24, 1991.

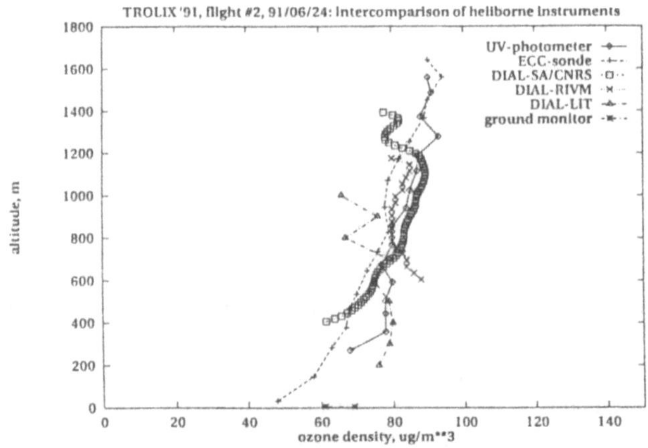

Fig. 5.8: Results from the helicopter flight on June 20, 1991.

Conclusions

The bad weather conditions during TROLIX were not in favor of lidar measurements and, in fact, quite untypical for situations to which tropospheric ozone research is focussed. As a consequence not all goals set for the campaign were reached. However, a unique data set was collected. An important conclusion was that all four tropospheric lidar systems that participated in this experiment, which were quite different in construction, had operated reliably. But also, in all participating tropospheric lidar systems some hardware related problems were detected that restricted the range of operation. On the basis of these results, new studies were initiated to improve the systems, mainly in the areas of:

- suppression of detector memory effects
- adjustment of overlap between transmitter and receiver
- data acquisition systems.

It was further concluded that for a homogeneous aerosol distribution, excellent accuracy of ozone lidar measurements was demonstrated. For inhomogeneous aerosol distributions large errors are introduced. The correction scheme has to be improved to yield accurate results under such adverse conditions and more work has to go into a sufficient characterization of aerosol properties. High temporal and spatial resolution was achieved for measurements of the vertical distribution of ozone. Further, *in-situ* measurements of vertical ozone profiles, at least when taken from a helicopter, have their own problems.

5.2.4 Comparison between lidar, ECC sondes, aircraft based measurements and point sampling monitors, 1991

T. Trickl

At IFU a KrF-excimer based tropospheric ozone lidar system was developed [21]. To validate the instrument, a total of four ECC sonde ascends were carried out at IFU in 1991. Due to transmission problems during the flight only two of those measurements were successful. After applying substantial corrections to the sonde data in order to reproduce the ozone density at ground level and the mountain-top monitoring stations of the institute, reasonable agreement with the lidar results was obtained.

Throughout 1991, lidar values were compared with those from the IFU monitoring stations in the valley and an the Wank and Zugspitze summits (up to 2960~m above sea level). Major discrepancies could only be seen in case of advection from the south, and were ascribed to wind inhomogeneities caused by the Alps.

Another comparison with an ozone profile determined by the IFU aircraft group also yielded good results. The airplane descended over Garmisch-Partenkirchen just 5 minutes before the start of one of the ECC sonde ascends. Thus, the comparison with the results from lidar and sonde sounding should be meaningful. The three simultaneously obtained ozone distributions are shown in Fig. 5.9 together with the corresponding densities recorded at the nearby IFU monitoring stations, demarcated in the figure by G, W and Z. The original ozone densities from the ECC sonde were multiplied by a factor of 0.72 in order to obtain good agreement with the monitoring stations. On the average, the differences stay within

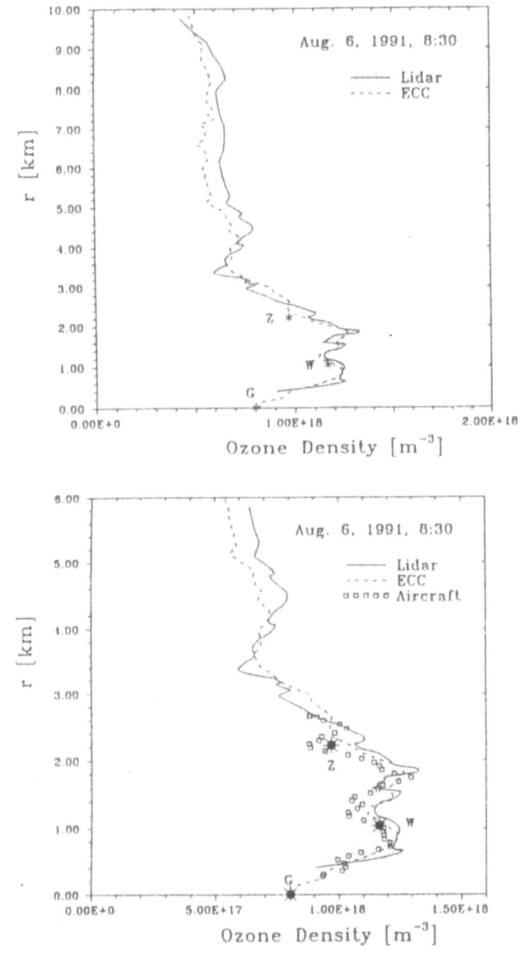

Fig. 5.9: Intercomparison of three vertical ozone profiles simultaneously recorded in Garmisch-Partenkirchen, Germany, on August 6, 1991, at about 08:30 CET. a) Full range of the lidar trace (only the lidar and ECC data are shown). b) Lower section of a), also including the results of the aircraft-based measurements. G, W and Z denote the half-hour averaged ozone densities from the IFU stations at Garmisch-Partenkirchen (G), on Wank (W) and Zugspitze (Z).

the ±5 % limit initially proposed for the IFU tropospheric ozone lidar system. A detailed discussion of the main error contributions is given in Kempfer *et al.* [21]. It should be mentioned here that the densities derived from the lidar measurements on that day (August 6, 1993) are substantially lower in the free troposphere than those from the nearby Hohenpeissenberg observatory. Since this has been the case also for other days more work will be needed to identify the source of this discrepancy.

Conclusions

The average reliability of the IFU system was estimated to lie within 4 μg/m^3 (3.1 ppb near the ground). If the detector gain and the position of the range-gate are not set with extreme care, problems with averaged single bit steps in the 277 nm backscatter profiles may occur in regions of low signal. These steps result in small, but distinct ozone density oscillations. The use of 12-bit digitizers together with the dithering technique (see, *e.g.*, [22]) will help to avoid these artifacts.

5.2.5 Verification measurement of the ARGOS system, June 1992

U.B. Goers, R. Baumgart and C. Weitkamp

Prior to the numerous measurements of tropospheric ozone distributions with the differential absorption lidar ARGOS developed at GKSS, a verification experiment was carried out. The main objectives were to verify the hardware and software performance, the concentration extraction algorithm and the validity of the ozone absorption cross sections used for the signal and for the reference wavelengths.

Strategy

In this experiment a UV-photometric analyzer was used as a reference. The detection limit and precision of the analyzer (Thermoelectron Model 49) as given by the manufacturer are 2 ppb or about 4 μg/m^3. The sampling system connected to the analyzer was mounted to a mast located in a forested area near Albsfelde, 40 km east of Hamburg. Only the uppermost sampling orifice at 47 m above ground level was well above canopy and accessible to the lidar. The azimuth and elevation angles of the lidar were 79° (east by north) and 3°, respectively. The lidar beam crossed an open field on its first 700~m and then passed closely above the treetops before it reached the area of interest at a distance of 1030 m.

Measurement conditions

The verification measurements were carried out on two sunny and very warm summer days. The lidar data were taken at intervals of 26 minutes in a depth bin of 150 m length centered at the position of the mast. This temporal and spatial resolution yielded a precision of the lidar data of ± 3 μg/m^3. However, for a few daytime measurements, when ozone concentrations exceeded 150 μg/m^3, light absorption by ozone was so strong and the visibility decreased so markedly due to photochemical smog that data somewhat closer to the lidar had to be used for the comparison. A 4 hour interruption in the sequence of lidar measurements occurred when the power generator broke down around noon on July 20, 1992.

Results

The results of the measurements are shown in Fig. 5.10. The agreement of the *in-situ* and the lidar measurements is excellent, particularly during the period of low turbulence between midnight and about noon on July 21, 1992. The Rayleigh scattering term in the determination of the concentration is 3 ppb for the wavelengths used. The particle extinction correction turned out to be of the same magnitude. No parameters of any kind were fitted in the evaluation of the data [23]; the ozone absorption cross sections used are those of Molina and Molina [24] and the sulfur dioxide cross sections necessary to avoid SO_2 interference were taken from Weitkamp *et al.* [25].

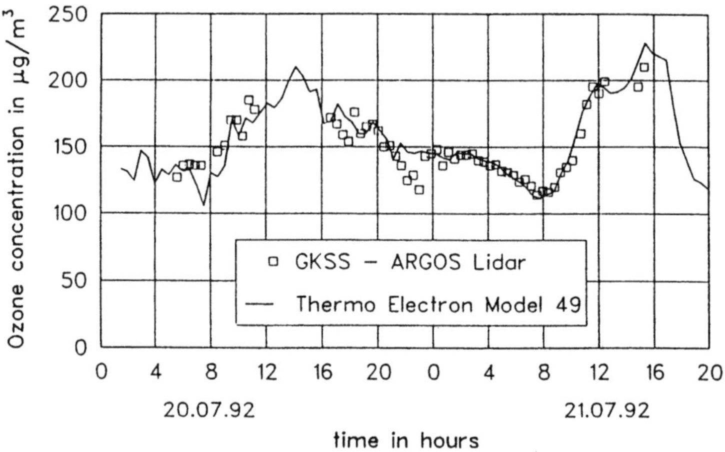

Fig. 5.10: Results from the verification measurement on June 20-21, 1992. The lidar depth bin width was 150 m. The number of laser pulse pairs was 15000. Measurement height of the *in-situ* device was 47 m.

To ascertain the explanation for the particularly good agreement during calm nighttime hours a plot of all four *in-situ* measurement time series is shown in Fig. 5.11. As the figure shows, periods of large scatter of the *in situ*-data coincide with periods of large scatter of the lidar results, and although the *average* agreement is still very good, individual data points differ somewhat more from one another than during the periods with little turbulence. Fig. 5.11 also shows that during periods with non-vanishing ozone concentration gradients, the lidar measurements agree more closely with the *in-situ* results from 47 m than with those taken at lower heights. This means that the (small) ozone concentration gradients that were determined from the *in-situ* data could equally well be measured with a lidar system.

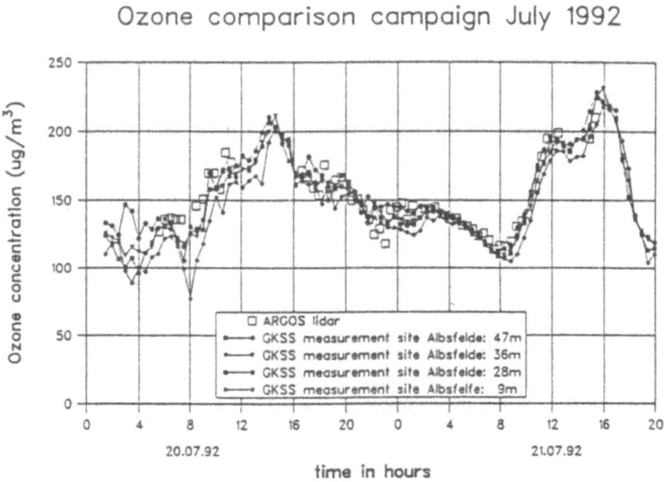

Fig. 5.11: Results from the verification measurement on June 20-21, 1992. The same results as in the previous figure are shown, together with additional measurements from the *in-situ* device at heights of 9, 28, 36 and 47 m.

Conclusions

The comparison of the lidar measurements with *in-situ* ozone concentration determinations with a well-calibrated UV-photometric analyzer have shown that there is no systematic deviation of the lidar data from the *in-situ* data within the $\pm 4\ \mu g/m^3$ and $\pm 3\ \mu g/m^3$ uncertainties of the UV-analyzer and the lidar system, respectively. This confirms the suitability of the lidar technique for tropospheric ozone concentration determinations, and that the comparison of the results from the two devices is a simple and efficient means for lidar verification measurements, at least in the horizontal operating mode.

The experiment described proves the accuracy of the lidar results *at one given distance*. A lidar result verification over the whole length of the lidar range would have to confirm the results presented point by point from minimum to maximum measurement distance. The tower at Albsfelde, however, would not have lent itself to such a measurement because of the comparatively low height and its forested surroundings which, in particular, prevent measurements from a distance closer than 1000 m.

5.2.6 Validation of infrared tropospheric ozone lidar, spring 1993

R. Barbini, F. Colao, A. Palucci and S. Ribezzo

An infrared CO_2 station has been installed at the ENEA Research Centre in Frascati, Italy, as a test site for studying laser remote sensing hardware and software in the 10~µm wavelength range in the framework of a more general program aimed at assessing the capabilities of IR lidars to selectively detect atmospheric gaseous pollution. The station has been fully operational since 1991 and has contributed greatly to the advances in understanding the performance of 10 μm lidar systems for remote monitoring of minor atmospheric constituents, infrared-absorbing gaseous pollutants, wind velocities and air temperatures [26, 27].

Results

Validation tests have been carried out using a mobile facility carrying standard *in-situ* instruments. The ENEA van was moved to a number of points along the horizontal line of sight of the lidar during a one week intercomparison campaign, held in June, 1993. In Fig. 5.12 the ozone profile obtained by lidar is compared to the *in-situ* data. Although the values from the lidar are close to its lower sensitivity limit (50 ppb at 200 m spatial resolution) reasonable agreement is found over a major part of the measurement range of 3 km.

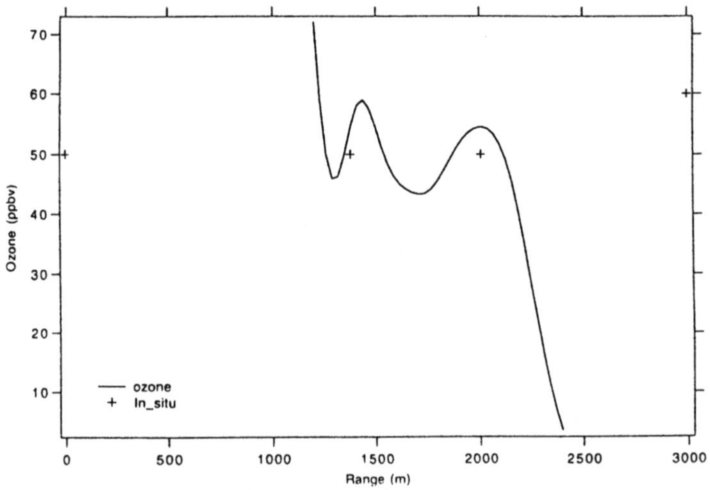

Fig. 5.12: Comparison between ENEA infrared ozone lidar and UV-photometric analyzer.

5.2.7 Simultaneous tropospheric ozone monitoring using lidar and DOAS systems, winter 1994

H. Edner, P. Ragnarson, S. Svanberg and E. Wallinder

Three remote-sensing systems for atmospheric trace gas monitoring have been constructed at the Lund Institute of Technology and have been applied simultaneously to monitor tropospheric ozone. An ozone DIAL system based on a Raman-shifted KrF excimer laser could be used for both vertical and horizontal profiling. A second DIAL system based on a Nd:YAG pumped dye-laser source was used for horizontal and slant-angle measurements. A differential optical absorption spectroscopy (DOAS) system, operated with high-pressure xenon lamps, was also used to measure the average ozone concentration over a horizontal path [28]. The scenario for the measurements is shown in Fig. 5.13. The measurement path of the excimer-based lidar system and the DOAS system were approximately the same. The path stretched 2 km right over the city of Lund. The measurement path of the Nd:YAG based system was chosen in the same horizontal plane, but perpendicular to the path covered by both other systems.

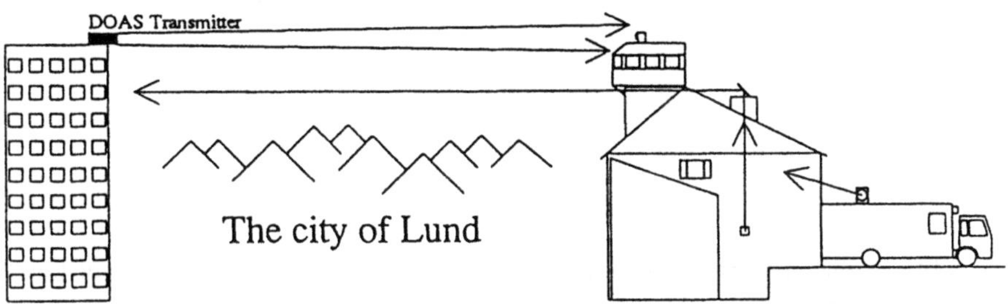

Fig. 5.13: Scenario for simultaneous measurements of tropospheric ozone using lidar and DOAS systems in Lund, Sweden, 1994.

Results

Simultaneous ozone measurements from the three systems obtained during one day are shown in Fig. 5.14, and Fig. 5.15 shows the cross correlation between the concentrations deduced from the DOAS and lidar systems over several days. The horizontal mode of operation was chosen for the lidar systems, in order to probe the same air mass, this, however was not achieved for the-dye-laser based system. A value of 13.4 $\mu g/m^3$ for the excimer-based system, and 7.5 $\mu g/m^3$ for the dye-laser-based system were subtracted from the ozone concentration values to correct for effects from molecular extinction. A correction for aerosol extinction amounting to 2-17 $\mu g/m^3$ and 1-8 $\mu g/m^3$, respectively, changing slightly during the days, was also subtracted.

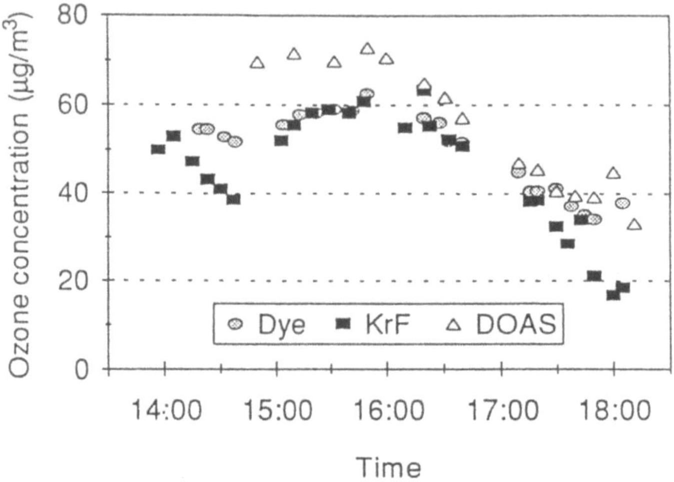

Fig. 5.14: Ozone concentrations measured over a horizontal path over the city of Lund, Sweden, 1994.

Discussion

It is interesting to compare the results of DIAL and DOAS measurements, since they both were made in a large air volume along the optical path. Both techniques utilize optical absorption, but where DIAL only employs two wavelengths with rather large separation the DOAS technique uses small higher order structures on the ozone absorption profile. The DOAS evaluation will normally not be affected by extinction due to aerosol scattering. In Fig. 5.15 the best linear fits of the correlation between the data sets are shown. The coefficients of these linear fits were 0.88 and 1.01 for the dye laser based and KrF laser based systems, respectively. The latter had a mean negative offset of 15 μg/m^3, but also a bit more scattered data. The difference between the DOAS and the dye laser based system might be explained by differences in the air masses probed.

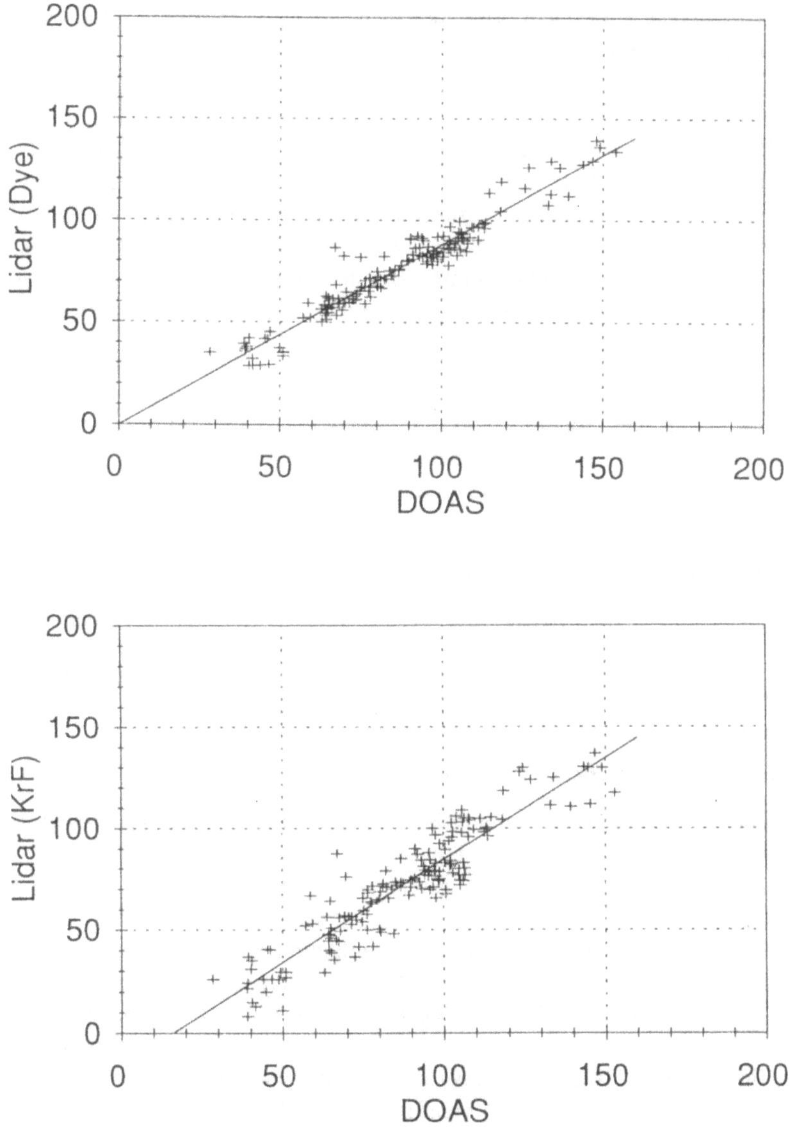

Fig. 5.15: Cross correlation between the concentrations deduced from the DOAS and lidar systems over several days, Lund, Sweden, 1994.

5.2.8 Comparison between lidar and tethered and free flying ECC sondes, summer 1994

J. Bösenberg, G.C. Grabbe, V. Matthias, and C. Senff

The MPI tropospheric ozone lidar has been used for extensive measurements of the vertical ozone distribution in the lower troposphere since 1992. In particular high resolution studies of the vertical transport in the boundary layer were conducted [29]. For assessing the accuracy of these high resolution measurements, it is particularly important when performing an intercomparison with other instruments that the same air volume is probed, since there is a rather large variability at small scales.

For this purpose, an intercomparison campaign has been organized in June 1994 at the Meteorologisches Obervatorium Lindenberg (MOL) to compare lidar retrievals to *in-situ* measurements of the vertical ozone distribution using ECC sondes carried by both tethered and free flying balloons.

The lidar system was installed close to the launch point of the tethered balloon on the MOL site, to make sure that the distance between the probed volumes remains small. The free flying balloon was launched at some distance from the lidar. Additional to the lidar measurements of ozone, ground level ozone was measured using a UV photometric ambient ozone analyzer (Environment S.A., Type $O_3$41M).

Comparison lidar - tethered balloon

For all intercomparisons between lidar and tethersondes it was attempted to keep the flight altitude of the balloon as constant as possible, with different heights selected according to the flight conditions. Due to dynamic forces, deviations less than 50 m occurred for most of the cases, for stronger winds deviations of more than 100 m were observed. During this campaign, seven intercomparisons with tethersondes were attempted, six of them were successful. Of these six intercomparisons, four could be extended for 30 to 90 minutes, two times the balloon was not sufficiently stable in height or did not reach the minimum height of the lidar (200 - 210 m) to allow simultaneous measurements for more than a few minutes. For the two short measurements only the absolute values of both instruments were compared, for the longer time series the trend in time and the variances were compared, too.

For one of these measurements, the results are shown in Fig. 5.16. The close agreement of both results is obvious. In Table 5.5 the results of all 6 intercomparison periods are summarized. Not only the mean values, but also the variances have been compared. The main result is that the mean values agree

Table 5.5: Mean values and standard deviations observed with DIAL and ECC sondes attached to a tethersonde during the intercomparison experiment in Lindenberg, 1994.

Day June '94	Time (UT)	Comp. height range / m	λ(lidar) / nm	Mean differ. μg/m³	Rel. differ. %	σ(diffs) μg/m³	σ_{O3}^{DIAL} μg/m³	σ_{O3}^{ECC} μg/m³
8th	03:32-04:21	270-330	268/292	<–0.1	<0.2	5.6	4.63	1.86
8th	08:34-10:10	270-330	268/292	2.3	3.8	2.6	2.72	1.50
8th	14:04-15:01	410-500	268/292	1.9	1.7	2.7	1.17	1.51
14th	14:44-15:19	310-400	277/313	5.8	6.0	2.1	1.26	0.82
14th	19:22-19:45	200-260	277/313	–6.7	–5.6	–	–	–
20th	15:47-15:52	210-300	268/292	–4.4	–3.6	–	–	–

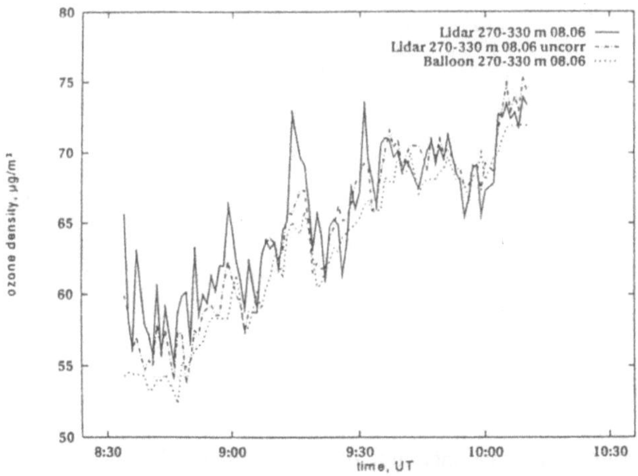

Fig. 5.16: Comparison between lidar and tethered balloon on June 14, 1994 in the altitude interval between 310 and 400 m.

within 2 to 6 μg m^{-3}. The variance of the lidar results for two cases is substantially larger than those measured with the ECC sonde, for the remaining two cases bot variances compare well. A more detailed analysis of the variance spectra shows that the differences are neither due to too slow response of the ECC sonde nor to large random noise in the DIAL data. The two cases with enhanced lidar derived variance correspond to cases, where the measurements were performed in a layer of relatively large aerosol gradients at the top of the mixed layer or at an internal boundary layer, repectively. It is therefore suspected that residual errors from the aerosol correction are responsible for the extra variance of the lidar results.

It may be noted here that the analysis of the variance spectra provides an excellent means to estimate the noise level of the DIAL results. Since consecutive lidar measurements are completely independent their noise spectrum has to be white. When the atmospheric variance is small at the high-frequency end of the spectrum the remaining spectral density can be used to estimate an upper limit for the random noise. Fig. 5.17 shows as an example a measured variance spectrum for a two-hours measurement on June 1, 1993. Measurement height was 525 m, resolution 10 s and 75 m in time and height, respectively. The indicated slope is that expected for isotropic turbulence in the so called inertial subrange, so up to about 0.005 Hz the spectrum appears to be dominated by atmospheric effects. Beyond that, the spectrum flattens out to a constant level, corresponding to a noise level of 1.8 μg m^{-3}. Similar values are mostly observed for high resolution boundary layer measurements, which is considered an excellent result.

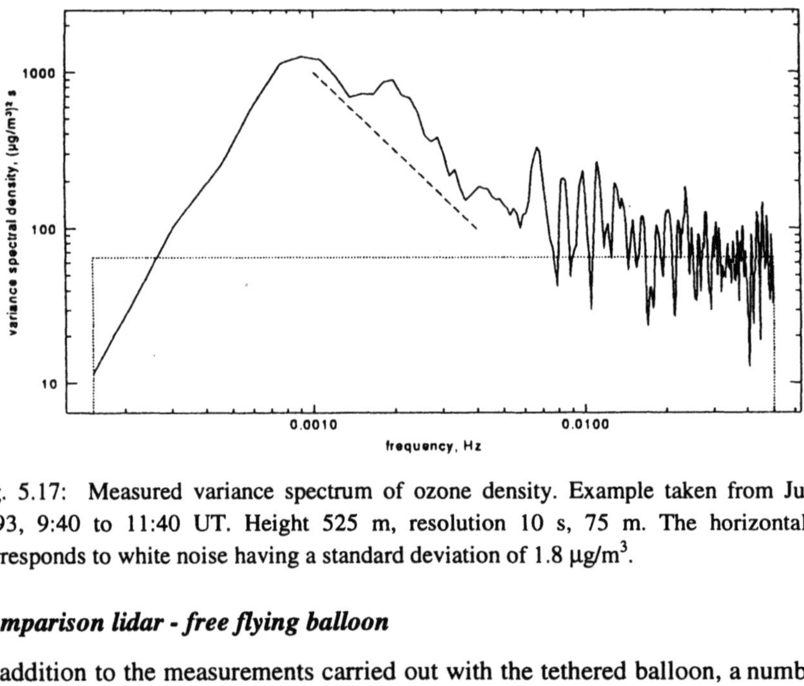

Fig. 5.17: Measured variance spectrum of ozone density. Example taken from June 1, 1993, 9:40 to 11:40 UT. Height 525 m, resolution 10 s, 75 m. The horizontal line corresponds to white noise having a standard deviation of 1.8 μg/m^3.

Comparison lidar - free flying balloon

In addition to the measurements carried out with the tethered balloon, a number of 10 free flying balloons were launched. In Fig. 5.18 the vertical ozone distribution is shown as measured by the lidar, the ECC sonde. The lidar was optimized for use in the altitude range between 1200 and 2600 m. The lidar values were averaged over 8 minutes and the vertical resolution was 90 m. There is was difference of 20 μg/m^3 between the ground level monitor, which was installed close to the lidar, and the first data point of the ECC sonde, which was launched at some distance away from the lidar. Data at the low end of the lidar range were inside the mixed layer and were close to the values from the sonde. At the transition from the boundary layer to the free troposphere deviations up to 10 μg/m^3 could be seen.

These differences were probably due to problems related to the retrieval of ozone in regions of spatially inhomogeneous aerosol, but also to the limited response of the sonde when flying through layers with large gradients in the ozone concentration. In the free troposphere between 1600 and 2300 m the differences between ECC sonde and lidar were less than 5 μg/m³. Above 2350 m both instruments showed an ozone concentration decreasing with altitude, but at slightly different altitudes. Since the free flying balloon had drifted about 3.5 km with the wind the difference could be ascribed to a difference in measurement volumes. The differences between the lidar and ECC values showed a standard deviation of 4.6 μg/m³ or 5.5 %.

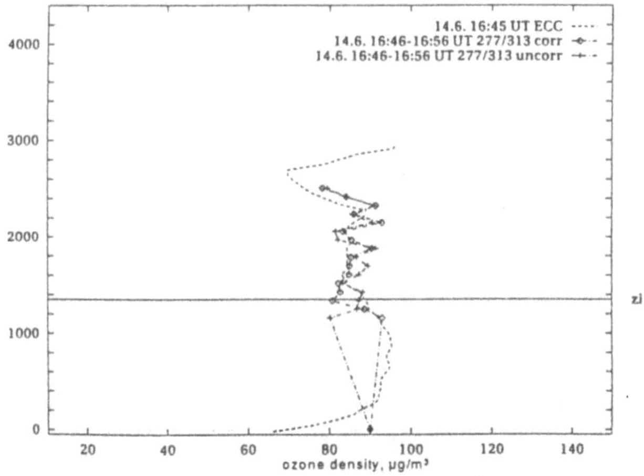

Fig. 5.18: Comparison between lidar and free flying balloon on June 14, 1994.

In Fig. 5.19 the vertical ozone distribution is shown as measured by the lidar, the ECC sonde and the ground level monitor. The lidar was optimized for use in the altitude range between 1000 and 1600 m during the ascent of the sonde. Before and after the launch of the balloon the lidar was optimized for measurements between 400 and 800 m. The sonde profile shows a sharp increase of ozone in the lowest 300 m. This is probably due to the stable nocturnal layer structure. In this layer ozone is depleted during the night. A residual layer started at about 300 m, which is representative for the planetary boundary layer from the previous day. The pronounced ozone minimum at 900 m altitude in the ECC measurements was not related to any recognizable boundary between atmospheric layers. A possible cause could be a high SO_2 concentration, which has a known lowering effect on the ozone values produced by the ECC sonde [30]. Possibly advection of SO_2 from a browncoal power plant about 50 km away from the MOL site had taken place. This presumption was supported by data from the radio sonde launched at 6:00 UTC, which showed wind speeds up to 6 m/s above the stable planetary boundary layer from the direction of the power plant. Since no measurements of

SO_2 or any other trace gases were made, the reason for the pronounced minimum could not be determined.

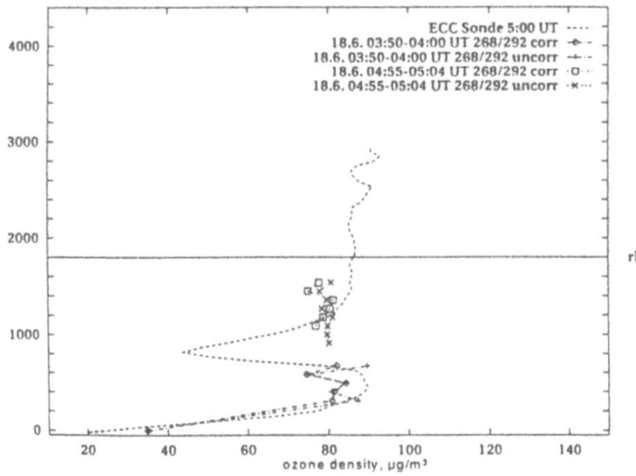

Fig. 5.19: Comparison between lidar and free flying balloon on June 18, 1994.

Discussion

Good agreement was found between lidar measurements and ECC sonde measurements of the vertical ozone distribution. The overall differences that were found amount to about 5 $\mu g/m^3$. For the cases where larger differences were found in certain altitude intervals plausible explanations could be found. For the comparison of variances the agreement is less favorable for cases, where the measurements were performed in layers of enhanced aerosol gradients. The additional variance of the lidar data is attributed to errors remaining from the aerosol corrections. In cases of a more homogeneous aerosol distribution, the agreement found in the measured variances is good. A more detailed analysis of these intercomparisons is found in [31].

5.2.9 Comparison between lidar and an ECC sonde, summer 1995

A. Apituley, E.P. Visser, M. de Winter and M.A.F. Allaart

At RIVM a tropospheric ozone lidar system based on two Nd:YAG lasers and stimulated Raman shifting in hydrogen and deuterium has been constructed [32]. Intensive measurements, on a daily basis, weather permitting, are being carried out since September 1994.

Since November 1993 the Royal Dutch Meteorological Institute (KNMI), located about 2 km away from the RIVM lidar site, performs balloon soundings of ozone using ECC sondes. Normally, these soundings are carried out once a week. It was decided to do lidar ozone measurements concurrently with the balloon soundings (weather permitting) so that an intercomparison on a routine basis would be possible. An example of a lidar and an ECC ozone profile, measured at about the same time, is given in Fig. 5.20. It should be noted that the sonde data are momentary values measured during the ascent of the balloon, while the lidar profile was contructed from merging four separate measurements of 15 minutes (9000 laser shots each), where signal blocking from close range by means of a mechanical chopper was varied between: no blocking, blocking up to 2 km, blocking up to 3 km and blocking up to 4.5 km. The air masses sampled by both techniques were not the same (in addition to the differences already present due to the geographical separation of the lidar system and the launch pad of the sonde.

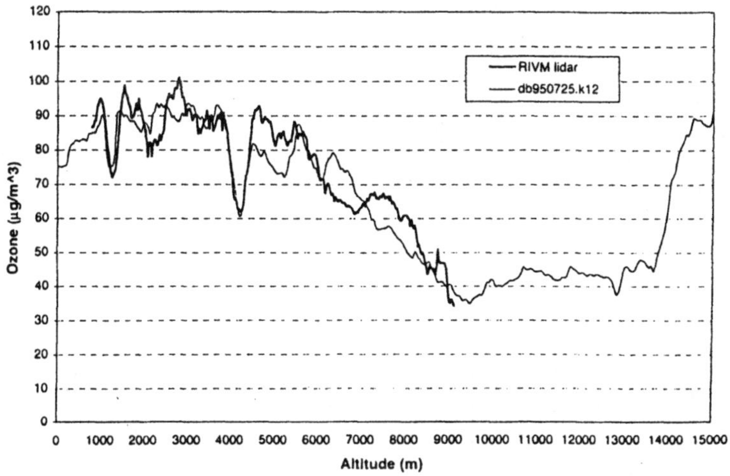

Fig. 5.20: Comparison between ozone profiles obtained by the RIVM tropospheric ozone lidar and a KNMI ECC balloon sounding, summer 1995.

5.3 Summary of results

In Table 5.6 the results of the system described in the above sections are summarized. For each lidar system a range interval can be defined where the instrument provides excellent results at high temporal and spatial resolution. This range interval is different for each system.

Intercomparisons with free flying ozone sondes are very often problematic. Not only are results from the lidar compared to the sondes difficult to interpret. Also, the results from different types of sondes are not always consistent; sometimes Brewer-Mast sondes show significantly lower ozone values than ECC sondes [3, 4], sometimes it is *vice versa* [5]. And further, comparisons of sondes with a UV-photometer show questionable results [7]. Therefore, the intercomparisons of tropospheric ozone lidar systems with ozone sondes should be treated with caution.

Table 5.6: Summary of the results from the comparative measurements presented in this chapter. In the column 'Wavelength Tuning', SRS indicates stimulated Raman scattering and Dye indicates the use of a dye laser. In the column 'Comparative Instrument', ECC indicates an ECC sonde, BM indicates a Brewer-Mast sonde and UV indicates a UV-photometric ozone analyzer. The operational mode, listed in the column 'Mode', was either horizontal (H) or Vertical (V). The values in the column 'Accuracy' give the absolute difference between the mean of the lidar measurements and the comparative measurements in the range interval and at the temporal resolution mentioned in the table. The precision given is the standard deviation of the lidar measurements. The CNRS lidar system was involved in several campaigns. This is indicated in the comparative instrument column. (a) indicates the campaign described in Sec. 5.2.1, (b) indicates the work described in Sec. 5.2.2.

Laser Source	Insititute	Wavelength tuning	Wavelengths / nm	Comparative instrument	Mode (H/V)	Range interval / km	Temporal resoln / min	Accuracy $\mu g/m^3$	Precision %
KrF	MPI	$SRSD_2$	268/292	ECC	V	0.2-.5	1	4-6	4-6
KrF	MPI	$SRSD_2$	268/292, 292/319	ECC	V	0.3-3.5	5-10	5	5
KrF	MPI	$SRSH_2$	277/313	ECC	V	1.0-2.5	5-10	3	3
KrF	IFU	$SRSH_2$	277/313	ECC/UV	V	0.4-10	12	6	—
KrF	LIT	$SRSH_2$	277/313	DOAS	H	0-2	5	15	5
Nd:YAG	LIT	Dye	278.7/286.4	DOAS	H	0-2	5	10	2
Nd:YAG	GKSS	Dye	280.91/282.72	UV	H	0.955-1.105	26	4	2
Nd:YAG	CNRS	$SRSD_2/H_2$	289/299	ECC(a)	V	6.5-9.5	30	8	-
Nd:YAG	CNRS	$SRSD_2/H_2$	289/299	BM(a)	V	6.5-9.5	30	11	-
Nd:YAG	CNRS	$SRSD_2/H_2$	289/299	ECC/BM/UV(b)	V	4.5-8	60	4	5
CO_2	ENEA	Grating	9504/9569	UV	H	1.25-2.5	2	4	—

For intercomparisons with lidars pointing horizontally, very good agreement is almost always obtained. There is less confusion about the air masses probed by the lidar and the other instruments (*e.g.*, UV-photometer or DOAS) and the horizontal structure of the atmosphere can generally be considered homogeneous (at least when averaging over several tens of minutes).

Again, the lidar systems described here are very different, measuring at different ranges with a very small region of overlap and using different measurement wavelengths. Also, the results presented were obtained at different locations, under varying circumstances. This makes comparison between the lidar systems among each other very difficult. However, the instrumental errors are more variable than the differences expected from lidar systems operating at different wavelengths. This implies that instrumental errors are important. Error sources of particular importance are the non-linearity of the detection (both in analog mode and photon counting mode), and optical alignment problems.

5.4 References

1. A. Brewer, J. Milford; The Oxford key ozone sonde, *Proc. Roy. Soc.* **256** (1960) 470.
2. W. Komhyr; Electrochemical concentration sondes for gas analysis, *Ann. Geophys.* **25** (1969) 203.
3. W. Attmannspacher, Dütsch; 2nd International ozone sonde intercomparison at the Observatory of Hohenpeissenberg, *Berichte des Deutschen Wetterdienst* **157** (1981) Deutschen Wetterdienst.
4. E. Hilsenrath, W. Attmannspacher, A. Bass, W. Evans, R. Hagemeyer, R. Barnes, W. Komhyr, K. Mauersberger, J. Mentall, M. Profitt, D. Robbins, S. Taylor, A. Torres, E. Weinstock; Results from the balloon ozone intercomparison campaign (BOIC), *J. Geophys. Res.* **91** (1986) 13137.
5. WMO; Third WMO intercomparison of the ozone sondes used in the global ozone observing system (Vanscoy, Canada, May 13-24, 1991), *World Meteorological Organization global ozone research and monitoring project, Report* 27, World Meteorological Organization, Geneva 1991.
6. M. Beekmann, G. Ancellet, D. Kley, G. Mégie, H. Smit; Intercomparison campaign for vertical ozone profiling in the troposphere at the Observatoire de Haute Provence, 1989: electrochemical sondes of ECC and Brewer-Mast type and a ground based lidar, *J. Atmos. Chem.* **19** (1994) 259.
7. M. Beekmann, G. Ancellet, D. Martin, C. Abonnel, G. Duverneuil, F. Eideliman, P. Bessemoulin, N. Fritz, E. Gizard; Intercomparison of tropospheric ozone profiles obtained by electrochemical sondes, a ground based lidar and an airborne UV photometer. *Atmos. Environ.* in press.
8. E.V. Browell, S. Ismail, S.T. Shipley; Ultraviolet DIAL measurements of O_3 profiles in regions of spatially inhomogeneous aerosols, *Appl. Opt.* 24 (1985) 2827.
9. D. Martin, Comparaison de profils d'ozone, Campagne OHP 91: Rapport final. *CNRS report*, CNRS programme environment, Paris 1992.
10. U. Platt, D. Perner, H. Platz; Simultaneous measurements of atmospheric CH_2O, O_3, and NO_2 by differential optical absorption, *J. Geophys. Res.* **84** (1979) 6329.

11. G. Ancellet, A. Papayannis, J. Pelon, G. Mégie; DIAL tropospheric ozone measurement using a Nd:YAG laser and the Raman shifting technique, *J. Atmos. Ocean. Tech.* **6** (1989) 832.

12. WMO; Report of the meeting of experts on sources of errors in detec- tion of ozone trends, *World Meteorological Organization global ozone research and monitoring project, Report* 12, World Meteorological Organization, Geneva 1982.

13. C. deJonge; Ground level measurements in the TROLIX'91 campaign. RIVM-report 222201008, RIVM, Bilthoven 1991.

14. C. deJonge; NO2 DIAL measurements in the TROLIX'91 campaign. RIVM-report 222201003, RIVM, Bilthoven 1991.

15. C. deJonge; Planetary boundary layer measurements in the TROLIX'91 campaign. RIVM-report 222201004, RIVM, Bilthoven 1991.

16. J. Bergwerff; Airborne ozone measurements in the TROLIX'91 cam- paign. RIVM-report 222201005, RIVM, Bilthoven 1991.

17. A. Apituley; RIVM Tropospheric Ozone Lidar Measurements during TROLIX'91. RIVM-report 222201007, RIVM, Bilthoven 1991.

18. G. Sonnemann, G. v. Cossart, J. Fiedler; Evaluation of ozone- (ECC-) and meteorological sonde data during TROLIX'91, *Technical report*, Institut für atmosphärenphysik Kuhlungsborn, Kuhlungsborn 1991.

19. A. Apituley; Comparison of the RIVM Tropospheric Ozone LIDAR to in situ measuring instruments using data acquired during TROLIX, *RIVM-report* 722701001, RIVM, Bilthoven 1995.

20. J. Bösenberg, G. Ancellet, A. Apituley, H. Edner, B. Galle, G. Sonnemann, *et al.*; Tropospheric ozone lidar intercomparison experiment, TROLIX '91: Field phase report, *MPI report* 120, MPI, Hamburg 1993.

21. U. Kempfer, W. Carnuth, R. Lotz, T. Trickl; A wide-range ultraviolet lidar system for tropospheric ozone measurements - development and application. *Rev. Sci. Instrum.* **65** (1994) 3145.

22. A. Langford, T. O'Leary, M. Profitt; Extending the dynamic range of differential absorption lidar measurements through large scale dithering, in: *Proc. 17th Int. Laser Radar Conf.* (XVII ILRC), Sendai, Japan 1994, p. 173.

24. L. Molina, M.J. Molina; Absolute absorption cross sections of ozone in the 185- to 350-nm wavelength range, *J. Geophys. Res.* **91** (1986) 14501.

25. C. Weitkamp, O. Thomsen, P. Bisling; Mess- und Vergleichswellenlangen zur Elimination von SO_2-Querempfindlichkeiten bei der Lidarfernmessung troposphärischen Ozons, *Laser und Optoelektronik* **24** (1992) 46.

26. R. Barbini, F. Coalo, T. Hermsen, A. Palucci, S. Ribezzo; Atmospheric water vapour measurements with the ENEA lidar station, *ENEA internal report* RT/INN/90/53, ENEA, Frascati, Italy 1990.

27. R. Barbini, F. Coalo, T. Hermsen, A. Palucci, S. Ribezzo; Preliminary measurements of vertical ozone profile with the ENEA lidar station, *ENEA internal report* RT/INN/91/21, ENEA, Frascati, Italy 1991.

28. H. Edner, P. Ragnarson, S. Spannare, S. Svanberg; Differential optical absorption spectroscopy (DOAS) system for urban atmospheric pollution monitoring, *Appl. Opt.* **32** (1993) 327.

29. C. Senff, J. Bösenberg, G. Peters, T. Schaberl; Remote sensing of turbulent ozone fluxes and the ozone budget in the convective boundary layer with DIAL and radar-RASS: A case study, *Contrib. Atmos. Phys.* in press.

30. A. Schenkel, B. Broder; Interference of some trace gases with ozone measurements by the KI method, *Atmos. Environ.* **16** (1982) 2187-2190.

31. G. Grabbe, J. Bösenberg, H. Dier, U. Gorsdorff, V. Matthias, G. Peters, T. Schaberl, C. Senff; Intercomparison of Ozone Measurements between Lidar and ECC sondes, *Contrib. Atmos. Phys.* in press.

32. J.A. Sunesson, A. Apituley, D. Swart; Differential absorp- tion lidar system for routine monitoring of tropospheric ozone. *Appl. Opt.* **33** (1994) 7045.

Chapter 6

Applications

6.1 The CNRS Contribution

Gerard Ancellet

Service d'Aéronomie du CNRS, Université Paris 6, Boite 102, 4 Place Jussieu, 75230 Paris Cedex 05, France

Within the TESLAS program, the CNRS/SA objectives were to improve the quality of ozone measurements both in the free troposphere (3-15 km) and in the boundary layer (0.4-3 km). The CNRS has developed compact and easy to operate lidar systems for implementation at the Observatoire de Haute Provence (OHP), which is part of the TOR network, and on an aircraft platform. Both systems are using a Nd-YAG laser and stimulated Raman shifting techniques for the generation of (up to 3) wavelengths in the near UV spectral range.

6.1.1 A 5-year lidar data base for the OHP station

The ozone lidar system developed for ozone monitoring in the free troposphere was installed at the OHP in August 1989 and is operated regularly since August 1990 [1]. However, the sampling was rather scarce at the beginning of the period which was devoted to the instrument optimisation (no analog detection before July 1990, change of the laser source in November 1992, manpower availability to run the lidar). The OHP lidar system is described in more detail in Section 3.6.1. The error terms for the ozone lidar are generally well understood, and some of them can be corrected when they are systematic. The main errors are the following:

i) statistical errors

Statistical errors are calculated from the raw signal assuming a Poissondistributed photon flux. This error remains too high for an integration time of 30 min and the nominal height resolution of the photon counting system of 150 m (18 % at 8 km). Therefore, a running digital low pass filter with an increasing window size is applied to the raw signal degrading the height resolution to 300 m at 6 km altitude and 1.5 km at 15 km. The digital filtering reduces the statistical error for nighttime measurements to an average value of 6 % in the height range 3.5 - 9.5 km (Table 6.1).

Table 6.1: OHP lidar error analysis.

Error	Accuracy			Precision		
	3-5 km	5-10 km	10-15 km	3-5 km	5-10 km	10-15 km
statistical				<5 %	6 %	15 %
methodology						
O_3 cross section		1 %				
Rayleigh extinction	<0.5 %			0.5 %		
Aerosol interference		<3 %			3 %	
instrumental						
misalignment	<10 %	<5 %		10 %	5 %	
signal induced noise		<4 %	<15 %		4 %	15 %
total	<10 %	<7 %	<15 %	11 %	9 %	20 %

At higher altitudes, the statistical error increases due to a decreasing signal to noise ratio. A critical value of the statistical error (15 %) was used as a criterion to determine the upper altitude limit of the ozone profiles. This limit is between 10 and 15 km, depending on the laser output energy, the optical transmission of the atmosphere and the level of the background sky light.

ii) errors due to interferences with atmospheric molecules and particles

An atmospheric model of the density and temperature is used in order to determine the variation in the Rayleigh extinction between the two wavelengths and the temperature dependence of the ozone absorption cross section. This model is derived from a temperature profile climatology used in an inversion algorithm for satellite measurements [2]. The error due the use of a model instead of the actual temperature and pressure profiles is less than 1 %. The aerosol load can be estimated from the backscatter signal at 299 nm [3]. The part of the profiles with cloud interferences or large desertic aerosol concentrations, is removed leading to a contribution to the precision and the accuracy which is better than 3 % for the ozone profiles included in the OHP database.

iii) instrumental errors

Two instrumental errors have been identified at the upper and lower limit of the measurement range: an electronic noise of the photomultiplier (upper altitude limit), and a misalignement of the optical system (lower altitude limit). They can be estimated from the analysis of the vertical derivative of each backscatter profile and a first order correction of each signal can be applied before the ozone calculation. The remaining error is then related to the uncertainty in this correction procedure. The effect of the misalignment of the laser beam with respect to the telescope axis is partially removed by using a modelled overlap function between the telescope and the laser beam in order to correct the measured signal. The misalignment is estimated by assuming that each slope of the raw signal at both wavelengths should correspond to the same ozone value in the range 3-4 km. The remaining error after this correction is of the order of 10 % below 6 km and 5 % above 8 km height (Table 6.1). In the upper measurement range, a spurious signal due to the response function of the photomultiplier tube to the large flux of photons from the low altitudes. This noise signal decreases more slowly than the atmospheric signal with height and is modelled by a 2nd order polynomial at altitude ranges where the atmospheric signal can be neglected (> 20 km). The accuracy of this correction is estimated by shifting the height window for the determination of the fitting polynomial from its optimal position. For night-time conditions, this accuracy is roughly a function of the statistical error as the induced signal becomes relatively more important when the atmospheric signal decreases. For example, for a statistical error of 7.5 % and 15 %, the accuracy due to the signal induced noise is respectively about 5 % and 18 %.

From this error analysis, we conclude that the best measurement range for the tropospheric system is between 5-10 km with an overall precision of about 10 % for an integration time of 30 min and a vertical resolution of 500 m. The lidar data base includes now 260 independent profiles (*i.e.*, with a time lag of more than 1 day) for the period 1989-1994. Average ozone values have been calculated in two 1 km deep layers for 5 periods of the year corresponding to the winter period (Nov./Feb.), the transition periods in March and October and the spring (Apr./Jun.) and summer (Jul./Sep.) periods. This is reported for the whole period 1989-1993 in Table 6.2, together with the seasonal variations derived from the Brewer-Mast (1988-1990) and the ECC (1991-1993) measurements performed at the OHP for the same time period. It shows a similar seasonal variation for the three data set with a spring/summer maximum with ozone values 13 ppb higher than the winter values. The differences between the ECC and the lidar in the spring period in the upper troposphere are may be linked to a different sampling of the meteorological situations. Indeed, the monthly averages of the vertical profiles of stability and humidity, which are calculated from the Nîmes sounding station, are quite different for the lidar and ECC measuring days. The low Brewer-Mast data in the upper troposphere during the winter are probably related to a difference in the influence of stratosphere/troposphere exchanges for this data base compared to the other two where the average tropopause level (8.5 km) from November to February was 1 km lower and the static stability in the 6-7 km is also higher. In order to obtain a good temporal coverage, one has to combine the different instrument data bases before studying an ozone trend or a seasonal variation of the ozone production processes.

Table 6.2: Ozone averages in ppb for the lidar (89-93), BM(88-90) and ECC(91-93).

Brewer	av	st	md	av	st	md		av	st	md	nb
		2-3 km			4-5 km				6-7 km		
1	41	8	40	40	9	37		38	12	36	18
2	51	10	53	49	13	53		52	15	50	10
3	53	8	52	54	10	55		59	17	56	19
4	55	9	56	56	13	56		58	15	52	16
5	42	12	43	44	12	43		47	20	41	9
ECC	av	st	md	av	st	md		av	st	md	nb
		2-3 km			4-5 km				6-7 km		
1	42	5	42	44	5	45		47	10	46	38
2	43	6	43	44	10	42		44	5	43	7
3	55	8	56	57	8	56		61	13	63	26
4	53	10	52	54	10	53		60	12	60	31
5	42	5	43	47	8	50		49	11	53	8
Lidar				av	st	md	nb	av	st	md	nb
					4-5 km				6-7 km		
1				44	8	44	26	44	13	42	46
2				41	13	43	3	52	15	51	10
3				53	16	53	11	70	20	67	35
4				60	14	58	18	59	15	60	50
5				53	64	53	5	53	15	50	10

1 = November/February
2 = March
3 = April/June
4 = July/September
5 = October

av = average
st = standard deviation
md = median value
nb = number of profiles

6.1.2 Stratosphere/troposphere studies from ground-based lidar measurements

The OHP system developed through the TESLAS program was also involved in case studies of stratosphere-troposphere exchange conducted within the European program CEC/TOASTE (Transport of Ozone And Stratosphere Troposphere Exchange) during 1990/1991. During one of these campaigns (19-30 Nov. 90), a tropopause fold connected to a developing cut-off low system was observed over Southern France [4]. Quasi-continuous lidar measurements at the OHP on November 23 represent a transverse cross section through an upper tropospheric frontal zone and make evident a tropopause fold with enhanced ozone concentrations (6-7 $\times$ 10^{10} molecules cm^{-3} compared to about 4 $\times$ 10^{10} molecules cm^{-3} outside of the fold) and sloping down to a 4.5 km altitude (Fig. 6.1).

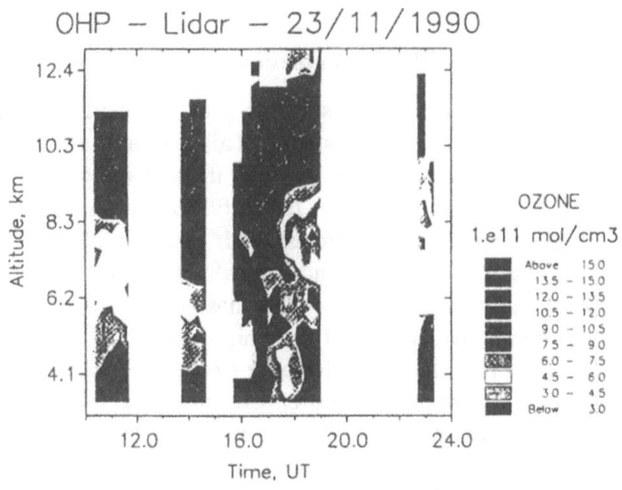

Fig. 6.1: Lidar ozone measurements in 10^{11} molecules cm^{-3}, at OHP on Nov. 23rd; a stratosphericintrusion during 12.00 to 18.00 h, descending down to 4.5 km is visualised by high ozone values, in dark colour.

The horizontal and vertical extension of the fold was determined by using PV-fields and by analysing radio soundings of the European network. Air-mass trajectory calculations show that the anticyclonic part of the fold is irreversibly transferred into the troposphere. Additionally, air masses on the cyclonic side of the fold are trapped by the circular motion around the cut-off low system and are irreversibly transferred into the troposphere in a second tropopause fold on the western flank of the cut-off low, 2 days later. The total air volume transferred to the troposphere is then approximately 1100 km $\times$ 500 km $\times$ 1.8 km and the ozone amount transferred to the troposphere about 6.5 $\times$ 10^{32} molecules. This estimation obtained during autumn, are at least three times lower than those obtained for spring [5-7], when most of the tropopause folds have been investigated up to now.

At a later stage of the evolution of the same cut-off low, lidar measurements made in the low pressure while it was stationary above the OHP, show a decrease of ozone concentrations in the layer just above the tropopause [4]. This corresponds to an ozone quantity transported into the troposphere of 6×10^{32} molecules, very similar to the value obtained for the tropopause fold observed at OHP some days before, showing the potential importance of cut-off lows for stratosphere-troposphere exchange processes. Although the exact mechanism of the observed cut-off low erosion is not clear, a vertical mixing coefficient K_z of 2.7 $m^2 s^{-1}$, comprising turbulent and convective mixing, can be calculated from the lidar measurements. Comparable values of K_z (1.5-2.2 $m^2 s^{-1}$) were obtained from ST-radar measurements in similar cut-off low systems, the erosion of the cut-off low being induced in one system by convective turbulence and in another by wave turbulence interaction [8].

6.1.3 The ALTO airborne lidar instrument

An airborne lidar has been designed using the same concept as above, but with new developments for aircraft implementation. This system is described in Section 3.6.10. The lidar instrument makes use of three simultaneous wavelength emissions at 266/289/316 nm. Although the 266/289 nm pair provides the best accuracy, the measurement range is limited with this pair (1 or 2 km), and for larger ranges, one must rely on the second pair which is unfortunately much more sensitive to aerosol interferences. The system has been used in a ground based mode during one month during a field campaign organized at a remote site in Brittany for boundary layer studies performing ozone profiles from 0.5 km up to 4 km leaving the lidar system almost unattended for several hours. An example of a one 50 min. averaged lidar profile is given in Fig. 6.2a showing the two different ozone profiles obtained with the two different wavelength pairs 266/289 and 289/316 nm. The deviation of the 289/316 nm below 1.5 km is due to the large difference in the divergence of the 316 nm laser beam compared to the 289 nm beam inducing a different lidar geometrical overlap function for the two wavelengths. The system was then recently tested in the French Research Aircraft (ARAT/Fokker 27) during two flights on February 22nd and 23rd. The lidar was pointing downward while the aircraft has flown along the same track back and forth at two different altitudes 4.5 km and 2.7 km. The airborne UV photometer ozone measurements were also compared to the lidar profile during the ascending and descending phase. Although the lidar derived ozone profile must be regarded as very preliminary (no optimization of the processing near the ground return, no aerosol corrections), one example is shown in Fig. 6.2b for a 30 s time averaging. It shows the good measurement quality for this relatively small integration time and the relatively good agreement with the *in-situ* measurement considering that the UV photometer data below 2.7 km correspond to a later part of the flight.

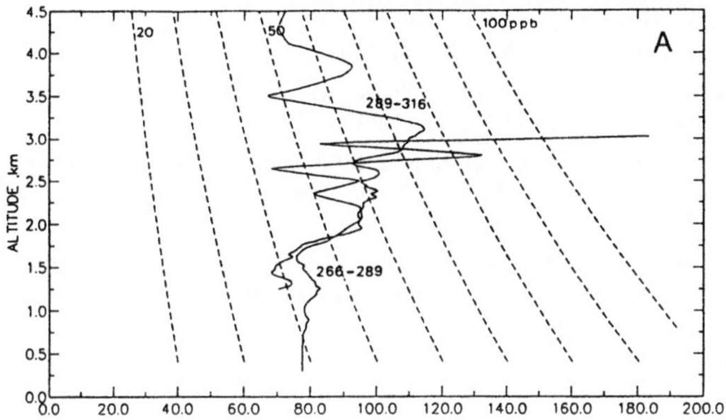

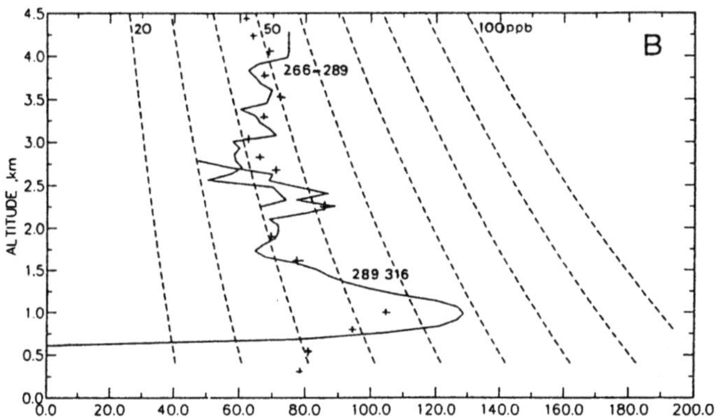

Fig. 6.2: a) Ozone profile obtained with the ALTO system used in a ground-based mode at the Porspoder Station using two wavelength pairs. b) Ozone profile obtained with the ALTO system mounted on the ARAT aircraft flying at 4.5 km (thick line) and ozone profile reconstructed from the measurements of the aircraft UV photometer (crosses).The dotted lines represents the constant O_3 mixing ratio profiles.

6.1.4 References

1. G. Ancellet, M. Beekmann; The Observatoire de Haute Provence, in: T. Cvitaš, D.Kley (eds.), *The TOR Network*, EUROTRAC ISS, Garmisch-Partenkirchen 1994, pp. 173-182.
2. A. Chedin, N.A. Scott, C. Wahiche, P. Moulinier; The improved initialization inversion method: a high resolution physical method for temperature retrievals from satellites of the TIROS-N serie, *J. Climat. Appl. Meteor.* **24** (1985) 2.
3. A. Papayannis, G. Ancellet, J. Pelon, G. Mégie; Multiwavelength lidar for ozone measurements in the troposphere and the lower stratosphere, *Appl. Opt.* **29** (1990) 467-476.

4. G. Ancellet, M. Beekmann, A. Papayannis; Ozone transport in a tropopause fold / cut-off low event, *J. Geophys. Res.* **99D** (1994) 3451-3468.
5. E. Danielsen; Stratospheric-tropospheric exchange based on radioactivity, ozone and potential vorticity, *J. Atmos. Sci.* **25** (1968) 502-518.
6. W. Viezee, W.B. Johnson, H.B. Singh; Stratospheric ozone in the lower troposphere - II. assessment of downward flux and ground level impact, *Atmos. Environ.* **17** (1983) 1979-1993.
7. G. Ancellet, J. Pelon, M. Beekmann, A. Papayannis, G. Mégie; Ground based lidar studies of ozone exchanges between the stratosphere and the troposphere, *J. Geophys. Res.* **96** (1991) 22401-22421.
8. F. Bertin, A. Cremieu, R. Ney, A. Desautez; Tropospheric-stratospheric exchange coefficients in cut-off lows estimated with a high resolution UHF-radar, *TOASTE report* (1993).

6.2 Application of a New Shot per Shot Methodology to Tropospheric Ozone Measurements with a Shot per Shot DIAL Instrument

Eric Durieux and Luca Fiorani

EPFL-LIDAR Group, Bellerive 34, CH-1007 Lausanne, Switzerland

A mobile lidar instrument has been built by the EPFL Lidar group since the TESLAS collaboration was joined in 1992. This system has been used to support the development of a new methodology for the characterization of tropospheric ozone measurements [1] and for the measurement of ozone fluxes [2]. It is based on the original recording and treatment of each successive Lidar signal, by opposition to the usual hardware averaging approach. A R4000 mastered VME

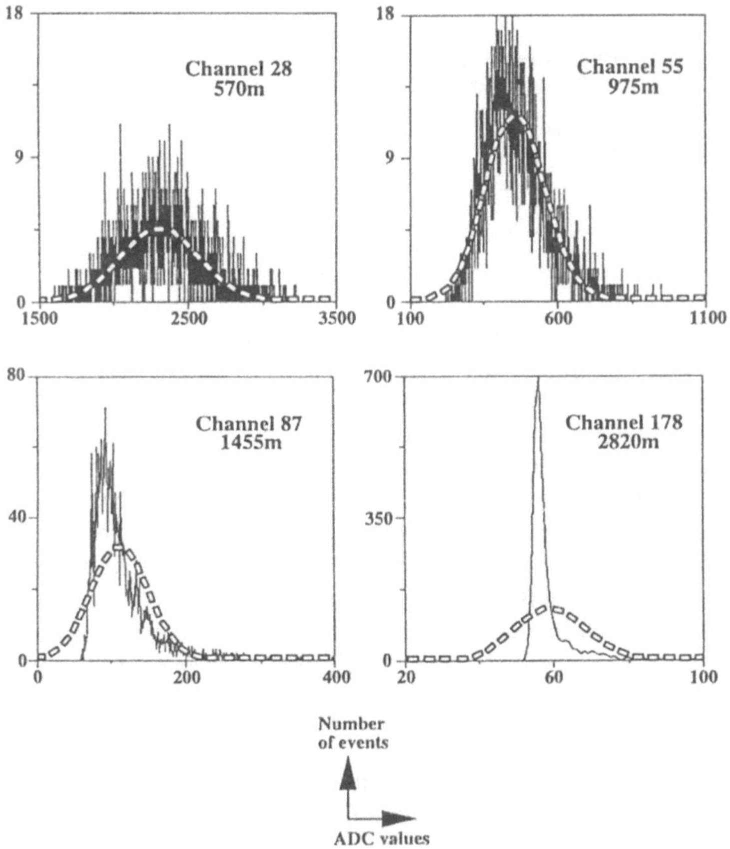

Fig. 6.3: Distributions of events at different distances for the ON wavelength.

architecture presented in Fig. 3.15 has been especially designed to ensure the synchronization of the experiment and the transfer of the sampled Lidar signals to the CPU [3]. Distributions of events are reconstructed for each sampling channel, where the number of Lidar signals for which a given ADC value is recorded, is plotted against the reachable ADC values. This is illustrated in Fig. 6.3 for four different channels of the 12 bits ADC used for the sampling of a Lidar return at the ON wavelength. A mean μ and a standard deviation σ are calculated with Gaussian estimators.

The resulting function is superimposed to the data in dashed line. The signal to noise ratio is thus precisely determined as shown in Fig. 6.4 where the σ_μ / μ ratios are represented for the two wavelengths.

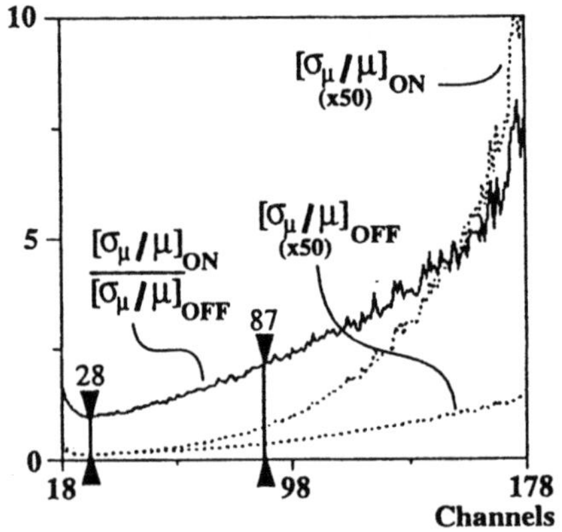

Fig. 6.4: Signal to noise ratios for the two wavelengths.

A phenomenological model has been proposed to represent its range and spectral evolutions [1]. The propagation of the measured standard deviations to the ozone concentrations leads to a precise determination of the statistical error on the results. They are shown to range from 3 % to 11 % in Fig. 6.5a for a typical uncorrected vertical profile as presented in Fig. 6.5b. The corrections on the concentrations for the atmospheric extinction independent of the ozone molecule are shown in Fig. 6.5c to raise the level of this uncertainty to about 16 %. This is mainly due to the presence of a numerical integration of the Lidar signal in the inversion scheme discussed in the methodology section of this document. Deviations of the distributions of events from a Gaussian shape are shown to appear at long distances in Fig. 6.3. At the present stage of the analysis, this is interpreted as a new signature for the SIB effect discussed in the instrument section of this document.

Correlations between successively acquired Lidar signals have already been studied to measure wind profiles with a fixed [4] or rotating [5] direction of the light beam. Fractional deviations from the mean Lidar signal have been reconstructed for the present fixed beam configuration of our instrument as shown in Fig. 6.6 for a telescope pointing at 30° from ground.

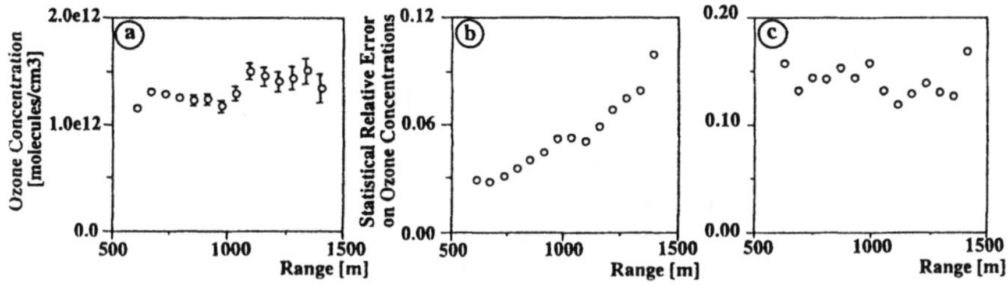

Fig. 6.5: a) A typical uncorrected ozone profile; b) statistical relative error on the uncorrected profile; c) statistical relative error on the corrected profile.

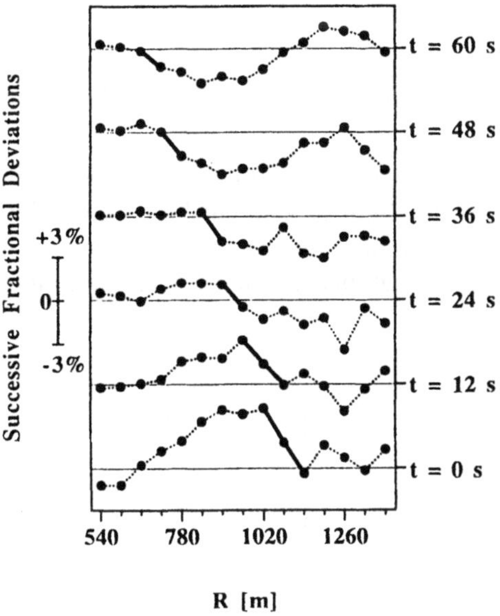

Fig. 6.6: Fractional deviations from mean for 6 samples of 500 lidar returns recorded with 10 s intervals.

The displacement of the marked pattern with time is due to a wind component along the beam. This study has led to a first combined measurement of ozone concentrations and wind vectors with a shot per shot DIAL instrument. This is illustrated in Figs. 6.7a and 6.7b where an ozone concentration and a wind speed along the beam axis are respectively represented.

The errors quoted on the wind profile are due to the ambiguities in its determination. They are preliminary at this stage of the analysis. A first measurement of an ozone flux with a single DIAL system is presented in Fig. 6.8.

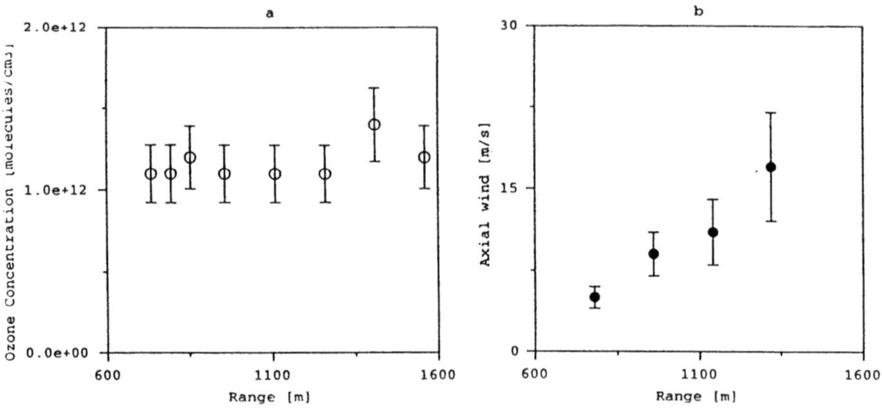

Fig. 6.7: a) An ozone profile obtained with a 30° inclination from ground of the light beam. b) A wind profile along the beam axis.

The first innovative outputs from the shot per shot methodology have been presented for tropospheric ozone measurements. Both experimental and processing developments will be pursued in order to optimize the extraction of as much information as possible from this approach, so that this system can be engaged into field campaigns in the best conditions.

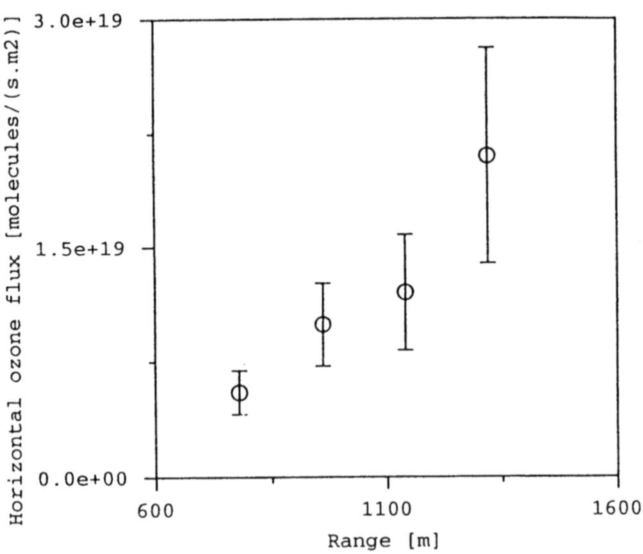

Fig. 6.8: Ozone flux along the beam axis.

References

1. E. Durieux, L. Fiorani, B. Calpini, M. Flamm, L. Jaquet, H. van den Bergh; Tropospheric ozone measurements over the Great Athens Area during the MEDCAPHOT-TRACE campaign with a new shot-per-shot DIAL instrument, *Atmos. Environ.* in press.
2. L. Fiorani, B. Calpini, L. Jaquet, H. van den Bergh, E. Durieux; A combined determination of wind velocities and ozone concentrations for a first measurement of ozone fluxes with a DIAL instrument, *Atmos. Environ.* in press.
3. E. Durieux, *et al.*; in: L.D. Laude (ed.), *Excimer Lasers*, Kluwer, Dordrecht 1994.
4. E.W. Eloranta *et al.*; *J. Appl. Meteorol.* 14, (1975) 1485.
5. J.L. Schols, E.W. Eloranta, *et al.*; *J. Geophys. Res.* **97** (1992) 18395.

6.3 The Multifrequency Multichannel Siberian Lidar Station for Sensing the Tropospheric-Stratospheric Ozone and Aerosol

V.V. Zuev and V.D. Burlakov

Institute of Atmospheric Optics, SB RAS, 1, Akademicheskii Ave.,Tomsk, 634055, Russia

In the vast territory of Siberia the city of Tomsk (56.5° N, 85.1° E) is a single point where regular lidar monitoring of stratospheric aerosol (since 1986) and ozone (since 1989) has been performed. Because of such geographic position, the data of long-term observations of ozone and aerosol obtained in Tomsk, can be of particular interest for the specialists in the field of regional and planetary climatology, ecology and atmospheric physics.

The multichannel multifrequency lidar station has been developed for complex measurements of vertical distribution of ozone and aerosol in the troposphere and stratosphere- Siberian Lidar Station (SLS). A detailed description of the procedure of lidar measurements at SLS is given in refs. [1, 2]. A simplified block-diagram of the station is presented in Fig. 6.9, the main parameters are given in Table 6.3. For sounding the stratospheric ozone we use the fundamental frequency of the excimer, XeCl laser radiation (308 nm) and first Stokes component of its stimulated Raman-scattering in hydrogen (353 nm). The simultaneously multifrequency aerosol sounding may be carried out at the wavelengths of 353 nm, 532 and 1064 nm of Nd:YAG - laser and at the 628 nm of gold - vapor laser. The lidar returns from the stratosphere were received in the focal plane of the receiving mirror 2.2 m in diameter. The multichannel operating conditions of the lidar are performed due to the laser beam tilt at small angle, of about 35'. For a sounding channel at the Nd:YAG and XeCl lasers wavelengths the spectral selection cell consisting of spectral dividing mirrors, interference filters, collimating and focusing lenses is located together with a photomultiplier immediately in a focal plane of a receiving mirror. At the sounding line of 628 nm the optical signal is transmitted from the receiving mirror focus to systems of spectral selection and recording with the use of a photon optically mounted with a fiber. Mechanical choppers operating synchronously with the lasers are used to cut off the signal from the near range. The signals are detected in the photon-counting regime. To detect the signals at the wavelength of 1064 nm, where the counting photomultipliers have low quantum efficiency, the photomultiplier PMT-83 is used, which is cooled down to the temperature –30 °C for decreasing the thermal noises. Because of technical and organization difficulties of regular conducting of multifrequency sounding at several wavelengths simultaneously such measurements are carried out irregularly. Extension of the bulk of measuring information due to sounding even at two wavelengths simultaneously makes it

Table 6.3: Technical parameters of the multifrequency lidar station.

TRANSMITTER				
Laser	λ / nm	E / mJ	P_{av} / W	f / Hz
Nd:YAG	1064	150		10
	532	60		10
XeCl	308	60		50
XeCl + Raman-cell(H_2)	353	30		50
Cold-vapor laser	628		2	2.5×10^3
Copper-vapor laser + β-BBO crystal:				
doubled yellow	289		0.5	7×10^3
sum frequency	271		0.5	7×10^3
*Beam divergence 0.1-0.2 mrad				

RECEIVER				
	Troposphere		Stratosphere	
	ozone	aerosol	ozone	aerosol
Telescope diameter / m	0.5	0.3	2.2	
Telescope focal length / m	1.5	1.0	10	
λ / nm	271, 289	532, 1064	308, 353	353, 532 628, 106

REGISTRATION

Photon counting regime
Detector: photomultipliers PMT - 83, PMT - 130, PMT - 142
Spatial resolution, m: 50 - 100 (troposphere), 100 - 500 (stratosphere)
Temporal resolution, min: 5 - 15 (troposphere), 15 - 30 (stratosphere)

possible to evaluate qualitatively the microphysical and optical characteristics of aerosol. Such sounding is conducted by us at wavelengths of 532 and 1064 nm. When tuning the lidar recording system for receiving the return signals of maximal range the lidar signals of molecular scattering at a wavelength of 532 nm are used for retrieving the temperature profiles in the 35 - 75 km altitude range, and at a wavelength of 353 nm - in the 20 - 40 km altitude range.

Sensing of stratospheric ozone is also possible when using a detached lidar with a 1m diameter receiving mirror. At present, the lidar facilities have also been

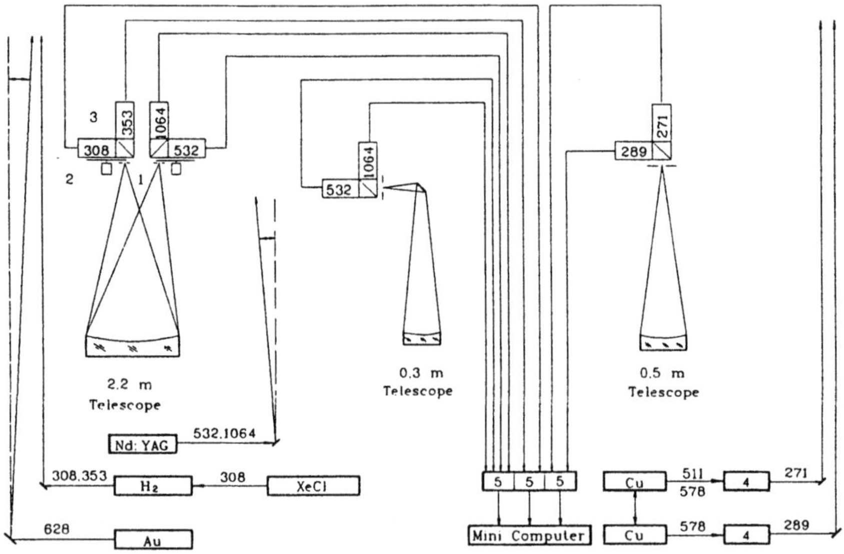

Fig. 6.9: The simplified block - diagram of the multifrequency lidar station: 1 - field diaphragm , 2 - chopper, 3 - cells of spectral selection with PMT, 4 - nonlinear crystals - BBO, 5 - photon counter.

developed for sensing the tropospheric ozone and aerosol. For aerosol sensing at wavelengths of 532 and 1064 nm a receiving telescope is used with a 0.3 m diameter mirror designed on the basis of the Newton scheme. The measurements of the vertical ozone distribution in the troposphere at wavelengths of 271 and 289 nm are carried out using a lidar with a 0.5 m diameter receiving mirror. The first test lidar measurements of ozone and aerosol in the troposphere started in 1994.

Some results of lidar measurements of ozone, aerosol and temperature are presented in other sections of this report

References

1. V.D. Burlakov, A.V. El'nikov, V.V. Zuev, *et al.*; A multichannel lidar for simultaneous measurements of the stratospheric ozone and aerosol with a primary mirror of 2.2 m in diameter, *Atmos. Ocean. Opt.* **5** (1992) 1022-1027.
2. V.V. Zuev, V.D. Burlakov, M.V. Grishaev, *et al.*; Siberian multifrequency lidar station for ground-based support measurements in lite experiment, *17th ILRCC Abstracts of Papers*, Sendai, Japan 1994, pp. 489-490.

6.4 The Aerosol Correction Method when Determining the Ozone Concentration using a Multifrequency Lidar

V.V. Zuev and B.S. Kostin

Institute of Atmospheric Optics, SB RAS, 1, Akademicheskii Ave.,Tomsk, 634055, Russia

The bulk of information obtained using a multifrequency lidar makes it possible to decrease the number of unknown values in the problem of ozone determination by the DIAL technique. This fact is confirmed by the examples given in Section 2.8 'Multiwavelengths techniques' (Figs. 2.10-2.13). However, in the above examples it is assumed that the aerosol coefficients are connected by the Ångström formulas. Here we consider the ozone laser sensing technique, which enables us to minimize a priori information.

When employing this technique a four-frequency lidar is used with the operating wavelengths of 0.277, 0.313, 0.532 and 1.064 μm. Three wavelengths are used for separating the scattering components. The wavelength of 0.313 μm serves for separating the backscattering molecular coefficient, that is, for determining the molecular component. The two other wavelengths of 0.532 and 1.064 μm define the aerosol component.

Particular emphasis has been placed on the determination of the aerosol component. The one-to-one correspondence exists between the spectral behavior of the two aerosol optical characteristics (the backscattering coefficient and the extinction coefficient). This correspondence is formally determined by the inverse operator introduced in the monograph by V.E. Zuev and I.E. Naats [1]. The characteristic property of the inverse operator is its independence of the kind of the particle size spectrum. For constructing the inverse operator a sphericity of scattering particles should be assumed as well as knowledge of spectral dependence of the refractive index of aerosol substance. The above-mentioned feature of the inverse operator is very important when developing the algorithm. Determination of certain optical characteristics through the other ones does not depended on the kind of the particle size spectrum.

Some principle features should be mentioned without a detailed description of the algorithm. First of all, note that the aerosol component is calculated using the model estimate method. This method is described in detail in the above-mentioned monograph by V.E. Zuev and I.E. Naats. The algorithm is iterational and converges fast. The number of iterations does not exceed five. The ozone concentration is determined by the standard DIAL technique after the work of the separation algorithm.

The lidar for tropospheric sensing is calibrated along a horizontal path. This is conducted by means of the screen with the known reflection coefficient. For the lidar calibration simultaneously with the calibration constants the aerosol light scattering coefficients are determined. The refractive indices of aerosol substances are calculated by means of a special program according to the values of these coefficients. The calculated values of refractive indices are taken as the initial conditions when interpreting the lidar signals.

The method was investigated using the numerical experiments performed for different models of molecular atmosphere. In these experiments the aerosol optical characteristics were calculated for a continental model of L. Deirmendjian. In the calculations we have used the different models of the refractive index of aerosol substance. The error obeying the Poisson law has been shown on thus calculated profiles of signals.

The results of the experiments have shown that in all the cases when a considerable variation of the refractive index of particles along the sounding path is lacking, the unknown profile of ozone density is exactly defined. In cases when the variation of the refractive index is large, the deviation of reconstructed ozone density from the exact value is also significant. The similar result is shown as a solid curve in Fig. 6.10.

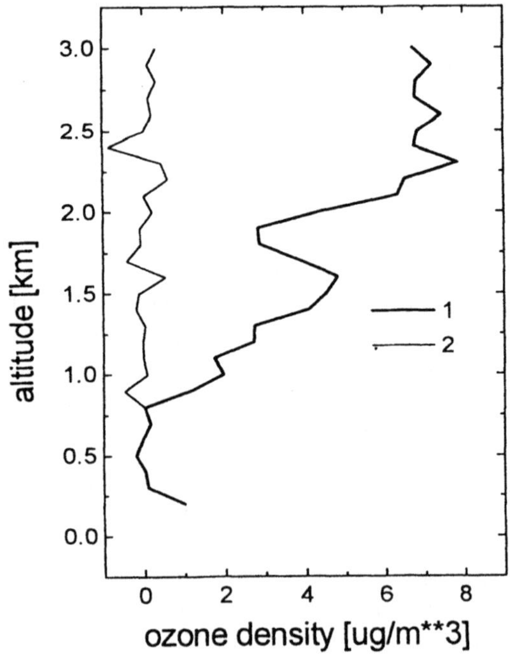

Fig. 6.10: Ozone density.

The divergence, being satisfactory in the beginning of the path, starts its increase from 0.9 km altitude. We dwell on this result in detail.

The situation, when the results of reconstruction are given in Fig. 6.10, is presented in Figs. 2.9 and 2.10 on pp. 27–28. To satisfy the requirements given in these figures, for the altitudes between 0.9 and 2.2 km the aerosol optical characteristics are calculated using the L. Deirmendjian model and the refractive index m = 1.47 - 0.022 i. For other altitudes the value of the refractive index is equal to m = 1.47 - 0.001 i. Because of the absence of the spectral model for the refractive index, it is selected to be constant for all the wavelengths. Now the result, given in Fig. 6.10, is clear. Before the 0.9 km altitude the exact value of the refractive index differs slightly from the specified one, which, in its turn, coincides with the initial value (m = 1.48). After the 0.9 km altitude the divergence became considerable.

To solve a given problem we propose simultaneously with the lidar sensing to measure at l = 0.532 m the signals of pure rotational Raman light scattering by molecules of nitrogen and oxygen which give the data on the air humidity profile. Because the maximal spectral shift of Raman scattering for a given wavelength is no more than 32 Å, the variation of optical characteristics may not be taken into account within this spectral range [2]. By separating the Rayleigh and Mie scattering components based on the use of our algorithm, the reconstructed profile of air density will depend on selected values of the complex refractive index of aerosol particles.

The true value of the complex refractive index can be found from the discrepancy of a measured signals of the pure rotational Raman scattering P_m and similar signals of P, calculated at different values of real and imaginary parts of the complex refractive index of aerosol particles. Fig. 6.11 presents the result referring to the two points of the sounding path of 0.9 km and 2.2 km.

This figure shows the relation $|P - P_m| / P_m$ depending on the real m and imaginary k parts of the refractive index of aerosol particles. One can see from the figure that in both cases the minimum of the relation is available. For the 0.9 km altitude the minimum is observed at m = 1.47 - 0.20 i (Fig. 6.11a). For the 2.2 km altitude the minimum is observed at 1.47 (Fig. 6.11b). These values of refractive indices are close to the true ones. The improved value of the refractive index over the whole sounding path makes it possible to calculate the refined ozone profile. A similar result is presented in Fig. 6.10 by the dashed line.

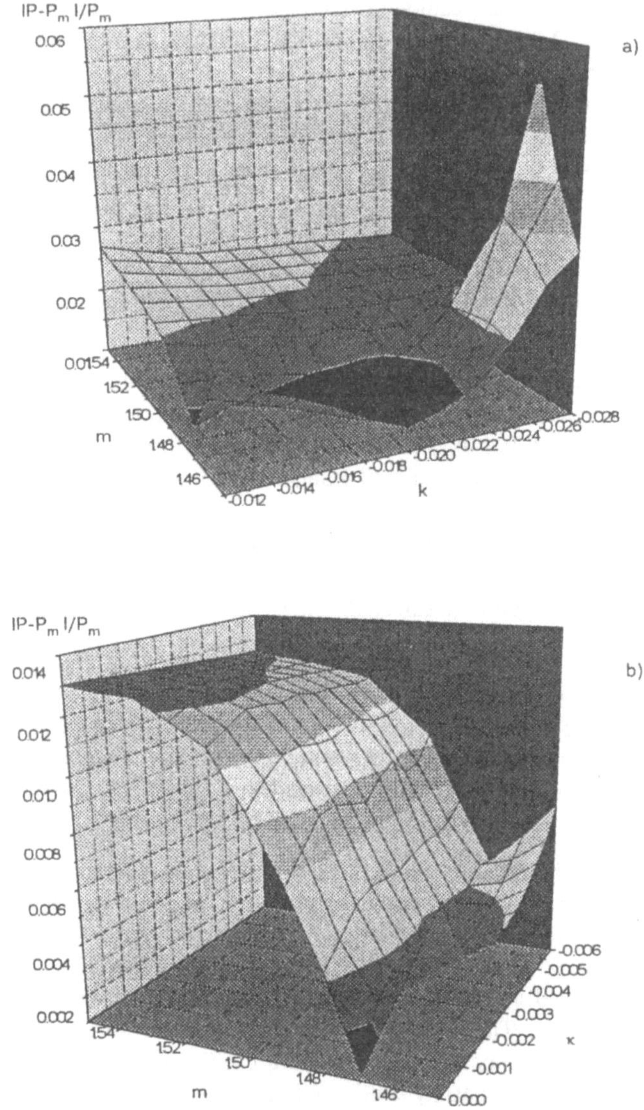

Fig. 6.11: The ratio of the Raman scattering signals as a function of imaginary and real parts of the refractive index of aerosol particles.

References

1. V.E. Zuev, I.E. Naats; Inverse Problems of Lidar Sensing of the Atmosphere, *Springer Series in Optical Sciences*, Vol. 29, Springer-Verlag, Berlin 1983, pp. 1-258.
2. S.N. Volkov, B.V. Kaul; Method for determination of back scattering and light attenuation coefficients in tropospheric aerosol layers using molecular and Raman scattering frequencies lidar, *Atmos. Ocean. Opt.* **7** (1994) 592-1602.

6.5 Vertical Soundings of Tropospheric Ozone with the IFU UV Lidar

W. Carnuth, U. Kempfer and T. Trickl

Fraunhofer-Intitut für Atmosphärische Umweltforschung,
D-82467 Garmisch-Partenkirchen, Germany

In January 1991, routine measurements with the new wide-range UV lidar were started at IFU [1-3]. A complete annual series of vertical ozone soundings with 580 measurements was achieved in 1991. However, only a limited number of files are available for the subsequent period due to the lack of funding. The data contain examples of ozone distributions obtained under a reasonable variety of conditions. The scientific results constitute a contribution to the TOR subproject, in which we have participated parallel to TESLAS. A short overview is given in the following.

6.5.1 Boundary layer (PBL)

The size and structure of the boundary layer is traced by evaluating the vertical distribution of the aerosol backscatter coefficients at 313 nm ("off" wavelength). For typical fair-weather conditions with moderate visibility next to the ground the aerosol backscatter coefficient is comparable with that for Rayleigh scattering and is, thus, clearly discernible.

The boundary-layer height in the afternoon of a typical summer day may easily exceed 3 km above the ground. As known from the literature (*e.g.* [4]) the night-time PBL shrinks to heights below 1 km. This behaviour shapes the vertical distribution of ozone. Enhanced O_3 densities are observed up to approximately the afternoon height of the PBL. During the night, the ozone density above 1 km does not change much, due to missing downward transport and deposition. Similarly, again because of the absence of convection, the aerosols at these heights do not completely disappear over night. There is a tendency for a two-step distribution of both ozone and the aerosols with reflects the separation into night-time PBL and residual layer. Two examples for the warm season can be seen in Fig. 6.12.

During the day, photochemically induced ozone production starts. According to Reiter *et al.* this enhancement is mainly caused by advection of precursors from outside the mountains and subsequent production above the valley [5]. We have seen production rates of 10 % and more per day, in good agreement with observations at rural sites (see, *e.g.* [5-7]).

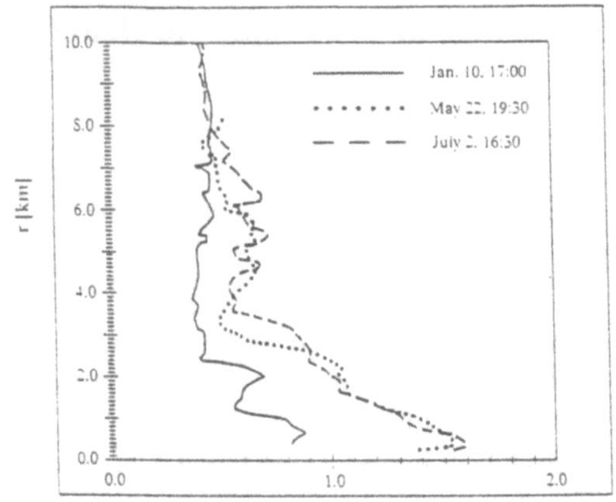

Fig. 6.12: Typical vertical ozone density distributions for different seasons in 1991; *r* is the height above the lidar (740 m a.s.l.).

Very astonishingly, such a two-step pattern of aerosols and ozone prevails also during high-pressure periods in winter (see January example in Fig. 6.12). It is less pronounced for the aerosols. The ozone-rich zone mostly exceeds 2.5 km above the ground, *i.e.* substantially above the inversion layer. An explanation of these findings is not easy and will be the subject of concentrated efforts in the future.

6.5.2 Free troposphere

Throughout the year, the fair-weather number densities of ozone in the free troposphere are roughly 50 % of the peak values in the PBL. On average the density is roughly independent of the height which means an increase of the mixing ratio towards the stratosphere. This might be ascribed to the natural input from the stratosphere. As in the boundary layer, a clear annual cycle is also observed (see Fig. 6.13), note the shifted concentration axis). Interestingly, such a behaviour is not seen near the coast where mostly clean marine air masses are analysed [8]. It is important to note that periodic diurnal density variations could not be found in the free troposphere, given our present accuracy.

As already indicated above the upward transport from the PBL is rather slow and cannot be discerned by measurements during a single day. Thus, the mechanisms governing the pronounced annual cycle of the ozone concentration in the free troposphere are not easily accessible. The observation of Grabbe *et al.* [8], of a missing summer maximum in the coastal free troposphere might indicate that most of the excess O_3 in summer is produced over the European continent rather than caused by upward transport on a hemispherical basis. On the other hand, some import of ozone from North America may be expected, which would take place in the free troposphere [9, 10].

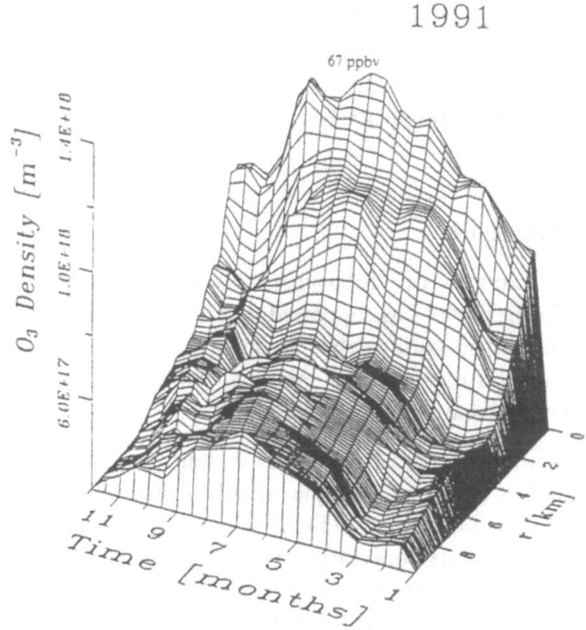

Fig. 6.13: Three-dimensional representation of the 1991 annual series (note the shifted density scale); reprinted from ref. [2] with permission of the American Institute of Physics (1996).

We expect convective transport above mountains or in clouds to dominate the anthropogenic contribution to the ozon concentration in the free troposphere. However, also a more quantitative estimate of the *in-situ* chemical production has to be derived in the future. Unfortunately, there is a lack of regular measurements of ozone precursors in the free troposphere. The available numbers indicate rather low densities.

6.5.3 Frontal passages

Several examples of frontal passages are available in the 1991 data. Frontal passages replace aged and polluted continental air by fresh air from outside the continent. As a consequence, initially low ozone concentrations are observed. Within a single day the typical two-step ozone distribution in the boundary layer may be restored. The observations indicate that mostly horizontal advection of ozone is responsible for this quick increase of concentration.

We expect that a study of frontal passages, combined with trajectory calculations to identify the origin of the air masses following the front, may help to distinguish the different contributions to the tropospheric ozone budget. Such investigations should focus on the free troposphere which contains information on the background ozone in the northern hemisphere.

Remote transport may be studied also in cases such as Föhn winds. Our results have shown that these winds, which frequently originate in Africa, contain substantially less ozone than observed for other wind directions. This is in good agreement by the more thorough analysis of the Zugspitze data by Scheel *et al.* [11].

6.5.4 Stratospheric ozone

The main natural source of tropospheric ozone is downward mixing of stratospheric O_3. Earlier work has suggested that intrusions of stratospheric air are particularly pronounced in spring when frequently cold-fronts sweep across our latitude [12, 13]. Recent experimental material within EUROTRAC points more towards a maximum in June and July [14, 15].

The IFU lidar has a range of 10 ± 2 km and is, thus, well suited for the study of such episodes. So far, only one example of subsidence of stratospheric air was found on July 24, 1993. July 24 was a highly clear day between two rain fronts on which particularly low ozone concentrations were registered below 4.5 km above the ground. Above 4.5 km a build-up of higher concentrations was observed. An analysis of radiosonde data for Central Europe confirmed our conclusion of the stratospheric origin of the enhanced concentrations in this altitude range [16].

In 1991, data were taken on a total of 80 days without revealing additional similar events. We do not exclude that this may in part be due to the fact that in this first year of system operation the selection of measurement days was not done in a sufficiently systematic way.

6.5.5 Conclusions

The first period of operation of our UV lidar has demonstrated the usefulness of this instrument for vertical sounding of ozone in the troposphere. The lidar constitutes a valuable extension of the mountain-top monitoring stations of IFU. Its high vertical resolution within the boundary layer and its short data- acquisition time are particularly suitable for the study of the relevant processes in the lower troposphere.

In addition to a complete coverage of the boundary layer, measurements in the entire free troposphere are possible. Due to a range up to 10 ± 2 km above the ground intrusions of stratopheric air can nicely be traced. Up to now, the accuracy above 6 km has been limited to about 10 % due to a signal-to-noise ratio quickly deteriorating with increasing height. The introduction of an additional wavelength channel (292 nm) and optimum daylight filtering, are expected to yield substantially higher accuracy, higher vertical resolution as well as a range extension to about 15 km. These modifications, planned in 1990, are currently under development under a new research contract.

6.5.6 References

1. U. Kempfer; Entwicklung und Anwendung eines differentiellen Absorptions-LIDAR-Systems zur Messung der troposphärischen Ozonkonzentration, *Dissertation*, Ludwig-Maximilians-Universität München, 1992.
2. U. Kempfer, W. Carnuth, R. Lotz, T. Trickl, *Rev. Sci. Instrum.* **65** (1994) 3145-3164.
3. W. Carnuth, U. Kempfer, T. Trickl, to be published.
4. R.B. Stull, *An Introduction to Boundary Layer Meteorology*, Kluwer Academic Dordrecht 1988.
5. R. Reiter, R. Sladkovic, H.-J. Kanter; *Meteorol. Atmos. Phys.* **37** (1987) 27-47
6. F.C. Fehsenfeld, M.J. Bollinger, S.C.L. Liu, D.D. Parrish, M. McFarland, M. Trainer, D. Kley, P.C. Murphy, D.L. Albritton; *J. Atmos. Chem.* **1** (1983) 87-105.
7. S.C. Liu, M. Trainer, F.C. Fehsenfeld, D.D. Parrish, E.J. Williams, D.W. Fahey, G. Hübler, P.C. Murphy; *J. Geophys. Res.* **92** (1987) 4191-4207.
8. G.C. Grabbe, J. Bösenberg, T. Schaberl; Ozone distribution in the lower troposphere investigated with the MPI ozone lidar, in: P.M. Borrell, P. Borrell, T. Cvitas, W. Seiler (eds.), *Proc. EUROTRAC Symp. '94*, SPB Academic Publishing, The Hague 1994, pp. 322-325.
9. D.J. Jacob, J.A. Logan, G.M. Gardner, R.M. Yevich, C.M. Spivakovsky, S.C. Wofsy, S. Sillman, M.J. Prather; *J. Geophys. Res.* **98** (1993) 14817-14826.
10. F.C. Fehsenfeld; Transport of O_3 and O_3 Precursors from anthropogenic sources to the North Atlantic, in: P.M. Borrell, P. Borrell, T. Cvitas, W. Seiler (eds.), *Proc. EUROTRAC Symp. '94*, SPB Academic Publishing, The Hague 1994, pp. 57-64, and references therein.
11. H.-E. Scheel, R. Sladkovic, W. Seiler; Ground-based measurements of ozone and related precursors at 47° N, 11° E, in: *EUROTRAC Annual Report '92*, Part 9, TOR, EUROTRAC ISS, Garmisch-Partenkirchen 1993, pp. 129-134.
12. E.F. Danielsen, V.A. Mohnen; *J. Geophys. Res.* **82** (1977) 5867-5877.
13. J.D. Mahlmann, W.J. Moxim; *J. Atmos. Sci.* **35** (1978) 1340-1373.
14. P. van Haver, D. de Muer; Seasonal and long term variation of stratospheric ozone intrusions into the troposphere and photchemical ozone production in the boundary layer at Uccle (Belgium), in: P.M. Borrell, P. Borrell, T. Cvitas, W. Seiler (eds.), *Proc. EUROTRAC Symp. '94*, SPB Academic Publ., The Hague 1994, pp. 335-340.
15. M. Beekmann, G. Ancellet, C. Mancier; Tropospheric ozone monitoring at the Observatoire de Haute Provence, in: P.M. Borrell, P. Borrell, T. Cvitas, W. Seiler (eds.), *Proc. EUROTRAC Symp. '94*, SPB Academic Publishing, The Hague 1994, pp. 367-372.
16. R. Sladkovic, private communication.

6.6 Remote Sensing of Turbulent Vertical Flux Profiles and the Budget of Ozone in the Convective Boundary Layer with DIAL and Radar-RASS

J. Bösenberg, C. Senff, and T. Schaberl

Max-Planck-Institut für Meteorologie, Bundesstr. 55, D-20146 Hamburg, Germany

6.6.1 Introduction

The high resolution DIAL system of the MPI in combination with a Radar-RASS wind profiler for measuring the vertical wind component has been used to measure vertical profiles of turbulent ozone flux under convective conditions. The fluxes are determined directly from the highly resolved ozone density and vertical wind speed measurements by using the eddy correlation technique. The time and height resolution of the combined DIAL/RASS system are 60s and 75m, respectively, which apparently is sufficient to resolve a major part of the contributions to the turbulent ozone flux in a convectively mixed boundary layer. In addition, the DIAL/RASS measurements were used to estimate the ozone production and destruction rate under the assumption that ozone advection is small compared to the other budget terms.

6.6.2 The method and the instruments

The critical point for the application of the eddy correlation technique is the capability of the measurement systems to achieve high temporal and vertical resolution in combination with high accuracy. For the ozone flux measurements described here, the KrF-laser based system was used with deuterium as the Raman shift gas, producing λ_{on} = 268 nm as online and λ_{off} = 292 nm as off-line wavelength. This offers a good compromise between high sensitivity to ozone absorption and attainable height range in the boundary layer. In addition, this is also an excellent choice for low sensitivity to aerosol interference, which is very important for high resolution measurements in the boundary layer. The repetition rate of the laser was 10 Hz, 100 shots were averaged online for subsequent evaluation. For the analysis, further averaging was performed, ending up with 60 s temporal and 75 m vertical resolution.

The Radar-RASS system of the Meteorologisches Institut der Universität Hamburg consists of a FM-CW (**f**requency **m**odulated - **c**ontinuous **w**ave) Radar operated at a center frequency of 1235 MHz and a sound source emitting acoustic waves parallel to the Radar beam. The Radar-RASS can be operated simultaneously in RASS and clear-air mode. In RASS mode, the sound velocity is retrieved from the Radar echoes scattered by the acoustic waves. In clear-air mode, the Radar signals, scattered by natural inhomogeneities of the refractive index of air, are used to derive the radial wind velocity component. During the ozone flux measurements the Radar-RASS system was pointing vertically in order to measure the vertical wind speed fluctuations. For details about the clear-air and RASS operation of an FM-CW Radar and the technical aspects of the Radar-RASS system see [1-2].

The resolution of the Radar/RASS system is 10 s temporally and 75 m vertically. Due to the horizontal distance of 25 m between DIAL and Radar-RASS and due to the largely different beam divergences, DIAL and Radar-RASS sample different measurement volumes. The geometry of the setup does not further restrict the size of the detectable turbulent eddies, since the horizontal scales of turbulent processes that can be sampled correctly with the system's time resolution of 60 s are larger than the system separation and the beam widths. Assuming a horizontal wind speed of 5 m/s, which is typical for a convectively mixed boundary layer, the smallest detectable eddy size is on the order of 600 m.

The height range of the combined DIAL/Radar-RASS system presently extends from 300 to a maximum of 800 m above the ground. Below 300 m ozone data have not been retrieved from the DIAL signals due to an incomplete overlap between the receiver telescope's field of view and the laser beam. The upper limit of the height range is determined by a low signal-to-noise ratio of either the DIAL or the Radar-RASS system.

The ozone flux measurements with DIAL and Radar-RASS were complemented by ground level measurements of ozone density, ozone flux and sensible heat flux. In addition, temperature, pressure, humidity, solar radiation, wind direction and wind speed at ground level were recorded.

Ozone budget. The mass budget of ozone for an air parcel in the atmosphere is described by the continuity equation. Assuming that the air in the boundary layer is incompressible, and that transport processes due to diffusion are negligible in the mixed boundary layer, the continuity reads [3]:

$$\frac{\partial \overline{\rho}_{O3}}{\partial t} + \sum_{i=1}^{3} \overline{u}_i \frac{\partial \overline{\rho}_{O3}}{\partial x_i} + \sum_{i=1}^{3} \frac{\partial \overline{u_i' \rho_{O3}'}}{\partial x_i} = \overline{S}_{O3} \tag{6.1}$$

where $\overline{\rho}_{O3}$ is the ozone density, x_i ($i = 1, 2, 3$) are the Cartesian coordinates, $\overline{u}_i$ is the wind speed, $\overline{u_i' \rho_{O3}'}$ is the turbulent ozone flux in the direction of the coordinate i, and $\overline{S}_{O3}$ is the source term of ozone. The overbar denotes temporal or spatial averaging of the respective variable.

Usually, for rather homogeneous terrain, the horizontal flux divergence can be neglected. With the x-axis aligned in the direction of the mean horizontal wind speed, we obtain

$$\overline{S}_{O3} - \overline{u} \frac{\partial \overline{\rho}_{O3}}{\partial x} - \overline{w} \frac{\partial \overline{\rho}_{O3}}{\partial z} = \frac{\partial \overline{\rho}_{O3}}{\partial t} + \frac{\partial \overline{w' \rho_{O3}'}}{\partial z} . \tag{6.2}$$

This means that the difference of the ozone source and the advection terms can be calculated for each height within the measurement range from the time derivative of the concentration and the vertical flux divergence. If the advection terms are small compared to the ozone source, it is possible to obtain estimates of the vertical distribution of the ozone production/destruction rate.

6.6.3 Results

The measurements were conducted at the TOR-site 'Hungriger Wolf', a flat site in North-western Germany with rather homogeneous rural environment. The weather situation on the four days of operation for this measurement campaign was characterized by clear summer weather with strong insolation and, therefore, strong convection and a high potential for photochemical reactions.

6.6.4 Ozone concentration

As an example, the results of the measurements taken on 30 June 1993 are presented here. On that day, simultaneous measurements with ozone DIAL and Radar-RASS were conducted from 10:30 to 14:32 UT and from 15:57 to 19:16 UT. Before and after, as well as inbetween these two measurement blocks, DIAL measurements of the ozone density in the free troposphere were performed. The ground data, including surface ozone density and flux, were recorded continuously during the whole time.

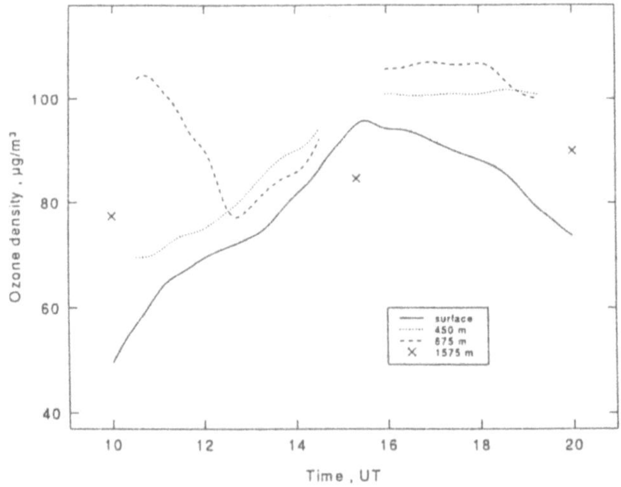

Fig. 6.14: Evolution of the ozone density on June 30 between 10:00 and 20:00 UT at the surface and at three selected heights.

In Fig. 6.14, the evolution of the ozone density on 30 June from 10:00 to 20:00 UT is shown at ground level and at three selected measurement heights: 450 m, 675 m, and 1575 m above the ground. For the ground, 450 m and 675 m, the time series of ozone density smoothed by applying a 30 min gliding average are plotted. The crosses in Fig. 6.14 represent the ozone measurements in the free troposphere at a height of 1575m, each averaged over 30 min timewise and 225 m in height. The evolution of the ozone density shows a typical behavior for a clear summer day under convective conditions: at the ground it increases steadily until the afternoon - from about 50 $\mu g/m^3$ at 10:00 UT to about 95 $\mu g/m^3$ at 16:00 UT - and then decreases again. At 450 m above the ground an almost identical behaviour is found except that the values are about 8 $\mu g/m^3$ higher than at the surface and that the ozone concentration does not drop after 16 UT. The same is true for the ozone density values measured at 675 m height after 12:30 UT. Before, the height of the boundary layer is less than 675 m, so that the measurements at 675 m height are performed in the residual layer, that is the remnant of the boundary layer of the previous day. Apparently the ozone in the residual layer is not destroyed during the night but stays at the high level it had reached the day before. This explains the high ozone density of 105 $\mu g/m^3$ measured at 10:30 UT at 675 m height. From 10:30 until 12:00 UT, the ozone concentration at this height decreases rapidly as the convective boundary layer, which at the beginning of the measurements contains about 30 $\mu g/m^3$ less ozone than the overlying residual layer, grows into the residual layer and quickly mixes the ozone-rich air downward. The ozone density in the free troposphere at 1575 m increases only slightly during the day from about 80 $\mu g/m^3$ to around 90 $\mu g/m^3$. The temporal evolution and vertical distribution of the ozone density in the lower troposphere shown in Fig. 6.14 seems to be typical for a clear summer day with strong convection. A number of very similar cases have been observed during routine ozone DIAL measurements [4].

6.6.5 Ozone flux

Fig. 6.15 shows ozone flux profiles as a function of height for six measurement intervals retrieved on 30 June with the DIAL/Radar-RASS combination.

The surface ozone fluxes measured with the fast *in-situ* sensors are also included. All heights are normalized with respect to the boundary layer height z_i. The averaging time for each ozone flux profile varies between 59 and 100 min. The fluxes measured in the boundary layer are mostly negative or near zero, and their magnitude is on the same order as the ground ozone fluxes. The exception to this are two uppermost measurements of the first and second flux profile. A detailed analysis shows, that in the presence of wavelike motions at the interface between two different airmasses the correlation of vertical wind and ozone density is very low. Consequently, the turbulent entrainment flux between residual and boundary layer cannot be sampled adequately with the eddy correlation technique. Therefore, the flux measurements near z_i will not be included in the following discussion of the measured flux profiles.

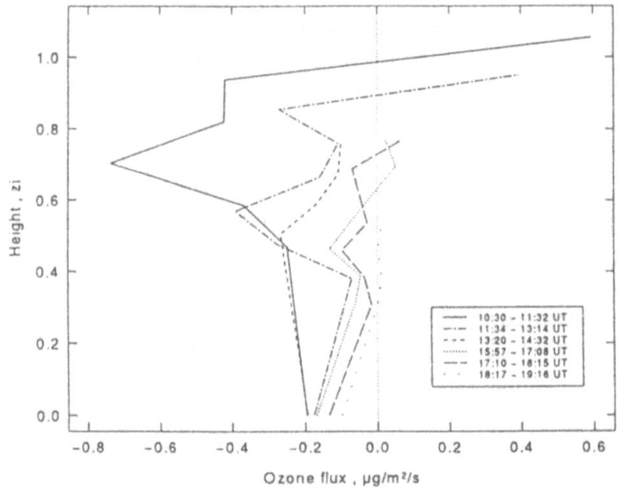

Fig. 6.15: Ozone flux profiles as a function of height normalized with respect to the boundary layer height z_i for six measurement intervals on June 30.

The first three intervals for which flux profiles have been measured (10:30-11:32 UT, 11:34-13:14 UT, 13:20-14:32 UT) are characterized by strong convection (surface sensible heat flux $\overline{w'T_s'} \approx$ 120 W/m^2) and a boundary layer rising from about 600m to 900m above the ground. For the first measurement interval, the downward ozone flux increases with height from around –0.2 µg/(m^2 s) at the ground and the lower portion of the boundary layer to about –0.6 µg/(m^2 s) at 0.8 z_i. Since between 10:30 and 11:32 UT most of the ozone-rich residual layer is mixed downward into the boundary layer (see fig. 6.), this strong negative flux in the upper part of the boundary layer and the negative vertical flux divergence at lower heights can be expected. After the residual layer has been mixed into the boundary layer, the flux profiles exhibit a rather constant flux with height of about –0.2 µg/(m^2 s). Measurement interval 4 and 5 (15:57-17:08 UT, 17:10-18:15 UT) are characterized by decreasing convective activity ($\overline{w'T_s'} \approx$ 50 W/m^2 and 25 W/m^2, respectively) and a stagnant boundary layer height at around 1000m. The downward fluxes are considerably lower than around noon and have values of about –0.05 µg/(m^2 s) in the lower part of the boundary layer. At around 0.7 z_i the ozone fluxes become zero, so the divergence is slightly positive. Zero fluxes in the middle of the boundary layer are plausible, since - as discussed above - the entrainment flux should be positive during this time of the day while the ground ozone flux is still negative. During measurement interval 6 (18:17-19:16 UT), which is about one hour before sunset, the convection breaks down completely ($\overline{w'T_s'} \approx$ 0 W/m^2), and a stably stratified surface layer starts to develop suppressing downward transport of ozone from the boundary layer to the sink at the surface. This explains why there is practically no more turbulent transport of ozone observed in the middle of the boundary layer. It should also be noted, however, that for the turbulence structure expected for this type of boundary layer, the resolution of the present system may not be sufficient, so that the transport of smaller eddies may not be detected. Further experiments with better resolution are needed to assess the validity of the results presented here for conditions of small convective activity.

6.6.6 Ozone budget

As explained above, the combination of ozone DIAL and Radar-RASS allows to estimate the ozone production/destruction rate under the assumption that the advection terms are small compared to the other ozone budget terms. For the measurements on 30 June, this assumption is most likely fulfilled.

The two ozone budget terms that can be inferred directly from the DIAL/Radar-RASS measurements - the local change of ozone density with time $\partial\bar{\rho}_{O3}/\partial t$ and the vertical ozone flux divergence $\partial\overline{w'\rho'_{O3}}/\partial z$ - and the sum of both, which is an estimate of the ozone production/destruction rate, are shown in Fig. 6.16 as a function of normalized height for measurement interval 2 (11:34 - 13:14 UT). In addition, the errors of the ozone production rate are plotted for each measurement height. The vertical bars indicate the rms errros of the ozone production rate. The

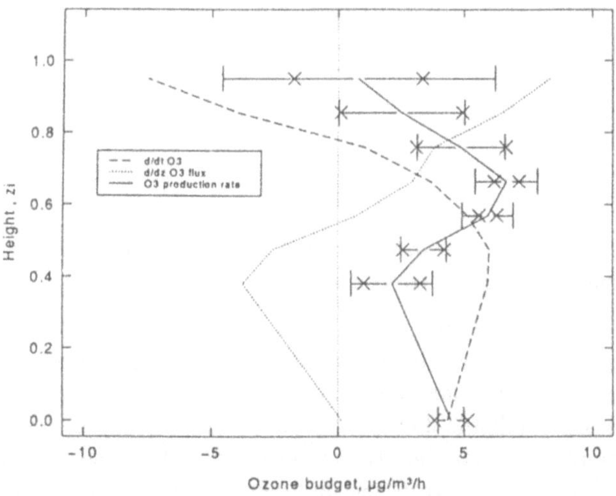

Fig. 6.16: Vertical profiles of the temporal derivative of ozone density, the vertical flux divergence, and of the ozone production rate as a function of normalized height for the measurement interval 11:34 to 13:14 UT on June 30. The vertical bars indicate the rms error of the production rate, the crosses mark the errors due to uncertainties in the aerosol correction.

crosses in Fig. 6.16 represent the systematic errors of the ozone production rate due to the aerosol correction errors of the ozone flux measurements. The errors caused by the flux sampling errors are not plotted since they are assumed to be significantly smaller than the other errors. The flux sampling errors should be correlated with height, because the large eddies that perform the bulk of the turbulent transport encompass the entire boundary layer. Therefore, the flux profile shape and, consequently, the vertical flux divergence do not change much when the sampling errors are regarded.

Within the margins of error, the ozone production rate for measurement interval 2 is about 4 μg/(m^3 h) and is fairly constant with height except for a slight maximum at 0.6 z_i of about 6 μg/(m^3 h). An ozone production rate constant with height is very plausible in a convectively mixed boundary layer since the ozone precursor gases should be well-mixed, and the photochemical reactivity should not exhibit any significant height dependence.

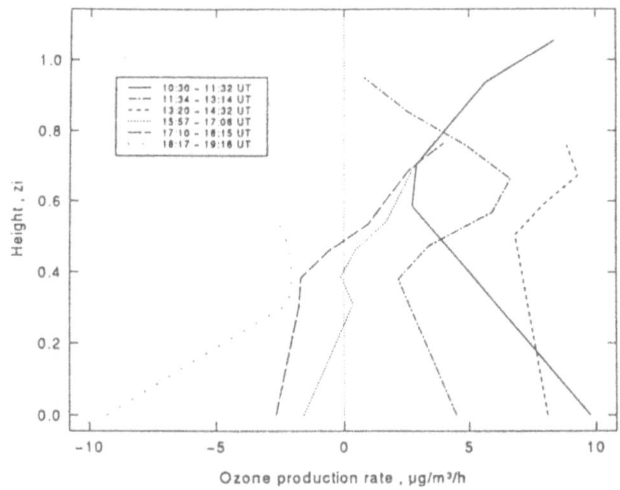

Fig. 6.17: Vertical profiles of the ozone production rate as a function of normalized height for the six measurement intervalls on June 30.

Fig. 6.17 depicts the ozone production/destruction rate as a function of normalized height for all six measurement intervals. For reasons of better legibility, error bars have been omitted. As stated above, the measurements near the top of the boundary layer are unreliable due to large statistical and aerosol correction errors. Below 0.8 z_i the rms errors of the ozone production/destruction rate are always less than 3.5 μg/(m^3 h), and the errors due to an erroneous aerosol correction are not larger than 1.7 μg/(m^3 h). The ozone production/destruction rate measured at ground level may be over- or underestimated, mainly because small-scale advection near the ground may not be negligible, even though the advection of ozone on a larger scale in the boundary layer is most likely small for the case of 30 June. Therefore, contributions to the surface ozone budget by small-scale advection could be misinterpreted as ozone production or destruction.

Due to the rather larger error margins, it is probably not justified to attempt a detailed interpretation of the height profiles. All of them appear to be compatible with the assumption of a height independent production/destruction rate, if the ground value is discarded as explained above. Regarding the mean ozone production rate in the boundary layer, it increases from about 4 $\mu g/(m^3 h)$ in the late morning (interval 1 and 2) to about 8 $\mu g/(m^3 h)$ in the early afternoon (interval 3). During the late afternoon (interval 4 and 5), the production rate drops significantly to around 1 $\mu g/(m^3 h)$ as can be expected since the intensity of the insolation and the photochemical reactivity also decrease. In the early evening (interval 6), slight destruction of ozone is observed in the lower boundary layer. The ozone production/destruction rates measured at ground level show a similar diurnal variation as in the boundary layer, but the actual values may be unreliable due to the above mentioned errors.

Due to the errors and the potential influence of advection effects, a more detailed analysis of the vertical distribution of the ozone production/destruction rate is not possible. But the results presented here demonstrate the potential of this method to yield time and height resolved measurements of the ozone budget terms in the boundary layer.

6.6.7 Conclusions

The measurements presented here demonstrate, that with the combination of a Radar/RASS and an ozone-DIAL the eddy correlation technique can be used to directly determine the major contribution to the vertical ozone flux, at least under conditions of fully developed convection in the central part of the boundary layer. The main advantage of this technique is, that vertical profiles of the flux are obtained, so that the flux divergence can be calculated, and that the measurements can be performed over extended periods of time with rather small operating costs.

The combination of concentration and flux profiles yields two terms of the ozone budget, so that in cases of no advection the ozone production/destruction can be estimated.

To obtain this estimate under more complex meteorological conditions the horizontal advection term has to be measured, too. This could be accomplished by deploying one or two additional ozone DIAL systems that would yield the horizontal gradients of ozone and, in combination with horizontal wind measurements, the ozone advection. Further improvement of the vertical and temporal resolution of both systems would make the technique applicable under a broader range of meteorological conditions.

6.6.8 References

1. G. Peters, D. Hasselmann, S. Pang; Radio acoustic sounding of the atmosphere using a FMCW Radar. *Radio Sci.* **23** (1988) 640-646.
2. G. Peters, H.J. Kirtzel, Complementary wind sensing techniques: sodar and RASS. *Ann. Geophysicae* **12** (1994) 506-517.
3. R.B. Stull; *An Introduction to Boundary Layer Meteorology*, Kluwer Academic Publisher, Dordrecht 1988.
4. G.C. Grabbe, J. Bösenberg, T. Schaberl; Ozone distribution in the lower troposphere investigated with the MPI ozone DIAL, in: *EUROTRAC Annual Report 92*, Part 9, TOR, EUROTRAC ISS, Garmisch-Partenkirchen 1993, pp. 156-161.

Chapter 7

TESLAS Publications 1988–1995

1988

Bisling, P., W. Lahmann, C. Weitkamp, W. Michaelis, W. Birkmayer; Argos: Advanced remote gaseous oxides, in: *Proc. 14th Int. Laser Radar Conference* , San Candido 1988, pp. 351-353.

Carnuth, W.; A differential absorption lidar for tropospheric ozone measurements, in: *Proc. 14th Int. Laser Radar Conference* , San Candido 1988, pp. 348-350.

Papayannis, A., G. Ancellet, J. Pelon, G. Megie; Tropospheric ozone measurements using a Nd-Yag laser and Raman shifters, in: *Proc. 14th Int. Laser Radar Conference* , San Candido 1988, pp. 472-475.

1989

Ancellet, G., A. Papayannis, J. Pelon, G. Megie; Dial tropospheric ozone measurement using a Nd-Yag laser and the Raman shifting technique, *J. Atmos. Ocean. Technol.* **6** (1989) 832-839.

Barbini, R., F. Colao, A. Palucci; Atmospheric backscattering measurements from the ground based CO_2 lidar station, in: *Proc. Int. Conf. on Laser and Electro-Optics*, Baltimore 1989, pp. 320.

Edner, H. P. Ragnarson, S. Svanberg, E. Wallinder; Vertical ozone probing with lidar, in: *Proc. Nordic Symp. on Atmospheric Chemistry*, Helsinki 1989.

Edner, H., S. Svanberg; Laser kartlägger luftföronreningar och nyttigt ozon, *NFR:s årsbok 1988/89; Forska på tvären,* Uppsala 1989, pp.159.

Svanberg, S.; Vertical ozone sounding with lasers, *Vad händer i luften över Europa? IVA-rapport* 366, Ingenjörsvetenskapsakademin, Stockholm 1989, pp. 46-47.

1990

Ansmann, A., M. Riebesell, U. Wandinger, C. Weitcamp, W. Michaelis; Raman lidar measurements of atmospheric aerosol extinction profiles, in: *Proc. 15th Int. Laser Radar Conference*, Part 1, Institute of Atmospheric Optics, Tomsk 1990, pp. 177-179.

Barbini, R., F. Colao, A. Palucci, S. Ribezzo, T. Hermsen, S. Orlando; Ozone and water vapour range resolved profiles from the ENEA LIDAR/DIAL station, in: *Proc. Laser '90 Conference,* San Diego 1990, pp. 320.

Barbini, R., F. Colao, T. Hermsen,, A. Palucci, S. Ribezzo; Misure di componenti atmosferici dalla statiozone LIDAR dell'ENEA, in: *Electtroottica '90*, Milano 1990, pp. 693.

Barbini, R., F. Colao, T. Hermsen,, A. Palucci, S. Ribezzo; DIAL measurements of atmospheric components, in: *Proc. 6th National Congress on Quantum Electronics and Plasma Physics*, Rome 1990, pp. 379.

Barbini, R., F. Colao, F. D'Amato, M. Giorgi, A. Palucci, S. Ribezzo, S. Orlando; Lidar cavity of the ENEA ground based station: DIAL and Doppler measurements, in: *Proc. 15th Int. Laser Radar Conference*, Part 1, Institute of Atmospheric Optics, Tomsk 1990, pp. 101-104.

Beekmann, M., G. Ancellet, J. Pelon, G. Megie, H.G.J. Smith, D. Kley; Intercomparison campaign between a ground based DIAL lidar and Balloon borne electrochemical sondes for tropospheric and lower stratospheric vertical ozone sounding, in: *Proc. 15th Int. Laser Radar Conference*, part 1, Institute of Atmospheric Optics, Tomsk 1990, pp. 210-213.

Carnuth, W., U. Kempfer, R. Lotz; Lidar measurements of tropospheric ozone, in: *Proc. 15th Int. Laser Radar Conference*, part 1, Institute of Atmospheric Optics, Tomsk 1990, pp. 202-205.

Milton, M.J.T., A. Sunnesson, J. Pelon, G. Ancellet, J. Bösenberg, W. Carnuth, H. Edner, L. Stefanutti; Raman shifted laser sources for DIAL measurements of ozone in the free troposphere, in: *Proc. 15th Int. Laser Radar Conference*, Part 2, Institute of Atmospheric Optics, Tomsk 1990, pp. 268-270.

Papayannis, A., G. Ancellet, J. Pelon, G. Megie; Multiwavelength lidar for ozone measurements in the troposphere and the lower stratosphere, *Appl. Opt.* **29** (1990) 467-476.

Pelon, J.; The joint European project TESLAS for lidar tropospheric ozone mesurement, in: *Proc. 15th Int. Laser Radar Conference*, Part 1, Institute of Atmospheric Optics, Tomsk 1990, pp. 121-124.

Riebesell, M., A. Ansmann, C. Weitkamp; Raman lidar measurement of the atmospheric aerosol extinction profile, in: *Proc. Topical Meeting on Optical Remote Sensing of the Atmosphere*, Lake Tahoe 1990, pp. TuD-1/308-311.

Sunesson, J.A., D.P.J. Swart; Tropospheric ozone lidar. feasibility and definition, *RIVM-Report* Nr. 222201002, Bilthoven 1990.

Sunesson, J.A.; Development of a DIAL System for Tropospheric Ozone Measurements, in: *Proc. 15th Int. Laser Radar Conference*, Part 2, Institute of Atmospheric Optics, Tomsk 1990, pp. 271-274.

Svanberg, S.; Atmospheric pollution monitoring using laser lidar, in: S. Martellucci, A.N. Chester (eds), *Optoelectronics for Environmental Science*, Plenum Press, New York 1990, pp. 3-13.

Svanberg, S.; Environmental monitoring using optical techniques, in: W. Demtröder, M. Ignuscio (eds), *Applied Laser Spectroscopy*, Plenum Press, New York 1990, pp. 417-434.

Thomsen, O; Messung des Absorptionsquerschnitts von Schwefeldioxid im Wellenlängenbereich von 265 bis 298 nm, *Report GKSS* 90/E/36, 117 pages, 1990.

Weitkamp, C., O. Thomson, P. Bisling; Sulfur dioxide absorption cross-section measurements between 265 and 298 nanometers, in: *Proc. Topical Meeting on Optical Remote Sensing of the Atmosphere*, Lake Tahoe 1990, pp. wD11-1/501-503.

1991

Ancellet, G., J. Pelon, M. Beekmann, A. Papayannis, G. Mégie; Ground-based lidar studies of ozone exchanges between the stratosphere and the troposphere, *J.Geophys. Res.* **96** (1991) 22401-22421.

Ansmann, A., M. Riebesell, U. Wandinger, C. Weitkamp, W. Michaelis; Combined Raman elastic-backscatter lidar for the independent measurement of backscatter and extinction profiles, in: *Proc. Second Symp. on Tropospheric Profiling: Needs and Technology*, Boulder 1991, pp. 19-20

Apituley, A.; RIVM tropospheric ozone LIDAR measurements during TROLIX '91, *RIVM report* 222201007, 1991

Barbini, R., F. Colao, A. Petri; Numerical analysis of a CO_2 SFUR Laser, *Il Nuovo Cimento* D **143** (1991) 143.

Barbini, R., F. Colao, S. Orlando, A. Palucci , S. Ribezzo; The ENEA lidar/dial facility, in: P. Borrell, P.M. Borrell, W. Seiler (eds.), *Proc. EUROTRAC Symp. '90*, SPB Academic Publishing bv, The Hague 1991, pp. 443-444.

Barbini, R., F. Colao, T. Hermsen, A. Palucci, S. Ribezzo; Preliminary measurements of vertical ozone profiles with the ENEA Lidar Station, *ENEA internal report* RTI/INN/91/21, 1991

Bergwerff, J.B.; Airborne O_3 measurements during TROLIX '91, *RIVM report* 222201005, Bilthoven 1991

Bösenberg, J., T. Schaberl; A differential absorption lidar system for tropospheric ozone profiling, in: *Proc. Second Symp. on Tropospheric Profiling: Needs and Technology*, Boulder 1991, pp. 5-6

Carnuth, W., U. Kempfer, R. Lotz, T. Trickl; Development and application of a tropospheric ozone lidar, in: P. Borrell, P.M. Borrell, W. Seiler (eds.), *Proc. EUROTRAC Symp. '90*, SPB Academic Publishing bv, The Hague 1991, pp. 445-447.

Carnuth, W., U. Kempfer, R. Lotz, T. Trickl; Ein troposphärisches Ozonlidar, Laser in der Umweltmesstechnik/Laser in remote sensing, Vorträge des 10. Internationalen Kongresses Laser 91, Springer Verlag, Berlin 1991, pp. 89-92.

Carnuth, W., U. Kempfer, R. Lotz, T. Trickl; Tropospheric ozone measurements with a UV lidar system. *Opt. Soc. Amer. Technical Digest Series* **18** (1991) 230-231.

Jonge, C.N. de; Ground based measurements during TROLIX '91, at Bilthoven, *RIVM Report* 222201008, Bilthoven 1991.

Jonge, C.N. de; Planetary boundary layer measurements during TROLIX '91, *RIVM Report* 222201004, Bilthoven 1991.

Jonge, C.N. de; NO_2-DIAL measurements during TROLIX '91, *RIVM Report* 222201003, Bilthoven 1991.

Krause, M.; Erzeugung frequenzverschobener Strahlung für ein Temperatur- und Feuchte-Lidar mit einem KrF-Laser unter Ausnutzung stimulierter Ramanstreuung, *GKSS Report* 91/E/26, 1991, p. 118

McIlveen, T.J., B.W. Jolliffe, M.J.T. Milton, N.R.S. Swann, P.T. Woods; A DIAL system for measurement of ozone and related photochemical pollutants in the free troposphere, in: P. Borrell, P.M. Borrell, W. Seiler (eds.), *Proc. EUROTRAC Symp. '90*, SPB Academic Publishing bv, The Hague 1991, pp. 449-450.

Morandi, M., V.M. Sacco, F. Castagnoli , L. Stefanutti; TODUL lidar system for tropospheric ozone measurements, in: P. Borrell, P.M. Borrell, W. Seiler (eds.), *Proc. EUROTRAC Symp. '90*, SPB Academic Publishing bv, The Hague 1991, pp. 455-458.

Pelon, J.; A joint European programme for lidar tropospheric ozone measurement: The TESLAS project, in: P. Borrell, P.M. Borrell, W. Seiler (eds.), *Proc. EUROTRAC Symp. '90*, SPB Academic Publishing bv, The Hague 1991, pp. 437-441.

Pelon, J., G. Ancellet , G. Mégie; Lidar development for tropospheric ozone, in: P. Borrell, P.M. Borrell, W. Seiler (eds.), *Proc. EUROTRAC Symp. '90*, SPB Academic Publishing bv, The Hague 1991, pp. 451-453.

Sonnemann, G., G. v. Cossart, J. Fiedler; Evaluation of ozone-(ECC-) and meteorological sonde data during TROLIX '91, *Internal Report*, Observatorium für Atmosphärenforschung, Kühlungsborn 1991.

Sunesson, J.A., A. Apituley; RIVM tropospheric ozone lidar, system description and first results, *RIVM Report* 222201006, Bilthoven 1991.

Sunesson, J.A., A. Apituley , D.P.J. Swart; Development of a Dial system for tropospheric ozone measurements, in: P. Borrell, P.M. Borrell, W. Seiler (eds.), *Proc. EUROTRAC Symp. '90*, SPB Academic Publishing bv, The Hague 1991, pp. 459-461.

Thomsen, O., P. Bisling, C. Weitkamp, W. Michaelis; Measurement of the sulfur dioxide absorption cross section in the solar-blind ultraviolet, in: P. Borrell, P.M. Borrell, W. Seiler (eds.), *Proc. EUROTRAC Symp. '90*, SPB Academic Publishing bv, The Hague 1991, p. 463.

Weitkamp, C., A. Ansmann, R. Riebesell, U. Wandinger, W. Michaelis; Wasserdampfmessungen mit einem Ramanlidar, in: *Laser 91, Kongreß Laser in der Umweltmeßtechnik*, Munich 1991, pp. 98-101.

Weitkamp, C., P. Bisling, J. Glauer, U.-B. Goers, S. Köhler, W. Lahmann, W. Michaelis; Das mobile ortsauflösende Schadgasfernmeßsystem ARGOS, in: *Laser 91, Kongreß Laser in der Umweltmeßtechnik*, Munich 1991, pp. 197-200.

1992

Apituley, A., J. Bösenberg, G. Ancellet, H. Edner,B. Galle, J.B. Bergwerff, G.V. Cossart, J. Fiedler, C.N. de Jonge, J. Mellquist, V. Mitev, T. Schaberl, G. Sonnemann, J. Spaakman, D.J.P. Swart, E. Wallinder; First results from TROLIX '91: An intercomparison of tropospheric ozone LIDARs, in: *Proc. 16th Int. Laser Radar Conf.*, Cambridge, Ma. 1992; *NASA Conf. Publ.* 3158, (1992) 53-55.

Barbini, R., F. Colao, A. Palucci, S. Ribezzo; The ENEA CO_2 lidar/DIAL station. Activity in a tropospheric remote sensing programm, in: *Proc. 16th Int. Laser Radar Conference*, Cambridge, Ma. 1992.

Barbini, R., F. Colao, A. Palucci, S. Ribezzo; Misure range resolved di ozono e vapor d'acqua in: *Proc. 2. Convegno Nazionale "Strumentazione e metodi di misura elettroottici"*, Firenze 1992, p. 263

Bisling, P., U.-B. Goers, J. Glauer, S. Kohler, W. Lahmann, C. Weitkamp, W. Michaelis; Luftschadstoffmessungen mit dem mobilen Lidarsystem ARGOS, in: *Optik Tagungsband 1992, Kolloquium "Theoria cum Praxi", 3. Berliner Optiktage*, Berlin-Adlershof 1992, Beitrag VO2, p. 10.

Bösenberg, J., G. Ancellet, A. Apituley, H. Bergwerff, H. Edner, B. Galle, C.N. de Jonge, V. Mitev, T. Schaberl, G. Sonnemann, E. Wallinder; First results from TROLIX '91: An intercomparison of tropospheric ozone LIDARs, in: *Proc. Quadrennial Ozone Symp.*, Charlottesville, Va. 1992.

Stefanutti, L., F. Castagnoli, M. Del Guasta, M. Morandi, L. Zuccagnoli, S. Godin, G. Mégie, J. Brechet, P. Therrier; Preliminary ozone measurements by means of DIAL in Dumont d'Urville, in: M. Colacino et al. (eds), *SIF Conf. Proc. 35, 4th Workshop Italian Research on Antarctic Atmosphere* (1992) 345.

Stefanutti, L., F. Castagnoli, M. Del Guasta, M. Morandi, V.M. Sacco, L. Zuccagnoli, S. Godin, G. Megie, J. Porteneuve; The Antarctic ozone lidar system, *Appl. Phys.* **B55** (1992) 3-12.

Weitkamp, C., O. Thomsen, P. Bisling; Meß- und Vergleichswellenlängen zur Elimination von SO_2-Querempfindlichkeiten bei der Lidar-Fernmessung troposphärischen Ozons, *Laser und Optoelektronik* **24** (2) (1992) 42-47; *Report GKSS* 92/E/24.

Weitkamp, C., U.-B. Goers, J. Glauer, W. Lahmann, P. Bisling, S. Köhler, W. Michaelis; Das mobile Lidarsystem ARGOS zur ortsaufgelösten Fernmessung gasförmiger Schadstoffe, in: *Umwelt-Meßtechnik, Tagung mit fachbegleitender Ausstellung zu den Themen Luft, Boden/Grundwasser, Oberflächengewässer*, Leipzig 1992, *VDI Tagungsbericht*, Düsseldorf 1992, pp. 105-114; *Report GKSS* 92/E/26 (1992)

1993

Ancellet, G., J. Pelon, G. Megie; Ozone lidar development for planetary boundary layer and free troposphere measurements, in: P.M. Borrell, P. Borrell, T. Cvitaš, W. Seiler (eds.), *Proc. EUROTRAC Symp. '92*, SPB Academic Publishing bv, The Hague 1993, p. 224.

Apituley, A., D.P.J. Swart, J.B. Bergwerft, C.N. de Jonge, J. Spakman; RIVM contribution to TROLIX 91: An intercomparison of tropospheric ozone lidars, in: P.M. Borrell, P. Borrell, T. Cvitaš, W. Seiler (eds.), *Proc. EUROTRAC Symp. '92*, SPB Academic Publishing bv, The Hague 1993, pp. 209-212.

Bisling, A., U.-B. Goers, J. Glauer, S. Köhler, W. Lahmann, C. Weitkamp, W. Michaelis; Luftschadstoffmessungen mit dem mobilen Lidarsystem ARGOS, in: *Beiträge zum Kolloquium Optik und Umwelt*, V 02, 12 S. GKSS 93/E75 (1993).

Bösenberg, J.; Measuring photo-oxidant profiles, in: P.M. Borrell, P. Borrell, T. Cvitaš, W. Seiler (eds.), *Proc. EUROTRAC Symp. '92*, SPB Academic Publishing bv, The Hague 1993, p. 69.

Bösenberg, J., G. Ancellet, A. Apituley, H. Bergwerff, G. V. Cossart, H. Edner, J. Fiedler, B. Galle, C.N. de Jonge, J. Mellquist, V. Mitev, T. Schaberl, G. Sonnemann, J. Spaakman, D.J.P. Swart, E. Wallinder; Tropospheric Ozone Lidar Experiment, TROLIX '91: Field Phase Report, *Report* Nr. 102, Max-Planck-Institut für Meteorologie, Hamburg 1993, pp. 1-239.

Bösenberg, J., T. Schaberl; First results from TROLIX '91 - a tropospheric ozone lidar intercomparison experiment, in: P.M. Borrell, P. Borrell, T. Cvitaš, W. Seiler (eds.), *Proc. EUROTRAC Symp. '92,* SPB Academic Publishing bv, The Hague 1993, pp. 208.

Carnuth, W., U. Kempfer, R. Lotz, T. Trickl; The first year of continuous ozone measurements with the IFU lidar, in: P.M. Borrell, P. Borrell, T. Cvitaš, W. Seiler (eds.), *Proc. EUROTRAC Symp. '92,* SPB Academic Publishing bv, The Hague 1993, pp. 225-226.

Edner, H., S. Svanberg, E. Wallinder; Evaluation of DIAL systems for tropospheric ozone measurements, in: P.M. Borrell, P. Borrell, T. Cvitaš, W. Seiler (eds.), *Proc. EUROTRAC Symp. '92,* SPB Academic Publishing bv, The Hague 1993, pp. 216-219.

Goers, U.-B., P. Bisling, J. Glauer, S. Köhler, W. Lahmann, C. Weitkamp, W. Michaelis; ARGOS: a differential absorption lidar for the depth-resolving measurement of sulfur dioxide, nitrogen dioxide and ozone, in: P. Zannetti, C.A. Brebbia, J.E. Garcia Gardea, G.Ayala Milian (eds), *Air Pollution*, Part II - *Analysis, Monitoring, Management and Engeneering*, Elsevier Sci. Publ., London 1993, pp. 487-494; *GKSS* 92/E/104

Papayannis, A., G. Ancellet, J.L. Conrad, C. Laqui, J. Pelon; Active optoelectronic system for stabilizing pulsed laser output energy at the 4th harmonic frequency, *Rev. Sci. Instrum.* **64** (1993) 612-617.

Ponomarev, Yu.N., I.S. Tryshkin; Measurements of absorptance and transmittance of atmospheric gases in the visible and UV spectral regions using a 10-metre-length multipass cell, in: *Proc. ASA workshop, Reims,* 1993, p. 91.

Sacco, V.M., F. Castagnoli, M. Del Guasta, M. Morandi, L. Stefanutti, S. Godin, G. Megie; The IROE-CNR and SA-CNRS ozone lidar in Antarctica, in: P.M. Borrell, P. Borrell, T. Cvitaš, W. Seiler (eds.), *Proc. EUROTRAC Symp. '92,* SPB Academic Publishing bv, The Hague 1993, pp. 227-230.

Swart, D.P.J., A. Apituley, J. Spakman, E.P. Visser, J.B. Bergwerff; RIVMs tropospheric ozone lidars for European and global monitoring networks, in: *Proc. Sixth Topical Meeting on Optical Remote Sensing of the Atmosphere*, Salt Lake City 1993.

Theopold, F.A., C. Weitkamp, W. Michaelis; Double-cavity étalon in the near infrared, *Optics Lett.* **18** (1993) 253-254; *GKSS* 93/E/15 (1993).

Theopold, F.A., J. Bösenberg; Differential absorption lidar measurements of atmospheric temperature profiles: Theory and experiment, *J. Atmos. Ocean. Technol.* **10** (1993) 165-179.

Wandinger, U., C. Weitkamp; Cloud measurements with a polarization Raman elastic-backscatter lidar, in: *Optical Remote Sensing of the Atmosphere, Technical Digest Series* Vol. 5, Washington 1993, pp. 126-128.

Weber, K., C. Weitkamp; Standardisierung auf dem Gebiet der optischen Fernmeßverfahren in Deutschland, in: C. Werner, W. Waidelich (eds), *Laser in der Umweltmeßtechnik - Lasers in Remote Sensing*, Springer-Verlag, Berlin 1993, pp. 201-205.

Weller, R., C. Weitkamp; Variationen der Ozonverteilung in der marinen und polaren Troposphäre, in: *Atmosphärisches Ozon - Prozesse und Wirkungen*, Arbeitsgemeinschaft der Großforschungseinrichtungen (AGF), Bonn 1993, pp. 9-11

1994

Ancellet, G., M. Beekmann, A. Papayannis; Impact of a cut-off low development on downward transport of ozone in the troposphere, *J. Geophys. Res.* **99**D (1994) 3451-3468.

Ancellet, G., M. Beekmann; Ozone stratosphere-troposphere exchange: Climatology and case studies, in: P.M. Borrell, P. Borrell, T. Cvitaš, W. Seiler (eds.), *Proc. EUROTRAC Symp. '94*, SPB Academic Publishing bv, The Hague 1994, pp. 812-819.

Beekmann, M., G. Ancellet, C. Mancier; Vertical ozone monitoring at the Observatiore de Haute Provence, in: P.M. Borrell, P. Borrell, T. Cvitaš, W. Seiler (eds.), *Proc. EUROTRAC Symp. '94*, SPB Academic Publishing bv, The Hague 1994, pp. 367-372.

Beekmann, M., G. Ancellet, G. Mégie, H.G. Smit, D. Kley; Intercomparison campaign for vertical ozone profiling in the troposphere at the Obervatoire de Haute Provence, 1989: Electrochemical sondes of ECC and Brewer-Mast type and a ground-based UV-DIAL lidar, *J. Atmos. Chem.* **19** (1994) 259-288.

Bösenberg, J.; Atmospheric processes and ozone profiles, in: P.M. Borrell, P. Borrell, T. Cvitaš, W. Seiler (eds.), *Proc. EUROTRAC Symp. '94*, SPB Academic Publishing bv, The Hague 1994, pp. 99-104.

Clappier, A., B. Calpini, E. Durieux, L. Fiorani, M. Flamm, H. van den Bergh; Numerical study of a pump and probe Lidar experiment for in situ tropospheric measurements of the OH radical, in: *Proc. 17th Int. Laser Radar Conf.*, Sendai 1994.

Clappier, A., B. Calpini, E. Durieux, L. Fiorani, L. Jaquet, H. Van den Bergh; Numerical simulation of a pump-probe LIDAR experiment, in: P.M. Borrell, P. Borrell, T. Cvitaš, W. Seiler (eds.), *Proc. EUROTRAC Symp. '94*, SPB Academic Publishing bv, The Hague 1994, pp. 949.

De Backer, H., E. P. Visser, D. De Muer, D.P.J. Swart; Potential for meteorological bias in lidar ozone data sets resulting from the restricted frequency of measurement due to cloud cover, *J. Geophys. Res.* **99** (1994) 1395-1401.

Durieux, E., B. Calpini, A. Clappier, L. Fiorani, M. Flamm, L. Jaquet, H. van den Bergh; An innovative high repetition rate shot per shot acquisition system and ist application to tropospheric LIDAR measurements, in: L.D. Laude (ed.), *Eximer Lasers*, Kluwer Academic Publishers, Dordrecht 1994, pp. 245-267.

Durieux, E., B. Calpini, L. Fiorani, L. Jaquet, H. van den Bergh; A shot per shot DIAL system for ozone measurements and for a possible determination of wind fields, in: *Proc. 17th Int. Laser Radar Conf.*, Sendai 1994.

Durieux, E., B. Calpini, L. Fiorani, L. Jaquet, H. van den Bergh; Atmospheric fluctuation tracking with a shot per shot LIDAR system, in: P.M. Borrell, P. Borrell, T. Cvitaš, W. Seiler (eds.), *Proc. EUROTRAC Symp. '94*, SPB Academic Publishing bv, The Hague 1994, pp. 957-962.

Edner, H., P. Ragnarson, S. Svanberg, E. Wallinder; Simultaneous tropospheric ozone monitoring using lidar and DOAS systems, *Lund Reports on Atomic Physics* LRAP-155, Lund Institute of Technology, 1994.

Fiorani, L., E. Durieux, B. Calpini, L. Jaquet, H. van den Bergh; Intercomparison of the analysis procedures for ozone LIDAR measurements, in: P.M. Borrell, P. Borrell, T. Cvitaš, W. Seiler (eds.), *Proc. EUROTRAC Symp. '94*, SPB Academic Publishing bv, The Hague 1994, pp. 950-956.

Kempfer, U., W. Carnuth, R. Lotz, T. Trickl; A wide-range ultraviolet lidar system for tropospheric ozone measurements: Development and application, *Rev. Sci. Instrum.* **65** (1994) 3145-3162.

Lehmann, S., V. Wulfmeyer, J. Bösenberg; A time dependent attenuator for dynamic range reduction of lidar signals, in: *Proc. 17th International Laser Radar Conf.*, Sendai 1994, pp. 189-190.

Matthias, V., J. Bösenberg, V. Wulfmeyer; Improvement of ozone measurements with DIAL by using an additional Raman channel, in: P.M. Borrell, P. Borrell, T. Cvitaš, W. Seiler (eds.), *Proc. EUROTRAC Symp. '94*, SPB Academic Publishing bv, The Hague 1994, pp. 326-329.

Matthias, V., V. Wulfmeyer, J. Bösenberg; Improvement of DIAL ozone measurements using an additional Raman channel, in: *Proc. 17th International Laser Radar Conf.*, Sendai 1994, pp. 453-455.

Sunesson, J.A., A. Apituley, D.P.J. Swart; Differential absorption lidar system for routine monitoring of tropospheric ozone, *Appl. Opt.* **33** (1994) 7045-7058.

Svanberg, S.; Differential absorption lidar (DIAL), in: M.W. Sigrist (ed), *Air Monitoring by Spectroscopic Techniques, Chem. Anal. Ser.* Vol. 127, J. Wiley, New York 1994, pp. 85-161.

Trickl, T.; Tropospheric ozone measurements: 1993 results and the future of the IFU lidar, in: P.M. Borrell, P. Borrell, T. Cvitaš, W. Seiler (eds.), *Proc. EUROTRAC Symp. '94*, SPB Academic Publishing bv, The Hague 1994, pp. 341-343.

1995 + in press

Apituley, A.; Comparison of the RIVM tropospheric ozone lidar to in situ measuring instruments using data acquired during TROLIX '91, *RIVM Report* 722701001, Bilthoven 1995.

Beekmann, M., G. Ancellet, D. Martin, C. Abonnel, G. Duverneuil, F. Eideliman, P. Bessemoulin, N. Fritz, E. Gizard; Intercomparison of tropospheric ozone profiles obtained by electrochemical sondes, a ground based lidar and an airborne UV-photometer, *Atmos. Environ.* (1995) in press.

Durieux, E. (ed.); *Lidar for Tropospheric Ozone Measurements, Proc. 12th TESLAS Workshop,* EPFL, Lausanne 1995.

Eisele, H., T. Trickl; The new design of the IFU tropospheric ozone lidar, *Opt. Soc. Amer. Technical Digest Series* **2** (1995) 150-151.

Grabbe, G.C., J. Bösenberg, H. Dier, U. Göersdorf, V. Matthias, G. Peters, T. Schaberl, C. Senff; Intercomparison of ozone measurements between lidar and ECC sondes, *Contrib. Atmos. Phys.* **69** (1996) 189-203.

Senff, C., J. Bösenberg, G. Peters, T. Schaberl; Remote sensing of turbulent ozone fluxes and the ozone budget in the convective boundary layer with DIAL and RADAR-RASS: a case study, *Contrib. Atmos. Phys.* **69** (1996) 161-176.

Völger, P., J. Bösenberg, I. Schult; Scattering properties of selected model aerosols calculated at UV-wavelengths: implications for DIAL measurements of tropospheric ozone, *Contrib. Atmos. Phys.* **69** (1996) 177-187.

Theses

M. Sc. / Diploma

Kempfer, U.; Entwicklung und Anwendung eines differentiellen Absorptions-LIDAR-Systems zur Messung der troposphärischen Ozonkonzentration, Ludwig Maximilian University of Munich, 1992.

Krause, M.; Erzeugung frequenzverschobener Strahlung für ein Temperatur- und Feuchte-Lidar mit einem KrF-Laser unter Ausnutzung stimulierter Ramanstreuung, University of Hamburg, 1990.

Lehmann, St.; Empfangssystem für Lidarsignale mit erweitertem Meßbereich, University of Hamburg, 1994.

Matthias, V.; Messung der atmosphärischen Ramanstreuung mit Hilfe eines UV-LIDAR-Systems, University of Hamburg, 1993.

Voelger, P.; Optische Eigenschaften von Aerosolpartikeln und deren Auswirkungen auf Ozonmessungen mit einem LIDAR, University of Hamburg, 1993.

Ph. D.

Grabbe, G.C.; Untersuchungen troposphärischen Ozons mit DIAL, University of Hamburg, 1995.

Goers, U.B.; Laserfernmessung von Schwefeldioxid und Ozon in der unteren Troposphäre mit Hilfe der differentiellen Absorption und Streuung unter den Bedingungen des mobilen Einsatzes und der besonderen Berücksichtigung des Einflusses von Grenzschicht-Aerosolen, University of Hamburg, 1994; GKSS-94/E/52.

Schaberl, T.; Messung des Ozonflusses in der unteren Troposphäre mit einem neuen Ozon-DIAL-System und einem Radar-RASS, University of Hamburg, 1995.

Wallinder, E.; Application of lidar techniques in environmental studies, *Lund Reports on Atomic Physics* LRAP-154, Lund Institute of Technology, 1994.

Appendix

Sets of Lidar Data used for the Intercomparison of Retrieval Algorithms

In the following the numerical values of the data sets which have been used for the intercomparison of retrieval algorithms are presented.

Table A1: Signals collected by the CNRS lidar system. Averaging over 52000 laser shots.

R / m	on	off	R / m	on	off
3855	59.049 81	53.098 71	9 555	0.587 805	1.444 105
4005	51.109 40	46.685 80	9 705	0.502 379	1.313 522
4155	44.484 31	41.368 89	9 855	0.424 552	1.196 924
4305	38.742 26	36.761 55	10 005	0.356 652	1.090 393
4455	33.802 55	32.797 06	10 155	0.301 399	0.982 218
4605	29.587 22	29.268 14	10 305	0.252 711	0.887 178
4755	25.920 04	26.258 86	10 455	0.220 832	0.814 480
4905	22.662 51	23.560 79	10 605	0.189 539	0.740 170
5055	19.827 87	21.132 22	10 755	0.159 999	0.672 100
5205	17.373 86	18.972 83	10 905	0.144 420	0.628 163
5355	15.249 53	17.088 72	11 055	0.127 292	0.589 782
5505	13.452 62	15.386 99	11 205	0.106 364	0.538 885
5655	11.902 63	13.914 06	11 355	0.086 0845	0.493 228
5805	10.545 03	12.606 24	11 505	0.071 321	0.452 501
5955	9.337 647	11.425 23	11 655	0.060 426	0.408 478
6105	8.271 733	10.365 29	11 805	0.050 061	0.367 177
6255	7.346 106	9.432 814	11 955	0.041 447	0.334 205
6405	6.511 194	8.590 814	12 105	0.033 867	0.305 615
6555	5.774 912	7.806 616	12 255	0.028 307	0.277 726
6705	5.121 319	7.086 160	12 405	0.024 756	0.248 976
6855	4.548 074	6.471 806	12 555	0.021 446	0.227 474
7005	4.049 449	5.901 170	12 705	0.018 229	0.210 689
7155	3.608 712	5.386 601	12 855	0.016 349	0.191 323
7305	3.224 250	4.926 263	13 005	0.014 314	0.175 515
7455	2.893 031	4.521 024	13 155	0.011 335	0.167 560
7605	2.614 432	4.144 959	13 305	0.010 365	0.158 949
7755	2.356 154	3.817 550	13 455	0.010 279	0.145 559
7905	2.129 844	3.521 914	13 605	0.007 3275	0.133 410
8055	1.924 384	3.251 821	13 755	0.005 9127	0.125 393
8205	1.746 584	3.006 607	13 905	0.006 2467	0.118 416
8355	1.583 104	2.781 272	14 055	0.006 1090	0.107 258
8505	1.416 304	2.565 777	14 205	0.004 383	0.100 056
8655	1.265 21	2.374 532	14 355	0.002 714	0.099 218
8805	1.130 03	2.190 309	14 505	0.004 343	0.094 843
8955	0.999 134	2.016 882	14 655	0.004 903	0.086 270
9105	0.881 727	1.854 386	14 805	0.003 069	0.077 387
9255	0.776 625	1.707 721	14 955	0.002 252	0.071 249
9405	0.680 209	1.577 722			

Table A2: Signals collected by the EPFL lidar system. Averaging over 5000 laser shots and 20 sampling channels. The altitude of the lidar system was of 491 m at the time of the data taking.

R / m	on	off	R / m	on	off
63	9.6863	6.9009	2223	0.0278	3.2961
183	6.1639	5.6004	2343	0.0154	1.5935
303	67.6228	60.5695	2463	0.0100	1.1769
423	95.8346	93.9900	2583	0.0105	1.1599
543	75.7652	90.5577	2703	0.0079	1.0971
663	50.4911	75.1144	2823	0.0060	0.6936
783	32.4220	62.1928	2943	0.0051	1.1350
903	20.4518	53.4497	3063	0.0062	2.1500
1023	12.4623	45.9682	3183	0.0052	2.0177
1143	7.0651	39.2587	3303	0.0029	1.5357
1263	3.5854	33.4689	3423	0.0022	1.0046
1383	1.5884	27.4225	3543	0.0019	0.4554
1503	0.6997	21.9741	3663	0.0020	0.1831
1623	0.3483	17.6161	3783	0.0021	0.0694
1743	0.1887	14.2431	3903	0.0006	0.0357
1863	0.1092	10.8155	4023	0.0028	0.0184
1983	0.0640	7.6199	4143	0.0013	0.0105
2103	0.0497	6.9457	4263	0.0001	0.0044

Table A3: Signals collected by the MPI lidar system. Averaging over 9000 laser shots and 20 sampling channels.

R / m	on	off	R / m	on	off
76.25	42.962	62.108	1876.25	8.8867	73.047
226.25	276.71	239.64	2026.25	5.3185	57.871
376.25	1653.3	1431.2	2176.25	3.3477	43.115
526.25	2224.4	2406.7	2326.25	2.4672	34.345
676.25	1238.6	1753.7	2476.25	5.9310	144.47
826.25	469.66	791.12	2626.25	3.4621	89.853
976.25	239.31	524.17	2776.25	1.2034	15.457
1126.25	129.78	359.04	2926.25	0.9794	11.482
1276.25	76.190	268.89	3076.25	0.7953	9.6716
1426.25	41.921	181.48	3226.25	0.6754	8.2766
1576.25	25.914	140.54	3376.25	0.5951	7.1789
1726.25	15.560	101.85	3526.25	0.4796	6.2536

Table A4: Signals collected by the RIVM lidar system. Averaging over many laser shots and 5 sampling channels. The original 30m resolution data set can be obtained under an electronic form from Arnoud.Apituley@rivm.nl.

R / m	on	off	R / m	on	off
65	19.115	13.643	2315	1.7520	1.7307
215	491.19	405.44	2465	1.3254	1.3716
365	1353.9	625.98	2615	1.0354	1.1145
515	1001.4	483.39	2765	0.834 40	0.871 67
665	614.30	310.01	2915	0.687 86	0.760 82
815	343.45	182.15	3065	0.541 25	0.647 32
965	195.02	111.03	3215	0.455 40	0.520 22
1115	111.37	66.838	3365	0.385 48	0.461 07
1265	65.124	41.918	3515	0.308 55	0.395 71
1415	37.169	25.520	3665	0.224 81	0.339 71
1565	23.394	17.402	3815	0.188 93	0.285 89
1715	14.481	11.754	3965	0.181 57	0.223 87
1865	9.0985	7.8767	4115	0.140 68	0.221 50
2015	5.6879	5.0523	4265	0.108 57	0.164 45
2165	2.7225	2.4950	4415	0.088 630	0.145 11

Transport and Chemical Transformation of Pollutants in the Troposphere

Volume 8, Part II

Joint European Development of Tunable Diode Laser Absorption Spectroscopy for Measurement of Atmospheric Trace Gases

David Brassington (London)
Coordinator

JETDLAG Steering Group

Horst Fischer	Mainz
Dieter Klemp	Jülich
Hélène Mac Leod	Paris
Maurus Tacke	Freiburg im Breisgau
Peter Werle	Garmisch-Partenkirchen

Chapter 8

General Scientific Report

David Brassington

Atmospheric Chemistry Research Unit, Imperial College, Silwood Park, Ascot, UK

8.1 Introduction

This report on the JETDLAG sub-project is divided into two parts: a general introduction to the TDLAS technique, and a summary of the aims and achievements of JETDLAG. Reports from individual investigators within the sub-project follow this overview report.

8.2 Review of Tunable Diode Laser Spectroscopy

8.2.1 Overview

The development of tunable lead salt diode lasers (TDLs) in the mid 1960s provided the first convenient high-resolution tunable source for laser spectroscopy throughout the mid-infrared. The first use of TDLs for atmospheric measurements employed a long open-path, with a retro-reflector to return the beam to the instrument [1, 2], and this technique is still used today. However the most important application of TDLs to atmospheric measurements has turned out to be their use in conjunction with a long-path cell to provide high sensitivity local measurements [3, 4]. This technique is commonly referred to as TDLAS (tunable diode laser absorption spectroscopy) and the development of this technique was the main aim of the JETDLAG subproject. This type of measurement was pioneered in the late 1970s [5-9] and has since developed into a very sensitive (down to 20 ppt for some species) and general technique for monitoring most atmospheric trace species. The only requirement is that the molecule should have an infrared line-spectrum which is resolvable at the Doppler limit. In practice this includes most molecules with five or fewer atoms together with some larger molecules [10].

The principle of TDLAS is absorption spectroscopy using a single isolated absorption line of the species. However typical line-strengths are such that, even for strongly absorbing species, a typical atmospheric concentration of 1 ppb produces an absorption of only 1 part in 10^7 over a 10 cm path-length. Conventional absorption spectroscopy techniques cannot measure such small absorptions. TDLAS overcomes this problem firstly by using a multi-pass cell to give path-lengths of 100 m or more. Such cells achieve the long path by using mirrors to fold the optical path, giving typically 100 passes of a 1 m base-length cell. Secondly various types of modulation spectroscopy are employed in which the diode-laser wavelength is modulated over the absorption linewidth at frequencies of anywhere between 100 Hz and 2 GHz. These modulation techniques allow absorptions as low as 1 part in 10^5 to be measured with a 1 Hz bandwidth. In combination these two techniques give detection limits of around 20 ppt for the most strongly absorbing species (*e.g.* NO_2, HCl, NH_3) and better than 1 ppb for almost all species of interest in atmospheric chemistry (provided they have resolvable lines).

TDLAS is typically used in a continuous sampling mode in which air is continually drawn through the multi-pass cell at a pressure of about 30 mbar (achieved with a pressure-reducing valve on the inlet). Operating at reduced pressure narrows the absorption lines so avoiding possible interferences from other species and also reduces the range over which the laser needs to be modulated, which in turn reduces the effects of laser noise.

The most critical component of a TDLAS system is the diode laser and the properties of these devices are considered in the next Section. The JETDLAG subproject was concerned exclusively with instruments using the lead-salt family of semiconductor lasers, which produce wavelengths between 3 and 30 μm. An individual laser of this type could typically be tuned over a 100 cm^{-1} range and would need to be operated at liquid nitrogen temperatures. TDLAS instruments can also use InGaAsP diode lasers (which do not need cooling) operating at wavelengths below 2 μm, but in this wavelength region absorption lines are due to overtone or combination bands with strengths typically two orders of magnitude lower than the fundamental bands. Thus these lasers cannot normally be used for measurements of trace species.

The main features of TDLAS which make it such a valuable technique for atmospheric measurements are as follows:

- As a high resolution spectroscopic technique it is virtually immune to interferences by other species - a problem that plagues most competing methods. This ability to provide unambiguous measurements leads to the use of TDLAS as a reference technique against which other methods are often compared.
- It is a general technique. The same instrument can easily be converted from one species to another by changing the laser and calibration cells. Similarly

it is easy to construct an instrument which will measure several species simultaneously by multiplexing the outputs of several lasers (commonly up to four) through the multi-pass cell.

- It offers automated measurements at time-constants of a minute or so. (This compares with the bubblers, diffusion, or denuder tubes which for many species are the main alternative and which require sampling periods of at least several hours.) The time-constant of TDLAS can be traded off against sensitivity and this allows flux measurements of relatively abundant species such as CH_4 and CO by the eddy-flux correlation technique. For this application time-constants of 70 ms have been achieved [11].

Against these advantages must be set the facts that TDLAS instruments are complex and expensive, and they currently require expert operators. The diode lasers themselves can be unreliable and each laser is unique, leading to the need to recharacterise the instrument whenever a new laser is installed.

8.2.2 Properties and operation of tunable diode lasers

Lead-salt tunable diode lasers are similar in principle to the more familiar GaAs semiconductor lasers used widely in communications. The simplest form of laser consists of a crystal of a lead-salt semiconductor such as $Pb_{1-x}Sn_xSe$ on which a p-n junction is formed by diffusion of a salt of different stoichiometry into the top surface. The crystal is cleaved to a chip about 300 μm long with front and rear faces about 100 μm square. These front and rear facets form the laser cavity and do not need any reflective coating as the Fresnel reflection is sufficiently strong. If electrodes are deposited on the top and bottom surfaces and a current of a few hundred mA passed through the junction, lasing action takes place at a wavelength determined by the semiconductor energy gap. This energy gap is temperature dependent and a typical laser can be temperature tuned over about 100 cm^{-1}. The energy gap is also dependent on alloy composition so that lasers of different wavelength can be obtained either by varying the stoichiometry of the salt (*i.e.* the value of x above) or, for wider tuning, by using different constituents *e.g.* PbSnTe or PbSSe. (For reviews of lead-salt TDLs see Wall [12], Preier *et al.* [13], Tacke [14], Feit *et al.* [15])

The simple laser just described is known as a diffused homojunction laser and has many disadvantages such as poor mode structure, low power, and the need to operate below liquid nitrogen temperatures. Lasers in current production have more sophisticated structure and, at the risk of some oversimplification, may be divided into those intended for operation below 77 K (Type I lasers) and those intended for operation above 77 K (Type II). Type I lasers require cooling in heavy, expensive and often unreliable closed cycle coolers, whereas liquid nitrogen dewars are sufficient for Type II lasers. Because of their simpler cooling needs Type II lasers are the first choice for any new TDLAS system but Type I lasers are still needed to fulfil some special requirements *e.g.* high power.

Type I lasers are based on PbSnTe, PbSSe, or PbSnSe. They are homojunction devices formed by a diffusion process but differ from the simple laser described above in using a mesa structure, which is typically 10 - 20 μm wide, to confine the active region laterally. This results in better mode structure and can give higher temperature operation. Type I lasers are available between 3000 and 400 cm^{-1} and operate at temperatures between about 20 and 70 K. Output powers vary between 100 μW and 2 mW.

Type II lasers are more complex devices. They are of double heterostructure (DH) or buried heterostructure (BH) construction and are formed by molecular beam epitaxy (MBE) with PbEuSeTe or PbSnTe active layers. The advantage of this structure is much better optical and carrier confinement in the active region compared to diffused homojunctions. The active layer typically is < 1 μm thick and < 10 μm wide. This results in good mode quality and high operating temperature. The highest reported temperature for c.w. operation of these lasers is around 200 K [12] although production lasers are normally only specified for operation up to about 120 K. Lasers of this type are available between 3050 and 900 cm^{-1} with single-mode operation available over much of this range. Powers are typically 200 μW but single-mode lasers with powers of 1 mW can be produced at most wavelengths. A single-mode output of 3.5 mW has been seen in one delivered laser [16].

Because of their small cross-sectional area TDLs produce a very divergent beam needing *f*/1 optics for collimation. Early lasers had poor beam quality with irregular near and far field distributions, sometimes with the main beam lobe emitted at an angle to the optical axis of the laser. However the improved optical confinement given by mesa construction or buried heterostructure design has resulted in improved beam quality [17].

Early studies with liquid helium cooled TDLs reported linewidths as low as 54 kHz [18] but recent studies [19, 20] suggest that 10 MHz is a more typical linewidth for current lasers. This is however adequate for TDLAS where absorption linewidths are around 200 MHz. Linewidth is found to vary considerably from laser to laser and also with operating conditions. Mode competition can increase the linewidth to several 100 MHz and linewidth is also increased by optical feedback into the laser due to scattering, or by vibrations from closed cycle coolers.

Laser noise is also found to be strongly laser and operating condition dependent [21, 22]. The trend is a $1/f$ dependence of noise power at low frequencies changing to a white noise spectrum due to shot noise at higher frequencies.

Compared to GaAs lasers lead-salt semiconductor lasers are at a relatively early stage of their development (due to the much smaller market and hence smaller development funding). Current development is aimed at better mode quality, higher power, and higher temperature operation (with the goal of allowing thermoelectric cooling). Various routes are being explored including buried quantum well (BQW) structures, and laser cavities employing distributed feedback (DFB) and distributed Bragg reflectors (DBR) . In the quantum well design the thickness of the active layer is reduced to ~0.1 μm resulting in quantization of the valence and conduction band energy levels. The lasers have predominantly single-mode characteristics and higher efficiency as a result of the quantum effects [12], and can have higher power and higher operating temperature than conventional structures. The DFB and DBR cavity designs use corrugated structures, either distributed along the entire cavity (DFB) or at each end (DBR). Due to the Bragg condition these structures act as distributed or discreet reflectors but only at specific wavelengths. They thus give additional control over the laser modes and can give predominantly single-mode operation [23].

Lead-salt TDLs are currently available from only two commercial manufacturers world-wide. This is a cause of some concern since with a rather small market the future availability of the lasers cannot be guaranteed.

TDLs are tuned by varying the temperature of the active region. This can be done either by varying the temperature of the cold stage on which the diode is mounted or by varying the laser drive current, which varies the ohmic heating of the active region. Varying the base temperature will tune the laser over its entire range but is slow. Varying the drive current gives a more restricted tuning range, of maybe 20 cm^{-1}, but allows high frequency modulation of the laser frequency. In practice the two methods are used in combination.

Fig. 8.1 shows typical spectra of a buried heterostructure laser at fixed base temperature and a series of increasing drive currents [14]. At a particular drive current the output consists of a series of longitudinal modes which are approximately contained within an envelope; the maximum of which tunes with the temperature-dependent energy gap of the semiconductor. The modes themselves tune at a different, slower, rate through the temperature dependence of the refractive index, *n*, and also through the dependence of *n* on carrier-density, which varies with the laser drive current.

Laser mode quality has an important influence on the usability of TDLAS instruments. On starting to use a new laser the first task is to find a combination of base temperature and drive current at which the laser produces a strong, preferably single mode, tuned to the absorption line being monitored. Assuming this can be accomplished (and it may take hours or days) then all is well until the laser is cycled to room temperature and recooled. It may then be found that the laser characteristics have changed and the process must be repeated. This problem of non-reproducibility of the mode structure is a major obstacle to TDLAS becoming a routine measurement technique, usable by non-expert operators. One of the aims of JETDLAG was to develop TDLAS into a technique suitable for routine use in the field and the lack reproducible mode behaviour has been a major problem in achieving this aim.

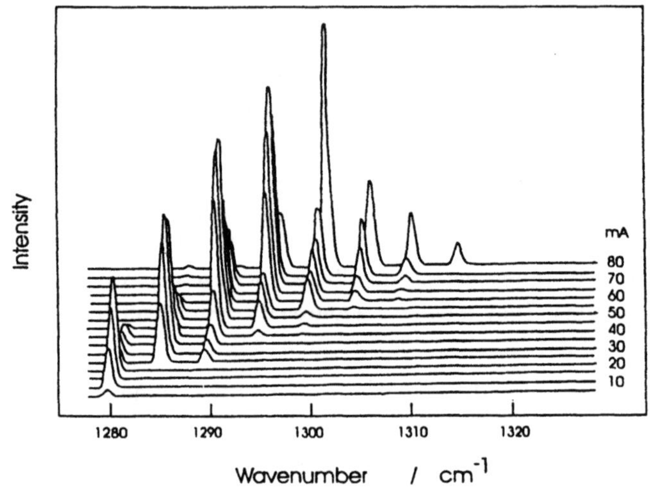

Fig. 8.1: Spectral plot of a buried heterostructure laser at fixed base temperature and a series of drive currents.

8.2.3 Spectroscopic principles of TDLAS

In order to achieve high sensitivity TDLAS uses modulation spectroscopy. This has two benefits. Firstly it produces a signal which is directly proportional to the species concentration, rather than the small change in a large signal which occurs with a conventional absorption measurement. This reduces drift. Secondly it allows the signal to be detected at a frequency at which the laser noise (one of the two main factors limiting sensitivity) is much reduced. There are two types of modulation spectroscopy employed in TDLAS systems: wavelength modulation spectroscopy (WMS) [24, 25] and frequency modulation spectroscopy (FMS) [26-27]. These terms are somewhat misleading since in both cases it is the wavelength (or optical frequency) of the laser that is modulated. The crucial difference is that for FMS the modulation frequency is equal to or greater than the absorption linewidth whereas for WMS the modulation frequency is much less than the linewidth. Thus FMS uses modulation frequencies of around 500 MHz and WMS frequencies of around 50 kHz. It has recently been appreciated that, since there is no fundamental difference between WMS and FMS, there can be advantages in operating at intermediate frequencies of between 1 and 50 MHz which have much of the performance advantage of FMS but with less of the experimental and cost disadvantages [25, 28]. This intermediate regime is known as high frequency WMS (HFWMS).

The experimental arrangement is essentially the same for both WMS and FMS and is shown in Fig. 8.2. The laser frequency is modulated at ν_m by modulating the drive current. In the presence of an absorption line the frequency modulation of the laser radiation produces an intensity modulation of the radiation transmitted through the cell and this is detected with a lock-in amplifier for WMS or a mixer for FMS. The intensity modulation contains components at multiples of the modulation frequency. For WMS, in the limit of small modulation, these components are proportional to the corresponding derivative of the lineshape, so that for example the component at ν_m is proportional to the slope of the line. Although TDLAS uses modulations which are comparable with the linewidth in order to give increased signal, the line shapes are still qualitatively similar to the shape of the derivatives. For WMS it is usual to detect at the second harmonic ($2\nu_m$) partly because this signal has a maximum at line centre. Line shapes for FMS are more complex but the principals of signal recovery are similar. One problem with FMS is that the high modulation frequencies demand expensive and fragile detectors. To overcome this a technique known as two-tone FMS (TTFMS) has been developed in which two modulations are applied, separated by typically 10 MHz. A signal is then produced at this difference frequency which is easier to detect.

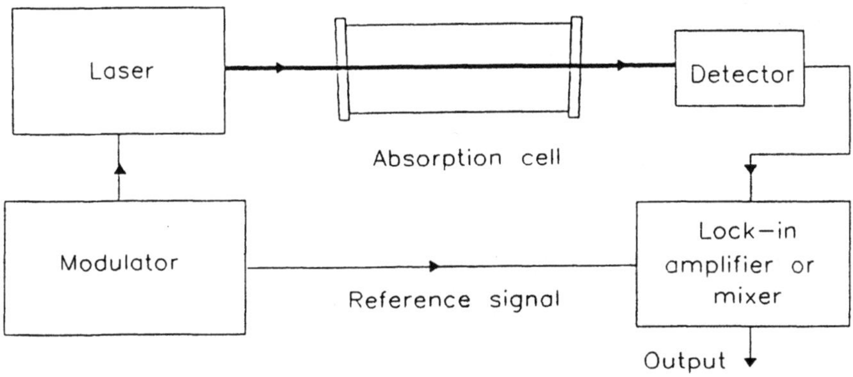

Fig. 8.2: Basic principle of TDLAS. In a real system the absorption cell would take the form of a multi-pass cell giving absorption paths of 100 m or more.

Historically WMS was developed first and is still the most common system. It is the only type of system which is commercially available. FMS is theoretically capable of two orders of magnitude better sensitivity than WMS, essentially because it operates at a frequency at which laser noise is negligible. However it has yet to show significantly better sensitivity in typical field measurements where a one minute or longer averaging period is used. An FMS system is significantly more expensive than WMS because the modulator, detector, and mixer all cost more than their lower frequency counterparts. At present it is an open question as to whether FMS will replace WMS for standard measurements of trace species. (However FMS does have a significant advantage where fast response ($\leq$ 1 s) is needed *e.g.* for eddy-correlation flux measurements.) Instruments developed within

JETDLAG have employed all four types of modulation (WMS, FMS, TTFMS and HFWMS)

8.2.4 Optical design

Fig. 8.3 shows a typical TDLAS system for detection of a single atmospheric species [29]. The beam from the TDL housed in the cold head is first collimated and then directed by a sequence of mirrors through the multi-pass White cell and onto the cooled detector. Part of the beam is split off and directed through a line-locking cell and also to a scanning Fabry-Perot interferometer which is used to give an on-line display of the laser mode-structure. A calibration cell can be inserted into the main beam under computer control. A visible He-Ne laser beam is combined with the invisible infrared beam to assist in alignment.

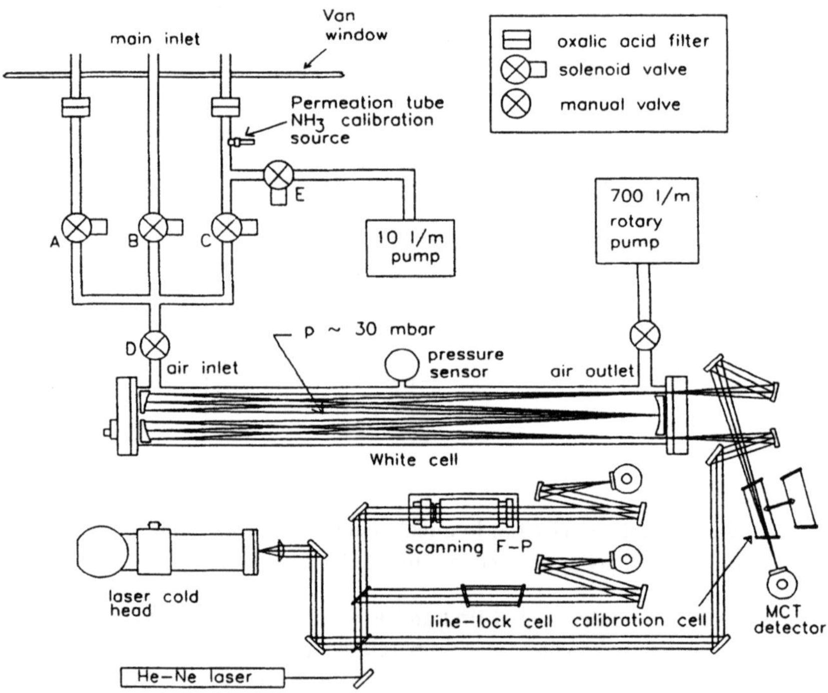

Fig. 8.3: Optical and sampling arrangement of a typical single-species TDLAS system.

The mirror arrangement of the White cell design of multi-pass cell [30, 31], as used in the system of Fig. 8.3, is shown in Fig. 8.4. As mirror B is rotated about an axis perpendicular to the plane of the paper the number of spots on the field mirror A increases. The figure shows the sequence of spots for 28 traverses of the cell. It can be seen that the total path (L) can be expressed as $4nS$ where n is an integer and S is the mirror separation. A typical cell of this type would have a 1 m mirror separation, a volume of 10 L and a path-length of 100 m. With gold coated mirrors an optical transmission of nearly 40 % can be achieved.

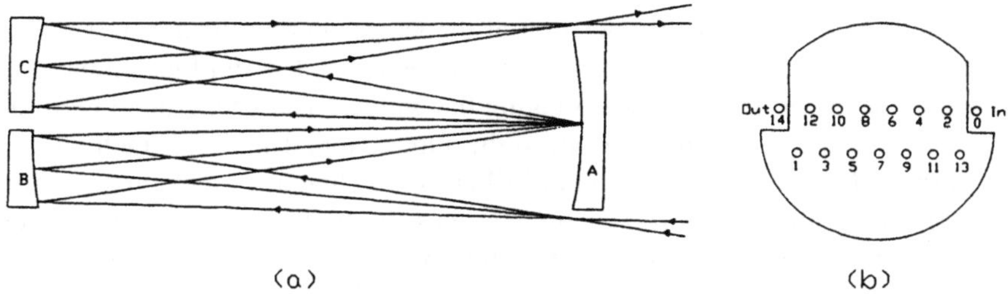

Fig. 8.4: Beam paths through a White cell. (a) shows the rays for the minimum number of passes (4). (b) shows the pattern of spots on mirror A for 28 passes.

For fast response measurements low-volume cells are needed. Various modifications to the design have therefore been introduced which better use the cell volume and these allow a 50 m path within a 1.4 L volume [32].

The White cell is increasingly being superseded in TDLAS systems by the Herriott cell [33-36] which offers advantages of low-volume, ease of alignment and reduced problems from etalon fringes. The Herriott cell consists of two spherical mirrors separated by nearly their radius of curvature. The optical beam is injected through a hole in one mirror and is reflected back and forth a number of times before exiting from the same hole. Unlike the White cell the beam remains essentially collimated throughout its traversals of the cell. In the original design the beam traces out elliptical paths on the two mirrors, however this does not give optimum use of the mirror area and thus a modified arrangement using slightly astigmatic mirrors often used. In this case the beam traces out a Lissajous figure.

Unlike the White cell a Herriott cell can support several independent optical paths, each with a separate entrance and exit hole in the input mirror [37]. This allows multi-species measurements with independent optical channels. Worsnop *et al.* [38] have described a Herriott cell, designed for TDLAS, with 76 mm diameter mirrors, 55 cm base length, a volume of 3 L and a total path-length of 100 m (182 passes).

Most TDLAS systems are limited in sensitivity not by laser or detector noise but by optical fringes superimposed on the measured spectrum. These result from unwanted etalons formed by reflections and scattering in the optical system. The fringes take the form of an approximately sinusoidal variation of the background signal with a period equal to the FSR of the etalon. This is true both for the direct absorption signal and for the various orders of modulation spectra [39]. These fringes can be reduced by careful optical design and adjustment, but it is usually difficult to reduce the fringe amplitude to a level much below that equivalent to an absorbance of 10^{-4} [28]. In order to achieve sensitivities in the 10^{-5} to 10^{-6} range some method of reducing the effect of the fringes must be found. These techniques can be categorized as follows: (i) mechanical modulation or dithering of the etalon spacing (ii) modified modulation schemes (iii) background subtraction and (iv) post-detection signal processing. The first two methods both rely on washing out the fringes by dithering the position of the fringes during the averaging of the spectrum. They thus work best for short period etalons. The third technique relies on the etalons being stable between taking the ambient and background (zero air) spectra so that they are cancelled out in the subtraction. The fourth method takes the form of analogue or digital filtering techniques including Fourier transformation. In practical TDLAS systems a combination of background subtraction with some form of post-detection processing is most commonly used. The fringe modulation techniques (i) and (ii) are rarely employed.

8.2.5 Multi-species instruments

Because of the limited tuning range of an individual laser multi-species operation usually requires a separate laser for each species. In the most common approach the collimated beams from these lasers are sequentially directed along the optical path through the instrument by either a rotating selecting mirror on a galvanometer drive [40] or by individual "pop-up" mirrors for each laser. Whilst this time multiplexing method does not provide true simultaneous measurement it is possible to cycle through four species within 10 s. Within the data-processing system the spectra of the individual species are accumulated separately over the required averaging period. However, because the duty cycle of each laser is reduced by multiplexing, a longer total period is needed to achieve a given sensitivity compared to the equivalent single-species measurement.

As an alternative to time multiplexing, the beams from the lasers can be combined along a common path using dichroic beamsplitters. By using different modulation frequencies the signals from the individual lasers can then be separated electronically. This known as frequency multiplexing [41].

Where two species have suitable lines within the current tuning range of a single laser (typically 20 cm^{-1}) both can be measured simultaneously using a technique known as jump scanning [42]. In this the normal low frequency ramp through the line is replaced by a ramp which first sweeps through the line of one species and then jumps discontinuously before sweeping through the line of the second species. The spectra of the two species are thus acquired simultaneously and appear as adjacent lines on the spectrum. These can be then analysed separately. The number of pairs of species which have suitably close lines is quite limited; examples are: HNO_3 and HCHO at 1720 cm^{-1}, NO_2 and NH_3 at 1625 cm^{-1} and CH_4 and C_2H_6. A further problem is that the laser must be able to access both lines within the same current scan and deliver strong, noise-free and preferably single-mode output at both. It is often difficult enough to find good laser operating conditions for one line let alone two.

If a Herriott cell is used as the multi-pass cell then its ability to provide several independent optical paths allows truly simultaneous measurements of several species by using separate optical channels for each. In such a system only the sampling system, Herriott cell, and data acquisition system are common to each species-measurement. The four-species ALIAS [37] and the proposed two-species ARGUS [43] instruments use this approach.

8.2.6 Sampling and calibration

The sampling system shown in Fig. 8.3 displays the main features needed. A high-speed rotary pump sucks air through the multi-pass cell at reduced pressure. This pressure is controlled by a valve on the cell inlet which is normally servo-controlled to maintain a fixed cell pressure. Computer controlled valves select one of three inlet sources: direct ambient air sampling (valve B), zero air (valve A), and calibration gas (valve C). The zero air is obtained in this system (which is for NH_3 monitoring) by passing ambient air through a scrubber consisting of an oxalic acid impregnated filter paper. The calibration gas is obtained using a permeation tube NH_3 source. A high speed (700 L/min) sampling pump is used because NH_3 posses severe sampling problems. For species such as NO_2 which pose no sampling problems a slower pump could be used. For stratospheric measurements from aircraft a pump can sometimes be dispensed with since the ambient pressure is already sufficiently low (below 100 mbar) and flow through the cell can be produced by the the forward movement of the aircraft [37].

Sampling problems with polar species such as HNO_3, HCl, and NH_3 are well known and not unique to TDLAS. These species are readily adsorbed onto the surfaces of the inlet system and multi-pass cell. This leads to memory effects due to subsequent outgassing, which can persist for days after high concentrations have been sampled, and also to a lower concentration of species in the multi-pass cell than in the air being sampled. Metals give the worst adsorption problems. PTFE and glass are the preferred materials for most species and these materials need to be used for the whole of the inlet system including the valves. The multi-pass cell also needs to be of glass or lined with PTFE.

For non-polar species where memory effects do not dominate, the instrument response time is controlled by the exchange time of the multi-pass cell. An approximate value for this can be found by assuming piston flow . Thus with a 10 L cell and a 600 L/min pumping speed the response time would be 1 s. Fast response is required for flux measurements and a 1/e response time of 70 ms has been achieved by using a 0.58 L White cell and a 1200 L/min pumping speed [44].

The best and most direct calibration method is to attach calibrant and zero air sources to the instrument inlet. This method has the advantage of calibrating the entire signal processing chain and also corrects for any loss of species due to surface adsorption (to the extent that this loss is proportional to concentration). The type of calibration source is species dependent but in general oven-stabilized permeation sources are preferred, although dilution of standard gas mixtures can be used. For polar species calibration concentrations should be comparable with levels being measured because adsorption losses can be non-linear.

Direct calibration can be time consuming especially for polar species with long equilibrium times. Because of this it is often convenient to use a sealed calibration cell as a transfer standard. This cell contains a sufficient concentration of the species being measured to give a strong signal when inserted into the optical path between the multi-pass cell and the detector (but low enough to still be in the linear response regime). The cell is filled with buffer gas to the same pressure as the multi-pass cell so that the lineshape is the same. The calibration is then in two stages. First the calibration cell is inserted in the beam and the signal compared with that from the primary calibration source. Next the atmospheric measurements are made and the signal compared with that obtained with the calibration cell inserted in the beam.

In principle TDLAS measurements can be calibrated absolutely using the known absorption line parameters, provided all the instrumental gains and modulation amplitudes are measured. In practice this is rarely attempted.

8.2.7 Signal processing

A TDLAS system can be operated either with the laser wavelength fixed at the point of maximum harmonic signal (often at line-centre) or with the laser repetitively scanning through the line and building up a spectrum by accumulating the scans in a signal averager. The first method theoretically gives the highest sensitivity for a given bandwidth because it gives maximum duty factor on the point of maximum signal. However unless fast response is needed (*e.g.* for flux measurements) the second method has overwhelming advantages because it avoids various systematic errors which can affect the fixed wavelength technique. These include etalon fringes which drift with time, and interferences from nearby absorption lines of other species. Scanning over the line also gives increased confidence in the measurement because the characteristic feature of the measured species is clearly seen and unwanted spectral features due to interfering species or etalon fringes can be identified. In this scanning method measurements of the ambient sample are typically interleaved with measurements of zero-air. Subtracting the zero-air spectrum from the ambient spectrum helps to remove etalon fringes and interferences. For polar species it also corrects for long term memory effects in the multi-pass cell. Fig. 8.5 shows an example of the signal processing of a measurement of 2 ppb of ammonia. The figure shows how the zero-air subtracted spectrum is fitted to the spectrum from the calibration cell, as described in the previous section. The ratio of these two spectra is directly proportional to species concentration. A simple regression fit is often adequate but more advanced techniques based on adaptive filters [45, 46] have been developed within JETDLAG.

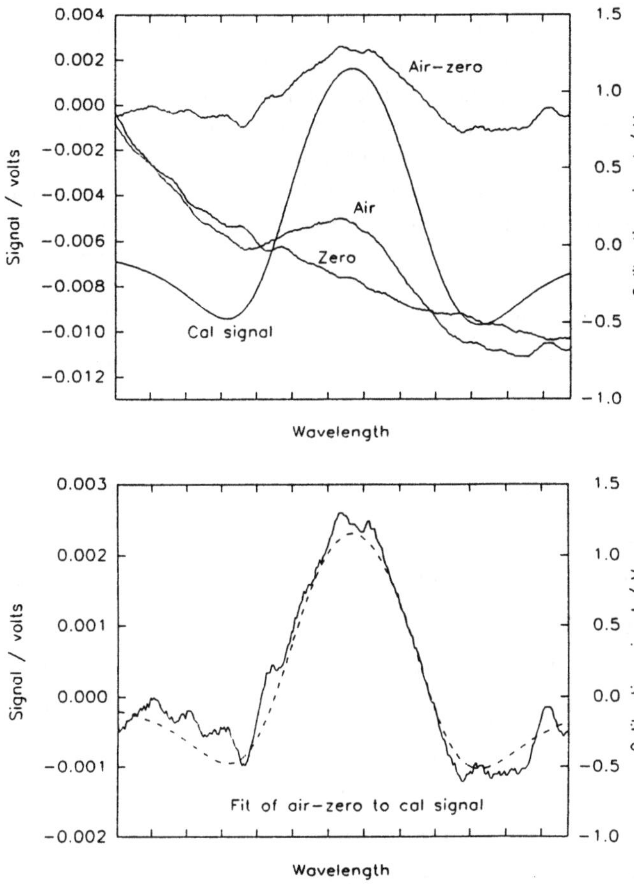

Fig. 8.5: Signal processing for a measurement of 2 ppb of NH_3. The top graph shows the air, zero air, and calibration spectra. The result of background subtracting the zero air spectrum from the air spectrum is also shown. The bottom graph shows the result of fitting the air - zero spectrum to the calibration spectrum (dashed line).

Where ambient and zero-air spectra are interleaved a single such pair of spectra can be obtained in about 30 s. However where ultimate sensitivity is not required the frequency of zero-air spectra can be reduced, allowing measurements every 10 s or less.

8.2.8 Sensitivity and precision limits

The sensitivity of TDLAS is normally limited either by laser noise or optical fringes, neither of which are easily predictable. Nevertheless experience shows that in field measurements, using alternate ambient and zero-air spectra, the absorbance sensitivity for a one minute measurement is typically between 5×10^{-6} and 10^{-5}. For a given path-length and absorbance sensitivity the minimum detectable

Table 8.1: Parameters of absorption lines suitable for TDLAS monitoring of a selection of species. The data correspond to a sampling pressure of 30 mbar and T = 296 K. The Minimum Detectable Concentration (MDC) (defined as that concentration giving a signal to r.m.s. noise ratio of 1) is also shown. This figure corresponds to a minimum detectable absorbance of 10^{-5} and a 100 m path-length. It is a realistic estimate of the performance of a typical TDLAS system using wavelength modulation spectroscopy and averaging over 1 minute. (Figures based on data from P. Werle and F. Slemr - unpublished.)

Species	M_r	λ / µm	ν / cm^{-1}	S / 10^{-19}cm	γ_v / 10^{-3} cm^{-1}	γ_v / MHz	σ / 10^{-17} cm^2	MDC / ppt
O_3	48	9.50	1053	0.42	2.64	79.2	0.53	260
N_2O	44	7.69	1301	1.56	2.77	83.1	1.92	70
CO	28	4.65	2151	1.89	3.85	115.5	1.85	73
CH_4	16	7.56	1322	0.53	3.25	97.5	0.60	225
NO	30	5.33	1876	0.34	3.26	97.8	0.39	350
SO_2	64	7.29	1372	0.49	3.86	115.8	0.41	330
NO_2	46	6.25	1600	2.18	2.90	87.0	2.70	50
NH_3	17	10.74	931	5.20	2.92	87.6	6.18	23
HNO_3	63	5.81	1722	0.72	3.75	112.5	0.25	218
HF	20	1.24	7856	0.76	12.09	362.7	0.27	500
HCl	36	3.40	2945	5.03	4.30	129.0	4.51	30
HBr	81	3.78	2649	0.45	2.90	87.0	0.57	240
HI	127	4.39	2278	0.02	2.21	66.3	0.03	4000
ClO	51	11.6	860	0.08	2.73	81.9	0.09	1500
OCS	60	4.87	2053	10.30	3.55	106.5	10.00	13
HCHO	30	3.56	2781	1.19	5.17	155.1	0.84	160
HCN	27	3.00	3337	3.69	5.97	179.1	2.33	58
CH_3Cl	50	3.29	3040	0.02	4.12	123.6	0.02	7800
H_2O_2	34	7.79	1284	0.45	3.52	105.6	0.43	320
C_2H_2	26	3.07	3260	2.29	5.43	162.9	1.65	83
H_2S	34	7.33	1365	0.01	4.89	146.7	0.007	20000
HCO_2H	46	8.98	1113	0.44	3.32	99.6	0.44	310
HO_2	33	7.08	1411	0.12	3.63	108.9	0.11	1200

concentration (MDC) for a particular species depends only on the strength (*S*) and halfwidth (γ) of the line being used. Table 8.1 gives spectral parameters for absorption lines suitable for TDLAS monitoring of a range of species. Also shown is the resulting MDC assuming an absorbance sensitivity of 10^{-5} and a 100 m path-length. The lines shown are not necessarily the strongest for that species nor do the wavelengths necessarily represent the only regions in which the species can be monitored.

By averaging over periods longer than 1 minute or by using longer path-lengths the MDCs shown in Table 8.1 can be improved upon. Thus the best sensitivity reported for NO_2 is 10 ppt using a 200 m path-length and a 7 minute total measurement time [47]. This same system gave an average sensitivity of 25 ppt during 11 flights of the CITE 2 measurement campaign.

Whilst Table 8.1 shows predicted sensitivities, based on known line strengths and the optical performance of a typical TDLAS system, Table 8.2 shows sensitivities actually achieved in field measurements for a range of species.

The above discussion relates to instruments using second harmonic WMS detection. It seems that the FMS technique can give a factor of 2 - 3 improvement in sensitivity compared to WMS systems, however these FMS systems are much more dependent on the quality of the laser than WMS systems, so this potential improvement can be difficult to realize reliably.

For measurements of trace species, sensitivity is usually more important than accuracy, and the 10 - 20 % accuracy which is typically attained without special precautions is normally adequate. However when monitoring the more abundant long-lived species such as CO, CH_4, and CO_2 high accuracy is often needed [48]. TDLAS accuracy depends firstly on the calibration procedure and secondly on instrument drift between calibrations. Calibration usually involves injecting a calibration mixture and also a zero air sample into the instrument inlet. The accuracy with which the species concentration in both these gases is known will limit TDLAS accuracy. Instrument drift between calibrations can be either baseline drift or gain drift. Baseline drift can be caused by time variation of optical fringes, and by interference from other non-constant atmospheric constituents with nearby lines. Gain errors can be caused by the temperature variation of the species absorption coefficient and by non-linearity of response. By analysing these factors and employing appropriate corrections it has been possible to obtain precisions of 0.5 % and accuracies of 2 % in measurements of CO and OCS [48, 49].

Table 8.2: Some examples of reported sensitivities achieved during field measurements of various species.

Species	Reference	Sensitivity / ppt	Time-constant
NO_2	[53]	25	70 s
	[70]	50	1 min
	[47]	25	10 min
	[61]	20	5 min
	[40]	150	3 min
	[72]	100	-
	[75]	25	150 s
HCHO	[60]	25	30 min
	[40]	300	3 min
	[62]	250	3 min
	[63]	100	30 min
	[53]	70	70 s
H_2O_2	[60]	75	30 min
	[68]	600	4 min
	[63]	150	60 min
	[53]	400	2 min
	[67]	50	60 min
HNO_3	[40]	300	3 min
	[72]	100	-
	[70]	300	1 min
	[63]	150	30 min
	[47]	75	10 min
NH_3	[57]	100	4 min
	[65]	100	5 min
HCl	[56]	50	1 min*
	[57]	100	1 s
SO_2	[70]	300	1 min
NO	[70]	300	1 min
O_3	[70]	300	1 min
N_2O	[78]	1.5 %	-

* 60 min equilibrium time

8.2.9 Use of TDLAS for atmospheric measurements

Concentration measurements

TDLAS has been used for atmospheric measurements of CH_4 [50, 51], CO [44, 49, 50, 52-55], HCl [56-59], HCHO [40, 51, 53, 60-66], H_2O_2 [51, 53, 60, 61, 63, 65-68], H_2O [42, 69], HNO_3 [10, 40, 47, 63, 65, 66, 70-74], NH_3 [6, 57, 65], NO [10, 42, 65, 70], NO_2 [9, 10, 40, 47, 51, 53, 61, 65, 66, 70, 72, 75, 76], N_2O [6, 77-80], O_3 [7], OCS [48], and SO_2 [5, 6, 8, 70]. Most of these measurements were at land sites but many were made with aircraft mounted instruments [44, 47, 49, 50, 51, 54, 55, 65, 66, 69, 75, 77, 79, 80] and several from ships [53, 56, 58, 61]. Intercomparisons with other measurement techniques were carried out in many of the studies [40, 51, 52, 60, 61, 64, 70-74, 76]. With the single exception of one of the H_2O measurements [69], which used sweep integration, all the above measurements used second-harmonic wavelength modulation spectroscopy.

Whilst TDLAS has mainly been used for tropospheric measurements interest in its use in the stratosphere is increasing. The earliest stratospheric measurements were made with the open-path baloon-borne BLISS instrument but more recently instruments using multi-pass cells have been developed for use on stratospheric research aircraft such as the NASA ER-2. Two such instruments have already flown. ATLAS [77], a single-species instrument which measures nitrous oxide as a tracer species [79, 80], and ALIAS [37] a Herriott cell based instrument with four independent optical channels, giving it the capability of simultaneous four-species measurements. Both ATLAS and ALIAS operate unattended in an unpressurized external pod of the ER-2.

Flux measurements

Because of its potential for fast response measurements TDLAS has been used for vertical flux measurements using the eddy-correlation method. In this technique the species concentration and the vertical wind velocity, *w*, are measured as a function of time. The flux is then given by [44]:

$$F = \overline{c'(w\rho)'} + \overline{w\rho}\ \bar{c}$$

where *c* is the mass mixing ratio of the species and ρ is the total air density. The primes indicate the instantaneous deviation from the mean of the quantity and the bars indicate a long-term average . Since it is normally the flux to the ground that is required, aircraft measurements are usually made at a range of heights and the flux extrapolated to ground level. The time resolution needed for c' and w' in order to achieve an accurate value for F depends on atmospheric conditions and lengthens with height, but ideally a bandwidth of about 10 Hz is required. This is equivalent to a 16 ms time-constant. However when making aircraft measurements Sachse *et al.* [50] found that at their 150 m minimum height most information was at frequencies < 1 Hz so that a 100 ms time-constant was adequate. Such a response time is easily achieved for the wind speed measurements by using sonic anemometers, the main difficulty is in achieving this response from the TDLAS concentration measurements.

Measurements so far made using this technique include airborne CO fluxes over the Amazon basin [44], and CH_4/CO fluxes over the Alaskan tundra [81], and ground based measurements of CH_4 flues over peatland [82]. Within JETDLAG Wienhold *et al.* have measured N_2O fluxes from fertilised grassland [83] using a TTFMS instrument. Anderson and Zahniser [84] used a TDLAS system with an open-path Herriott cell for flux measurements of CH_4 and O_3. An open path eliminates the response time of a closed multi-pass cell.

8.3 The JETDLAG subproject

8.3.1 Aims and history

The JETDLAG subproject was approved by the IEC in November 1987. The subproject had four specific objectives:

A. To develop and test a standard ground based TDLAS instrument capable of unattended field operation. The intention was to develop an instrument which would then be commercialized and thus be available for use by other EUROTRAC participants.

B. To construct special purpose instruments for other EUROTRAC programs such as TOR, GCE, TRACT, ALPTRAC, and BIATEX.

C. To develop new TDLAS components and techniques. In particular high temperature diode lasers, fibre optic components, high speed detectors, FM spectroscopy and signal processing.

D. To obtain high resolution IR spectral data for atmospheric species where these were missing or unreliable.

We shall see in the following sections that with the exception of objective A all these aims have been achieved. As approved in 1987 JETDLAG had 26 proposed projects and represented 26 separate institutions. Inevitably many of the proposed projects had funding problems. By February 1990 only eight projects had received funding with a further 12 projects still expecting to receive funds. However by May 1992 work was actually in progress on 17 projects. Thus after a slow start JETDLAG eventually developed into a substantial subproject.

The fact that objective A, the development of a standard instrument suitable for commercial production, could not be achieved was due largely to problems in interfacing between commercial organisations and the independent research groups within JETDLAG. Inconsistency of funding of the various projects which were between them all necessary to the development of the instrument was a further factor. In retrospect in seems inevitable that there should have been problems with this objective. It is worth noting that over the eight year course of JETDLAG commercial instruments have become available, developed entirely by the companies concerned. However these are still bespoke systems which require some operator expertise. A standard off-the-shelf instrument is still not available.

When it became clear in 1992 that it would not be possible to achieve the aim of developing a standard instrument, the SSC decided to review JETDLAG. As a result of this review JETDLAG was reorganised into a much smaller subproject with only six individual contributions, all except one focused on using TDLAS for field measurements. The exception was a contribution aimed at further development of the diode lasers. Because of this reorganisation the individual final reports largely cover only those contributions which survived the reorganisation and so do not give a true picture of the totality of the work of JETDLAG.

8.3.2 Laser development

Diode laser quality is critical to many aspects of TDLAS performance. Laser noise, power, mode structure, and beam quality can all directly affect instrument sensitivity. Also the difficulty of producing instruments which can be easily used by non-expert operators is mainly due to non-reproducibility in the laser behaviour. If lasers could be produced that would reliably give a single mode output at a known wavelength when operated at a given set of operating conditions the usability of TDLAS instruments would be transformed. The ability to operate at higher temperatures would also reduce the complexity of instruments. Because of this, laser development was an important component of JETDLAG.

The group at the Fraunhofer Institute (IPM) in Freiburg undertook this work. They developed test methods for studying laser beam patterns, mode behaviour and linewidth and improved their techniques, including MBE, for fabricating buried heterostructure and double heterostructure lasers using PbEuSe, PbSe, PbSrSe, and PbSnSe materials [85-88]. The test instrumentation allowed studies of the properties of these lasers [17, 89]. Some work on DBR and DFB lasers was undertaken [90] but the main project to develop these types of laser cavity remained unfunded. Overall the work at IPM has resulted not only in improved lasers with higher temperature operation, better mode patterns and beam quality, but also in an increased understanding of the laser physics and of the fabrication processes. This will provide a basis for future laser development.

8.3.3 High resolution spectroscopy

High resolution spectroscopic studies, including measurements of line strengths, pressure broadening coefficients and pressure shifts, were undertaken by groups at Ispra and at Université Libre de Bruxelles. TDL spectroscopy was used at Ispra [91-95] to study NH_3, N_2O, CO_2, CH_3Cl, and HCOOH. FTIR measurements at Bruxelles [96-100] included HONO, C_2H_2, NO, NO_2 , N_2O_4 and C_2H_6, C_3O_2 and CH_3CHO.

TDL spectroscopy was also used as a tool in laboratory studies of atmospheric chemistry [101] mainly in support of the LACTOZ subproject.

8.3.4 Component development

TDLAS systems are difficult to align and once aligned are susceptible to thermal and mechanical drift. Use of fibre optic components could have benefits both in overcoming these alignment problems and also in achieving more compact and robust systems by replacing lenses mirrors and beamsplitters by fibre components. This was investigated by a project at the Fraunhofer Institute (IPM) in Freiburg. Zirconium fluoride fibres, with good transmission at 4 μm, were developed in conjunction with a commercial company and evaluated for use with TDLs [102]. Results were encouraging but more development would be needed before these components could be used in a TDLAS system.

Signal processing forms an important aspect of any TDLAS system but some groups paid special attention to novel or improved techniques. These included studies of improved signal averaging and fitting methods [45, 46] optimum strategies for increasing sensitivity by long term averaging [103], and comparisons of different techniques for multiplexing several lasers [41]. Attention was also paid to the important topic of quality assurance of TDLAS measurements [104].

At the start of JETDLAG high frequency modulation techniques (see Section 8.2.3) were just beginning to be applied to TDLAS. They appeared to offer significantly improved sensitivity and faster response. A project at the Fraunhofer Institute in Garmisch-Partenkirchen (IFU) was aimed specifically at developing such an FM instrument. An instrument was developed using single-tone FM spectroscopy with a modulation frequency of 100 MHz [105, 106]. This achieved a sensitivity of around 1.7×10^{-6} with a path length of 36 m at a 6 Hz bandwidth. This is theoretically equivalent to 7×10^{-7} over a 60 s average. Compared to the best WMS systems using 100 m path-lengths this represents an improvement of nearly three in sensitivity. However such performance proved difficult to obtain reliably under field conditions. As part of this program significant contributions were made to understanding the influence of high frequency laser noise on TDLAS performance [22] and quantum limited performance was demonstrated [107] (although not in a field usable system).

High frequency instruments were also developed by the group at the Max Planck Institute in Mainz and these are mentioned in the next Section.

8.3.5 Instrument development

The development and application of instruments for field measurements of atmospheric species was the central aim of JETDLAG. Five groups contributed to this: Fraunhofer Institut für Atmosphärische Umweltforschung, Garmisch-Partenkirchen (IFU); Max Planck Institut, Mainz (MPI); Forschungszentrum Jülich (KFA); Service d'Aéronomie, Paris; and Imperial College, London (ACRU).

The work of the Garmisch group on developing FM techniques has already been mentioned. The instrument was initially used for ambient NO_2 measurements [103] (30 ppt sensitivity) and was further developed to allow simultaneous measurements of two species using frequency modulation multiplexing of the two lasers. In this form the instrument made measurements of background levels of HCHO and H_2O_2 on the MV Polarstern during a 1994 cruise in the Atlantic Ocean. The group has also carried out intercomparison studies of TDLAS NO_2 and HCHO measurements [104]. For NO_2 the TDLAS was compared with DOAS and Luminox methods, for HCHO the comparison was with the enzymatic/flourimetric method. In both cases convincing agreement was obtained. The possibility of measuring HO_2 radical by TDLAS was investigated [108]. This is considerable challenge since sensitivities of better than 10^{-7} would be needed. Although not yet achieved, studies indicated that a sensitivity of 4×10^{-8} might be achievable, corresponding to 10 ppt of HO_2.

The main instrument developed by the MPI group was a four laser system (FLAIR) which was used both for ground ship and aircraft measurements. The instrument as originally developed used a 1.5 m baselength White cell with 28 L volume. A later version was made more compact and rugged for aircraft studies. This used a 1 m White cell with 7 L volume and 126 m total absorption path. Both systems used time division multiplexing between the lasers in order to monitor the four species. The instrument has recently been further improved to allow HFWMS spectroscopy at MHz frequencies. FLAIR was used in a variety of measurement campaigns. In its original form it was deployed on the FS Meteor and RV Polarstern research vessels during three campaigns during 1987-8 [53, 56, 61 [53, 56, 61]. During TROPOZ II in 1991 it made airborne measurements involving 28 flights measuring vertical profiles of HCHO, H_2O_2, NO_2 and CO. In 1992 FLAIR was used on a DC3 for airborne measurements in the SAFARI 92 campaign studying biomass burning. Species measured were NO_2, HCHO, N_2O and CH_4 . Since then a series of ground based campaigns have been undertaken including two FIELDVOC campaigns and an OCTA campaign at the Izana TOR station.

In addition to FLAIR the MPI group have constructed a fast response instrument using TTFMS (see Section 0) which has been used for eddy-flux measurements of N_2O and CO [83, 109]. A further instrument has been constructed to measure isotope ratios ($^{12}C/^{13}C$ and H/D) in atmospheric CH_4 [110].

At KFA an instrument was developed specifically for HCHO measurements as part of the TOR project. It is housed in a van and uses a 1 m White cell. Sensitivity was 300 ppt for a 1 minute measurement. The instrument has undertaken a range of field studies with particular emphasis on the influence of HCHO from car exhausts on photo-chemical ozone production.

The work at Service d'Aéronomie was directed towards developing an instrument for eddy-flux correlation measurements of O_3, CO, and CH_4 on board a Fokker 27 research aircraft [111]. The instrument was developed in collaboration with the Société Bertin with the Fraunhofer Institute Freiburg providing the diode lasers and electronics. It uses wavelength modulation spectroscopy and the beams from the three lasers are combined using dichroic beamsplitters. Different modulation frequencies allow the signals from the three lasers to be separated in the electronics. The White cell exchange time is < 1 s so allowing measurements at a 1 Hz rate. The instrument has successfully measured CO and CH_4 profiles and development continues.

The group at Imperial College developed an instrument which was originally employed on studies of the transformation of fossil-fueled power station emissions. It was housed in a van and capable of mobile operation. HCl was measured in a power station plume and NH_3 measurements were made during two cloud chemistry experiments at Great Dun Fell [57]. Originally a single species instrument, it was later modified to measure two species using time multiplexing and in this form was employed for simultaneous measurements of HNO_3 and NH_3. The instrument is of conventional design using a 1 m baselength White cell and 50 kHz modulation frequency. A high pumping speed (20 slpm at the inlet) is used to reduce sampling problems which are known to be severe for all the three species studied. Further, airborne, instruments are currently under development for tropospheric and stratospheric measurements.

8.3.6 Photoacoustic and intracavity spectroscopy

JETDLAG provided a home for projects aimed at developing two non-TDLAS laser spectroscopic techniques, which offer promise for measurements of trace species.

Photoacoustic spectroscopy uses a sensitive microphone to detect the sound pulses produced by the absorption of a pulsed or chopped laser beam, tuned to the species absorption, in a cell containing the atmospheric sample [112]. Two institutes in the Netherlands (Agricultural University Wagenigen and Catholic University Nijmegen) collaborated on developing instruments aimed mainly at NH_3 measurements using a CO_2 laser as the source [113-116]. The development work included optimisation of the cell design and use of Stark modulation to improve sensitivity and selectivity. A sensitivity of 0.5 ppb for NH_3 measurements was achieved. Related work at the Hungarian Academy of Sciences concentrated on optimisation of photoacoustic cell design [117].

TDLs are not usually sufficiently powerful to act as sources for photoacoustic spectroscopic measurements of trace species. A group at Cambridge University investigated possible improvements to the technique which would allow TDLs to be used. They achieved ppm sensitivity for SO_2 but were not able to reach the ppb sensitivities necessary for most atmospheric species [118].

As the name implies intra-cavity spectroscopy places the sample within the laser cavity. Because the light within the cavity makes many passes very high effective path lengths and hence absorption sensitivities can be achieved. With a dye laser an absorption sensitivity of 3×10^{-10} cm^{-1} has been seen. Work at Hamburg University attempted to extend intra-cavity spectroscopy to TDLs [119, 120]. This involves anti-reflection coating the lasers so that they may be operated in an external cavity. Using GaAlAs lasers absorption sensitivities of better than 10^{-6} cm^{-1} were obtained using linear and ring cavities. Use of a fibre-laser allowed sensitivities of 10^{-7} cm^{-1}. Thus the technique shows promise but has not yet produced a usable instrument.

8.3.7 Achievements

At the start of the JETDLAG sub-project TDLAS was just beginning to make an impact on measurements of trace atmospheric species. As can be seen from Sections 8.2.8 and 8.2.9 above the TDLAS technique has now developed to a point where it is extensively used to measure the concentrations, and increasingly also the vertical fluxes, of a wide range of species. The JETDLAG sub-project, representing as it does most of the European research in this field, has made major contributions in advancing TDLAS to this point. The only disappointment is that TDLAS has not developed into a routine instrument which can easily be operated by non-specialists. The reason for this lies mainly in the lasers which are still non-reproducible in their characteristics so that whenever a laser is changed or cycled to room temperature the instrument must be recharacterised. This problem is unlikely to be overcome in the near future since the market for these lasers is not large enough to support the development needed.

Specific areas in which JETDLAG has made important contributions to TDLAS development include:

i. Improvements to lasers, including higher temperature operation, better mode quality and improved beam patterns. The JETDLAG work in this area represents a major part of the world-wide effort.

ii. Developments of FM spectroscopic methods for higher sensitivity. The development of three field usable FM systems within JETDLAG is at the forefront of research in this area. The use of an FM system for flux measurements is also a significant achievement.

iii. Deployment of multi-species instruments using time and frequency multiplexing methods. This has resulted in instruments able to monitor up to four species simultaneously.

iv. Construction of robust and compact instruments capable of reliable operation in aircraft and ships.

v. Spectroscopic studies of a range of species. Spectroscopic data is essential to the optimum design of TDLAS systems.

vi. Development of improved signal processing strategies leading to increased sensitivity and better quality assurance.

vii. Demonstration, through a large array of contributions to field measurement campaigns both ground, ship and aircraft based, of the usefulness of TDLAS in atmospheric chemistry studies.

Thus JETDLAG can be considered a success and the techniques developed within it will certainly find continuing application with the projects of EUROTRAC 2.

8.4 References

1 R.T. Ku, E.D. Hinkley, J.O. Sample; *Appl. Opt.* **14** (1975) 854-861.

2. E.D. Hinkley; *Opt. Quantum Electron.* **8** (1976) 155-167.

3. D.J. Brassington; Tunable diode laser absorption spectroscopy for the measurement of atmospheric species, in: R.J.H. Clark, R.E. Hester (eds.), *Spectroscopy in Environmental Science*, J. Wiley, New York 1995, pp. 85-148.

4. H.I. Schiff, G.I. Mackay, J. Bechara; The use of tunable diode laser absorption spectroscopy for atmospheric measurements, in: M.W. Sigrist (ed.), *Air Monitoring by Spectroscopic Techniques*, J. Wiley, New York 1994, pp. 239-333.

5. J. Reid, J. Shewchun, B.S. Garside, A.E. Ballik; *Appl. Opt.* **17** (1978) 300-307

6. J. Reid, B.K. Garside, J. Shewchun, M.El-Sherbiny, E.A. Ballik; *Appl. Opt.* **17** (1978) 1806-1810.

7. M. El-Sherbiny, E.A. Ballik, J. Shewchun, B.K. Garside, J. Reid; *Appl.Opt.* **18** (1979) 1198-1203.

8. J.Reid, B.K.Garside, J.Shewchun; *Opt Quantum Electron.* **11** (1979) 385-391.

9. J. Reid, M. El-Sherbiny, B.K. Garside, E.A. Ballik; *Appl. Opt.* **19** (1980) 3349-3354.

10. H.I. Schiff, D.R. Hastie, G.I. Mackay, T. Iguchi, B.A. Ridley; *Environ. Sci. Technol.* **17** (1983) 352-364.

11. J.A.Ritter, D.H.Lenschow, J.D.W.Barrick, G.L.Gregory, G.W.Sachse, G.F.Hill, M.A.Woerner; *J. Geophysical Res.* **95D** (1990) 16875-16886.

12. D.L. Wall; Advances in tunable diode laser technology for atmospheric monitoring applications, in: *Measurement of Atmospheric Gases, Proc SPIE* **1433** (1991) 94-103

13. H. Prier, Z. Feit, J. Fuchs, D. Kostyk, W, Jalenak, J. Sproul; in: R.Grisar, G. Schmidke, M. Tacke, G. Restelli (eds.), *Monitoring of Gaseous Pollutants by*

Tunable Diode Lasers, Proc. Int. Symp., Freiburg 1988, Kluwer Academic Publishers, Dordrecht 1989, pp. 85-102.

14. M. Tacke; in: R.Grisar, H. Bottner, M. Tacke, G. Restelli (eds.), *Monitoring of Gaseous Pollutants by Tunable Diode Lasers, Proc. Int. Symp., Freiburg 1991*, Kluwer Academic Publishers, Dordrecht 1992, pp. 103-118.
15. Z. Feit, D. Kostyk, R.J. Woods, P. Mak in: R.Grisar, H. Bottner, M. Tacke, G. Restelli (eds.), *Monitoring of Gaseous Pollutants by Tunable Diode Lasers, Proc. Int. Symp., Freiburg 1991*, Kluwer Academic Publishers, Dordrecht 1992, pp. 105-114.
16. P. Werle; private communication.
17. M Agne, U. Schiessl, A. Lambrecht, M. Tacke; in: R.Grisar, H. Bottner, M. Tacke, G. Restelli (eds.), *Monitoring of Gaseous Pollutants by Tunable Diode Lasers, Proc. Int. Symp., Freiburg 1991*, Kluwer Academic Publishers, Dordrecht 1992, pp. 93-101.
18. E.D. Hinkley, C. Freed; *Phys. Rev. Lett.* **23** (1969) 277.
19. J. Reid, D.T. Cassidy, R.T. Menzies; *Appl. Opt.* **21** (1982) 3961-3965.
20. G. Spilker, R Daddato, U. Schiessl, A. Lambrecht, M. Tacke; in: R.Grisar, H. Bottner, M. Tacke, G. Restelli (eds.), *Monitoring of Gaseous Pollutants by Tunable Diode Lasers, Proc. Int. Symp., Freiburg 1991*, Kluwer Academic Publishers, Dordrecht 1992, pp. 85-91.
21. H. Fischer, M. Tacke; *J. Opt. Soc. Am.* **B8** (1991) 1824-1830
22. P. Werle, F. Slemr, M. Gehrtz, C. Brauchle; *Appl. Opt.* **28** (1989) 1638-1642
23. A. Fach, H. Bottner, M. Tacke; in: R.Grisar, H. Bottner, M. Tacke, G. Restelli (eds.), *Monitoring of Gaseous Pollutants by Tunable Diode Lasers, Proc. Int. Symp., Freiburg 1991*, Kluwer Academic Publishers, Dordrecht 1992, pp. 63-68.
24. J. Reid, D. Labrie; *Appl.Phys.* **B26** (1981) 203
25. J.A. Silver; *Appl. Opt.* **31** (1992) 707-717
26. W. Lenth; *IEEE J. Quant. Electron. QE*-**20** (1984) 1045-1050
27. D.E. Cooper, R.E. Warren; *J. Opt. Soc. Am.* **B4** (1987) 470-480.
28. D.S.Bomse, A.C. Stanton, J A. Silver; *Appl. Opt.* **31** (1992) 718-731.
29. D.J. Brassington; *EUROTRAC Annual Report '90*, Part 7, JETDLAG, EUROTRAC ISS, Garmisch-Partenkirchen 1991, pp. 25-28.
30. J.U. White; *J. Opt Soc Am.* **32** (1942) 285-288.
31. H.J. Bernstein, G. Herzberg; *J. Chem Phys.* **16** (1948) 30-39.
32. S. Nadler, D. Karecki, G. Mackay, H. Schiff; in: *Optical Methods in Atmospheric Chemistry, Proc SPIE* **1715** (1992) 194-199.
33. D. Heriott, H. Kogelnik, R. Kompfner; *Appl. Opt.* **3** (1964) 523-526.
34. D.R. Herriott, H.J. Schulte; *Appl. Opt.* **4** (1965) 883-889.
35. R. Altmann, R. Baumgart; C. Weitkamp; *Appl. Opt.* **20** (1981) 995-999.
36. W.R. Trutna, R.L. Byer; *Appl. Opt.* **19** (1980) 301-312.
37. C.R. Webster, R.D. May, C.A. Trimble, R.G. Chave, J. Kendall; *Appl. Opt* **33** (1994) 454-472.
38. D.W.Worsnop, D.D. Nelson, M.S. Zahniser; *Optical Methods in Atmospheric Chemistry, Proc SPIE* **1715** (1992) 18-33.

39. C.R. Webster; *J. Opt Soc Am.* **B2** (1985) 1464-1470.
40. G.I. Mackay, H.I. Schiff, A. Wiebe, K. Anlauf; *Atmos. Environ.* **22** (1988) 1555-1564.
41. R. Mücke, P. Werle, F. Slemr, W. Prettl; Comparison of time and frequency multiplexing techniques in multicomponent FM spectroscopy, *Proc. SPIE* **1433** (1991) 136-144.
42. A. Fried, B. Henry, J.R. Drummond; *Appl. Opt.* **32** (1993) 821-827.
43. H.J. Kramer; *Observation of the Earth and its Environment, Survey of Missions and Sensors*, Springer-Verlag, Berlin 1994.
44. J.A. Ritter, D.H. Lenschow, J.D.W. Barrick, G.L. Gregory, G.W. Sachse, G.F. Hill, M.A. Woerner; *J. Geophys. Res.* **95D** (1990) 16875-16886.
45. P. Werle; Signal processing strategies for tunable diode laser spectroscopy, in: *Proc. Int. Symp. Optical Sensing for Environmental Monitoring*, October 1993, *Proc. SPIE* **2112** (1994)
46. P. Werle, B. Scheumann, J. Schandl; Real time signal processing concepts for trace gas analysis by TDLAS, *Opt. Eng.* **33** (1994) 3093-3105.
47. H.I. Schiff, D.R. Karecki, G.W. Harris; *J. Geophys. Res.* **95D** (1990) 10147-10153.
48. A. Fried, J.R. Drummond, B. Henry, J. Fox; *Appl. Opt.* **30** (1991) 1916-1932.
49. G.Sachse, G.F. Hill, L.O. Wade, M.G. Perry; *J. Geophys. Res.* **92D** (1987) 2071-2081.
50. G.W. Sachse, J.E. Collins, G.F. Hill, L.O.Wade, L.G. Burney, J.A. Ritter; in: *Measurement of Atmospheric Gases, Proc SPIE* **1433** (1991) 157-166.
51. G.I. Mackay, D.R. Karecki, H.I. Schiff; in: *Measurement of Atmospheric Gases, Proc SPIE* **1433** (1991) 104-119.
52. A. Fried, B. Henry, D.D. Parrish, J.R. Carpenter, M.P. Buhr; *Atmos. Environ.* **25A** (1991) 2277-2284.
53. G.W. Harris, J.P. Burrows, D. Klemp, T. Zenker; in: R. Grisar, G. Schmidke, M. Tacke, G. Restelli (eds.), *Monitoring of Gaseous Pollutants by Tunable Diode Lasers, Proceedings of the International Symposium, Freiburg 1988*, Kluwer Academic Publishers, Dordrecht 1989, pp. 68-73.
54. G.W. Sachse, R.C. Harriss, J. Fishman, G.F. Hill, D.R. Cahoon; *J. Geophys. Res.* **93D** (1988) 1422-1430.
55. R.C. Harriss, G.W. Sachse, G.F. Hill, L.O.Wade, G.L. Gregory; *J. Geophys. Res.* **95D** (1990) 16927-16932.
56. G.W. Harris, D. Klemp, T. Zenker ; *J. Atmos. Chem.* **15**, 327-332 (1992)
57. D.J. Brassington; Measurements of atmospheric HCl and NH_3 with a mobile tunable diode laser system, in: R. Grisar, G. Schmidke, M. Tacke, G. Restelli (eds.), *Monitoring of Gaseous Pollutants by Tunable Diode Lasers, Proceedings of the International Symposium, Freiburg 1988*, Kluwer Academic Publishers, Dordrecht 1989, pp. 16-24.
58. P. Pokrowski, W Herrmann; in: *Laser Spectroscopy for Sensitive Detection, Proc SPIE* **286** (1981) 33-38.
59. A. Fried, R. Sams, W.W. Berg; *Appl. Opt.* **23** (1984) 1867-1880.
60. G.I. Mackay, D.R. Karecki, H.I. Schiff; in: *Optical Methods in Atmospheric Chemistry, Proc SPIE* **1715** (1992) 126-137.

61. G.W. Harris, D. Klemp, T. Zenker , J.P Burrows, B. Mathieu; *J. Atmos. Chem.* **15** (1992) 315-326.

62. G.W. Harris, G.I. Mackay, T. Iguchi, L.K. Mayne, H.I. Schiff; *J. Atmos. Chem.* **8** (1989) 119-137.

63. J.W. Drummond, H.I. Schiff, D.R. Karecki, G.I. Mackay; in: *Proc. APCA '89*, June (1989)

64. T.E. Kleindienst *et al.*; *Atmos Environ.* **22** (1988) 1931-1939.

65. H.I. Schiff, G.W. Harris, G.I. Mackay; in: R. Grisar, H. Preier, G. Schmidke (eds.), *Monitoring of Gaseous Pollutants by Tunable Diode Lasers, Proc. Int. Symp., Freiburg 1986,* Reidel Publishing Co. 1987, pp. 4-16.

66. H.I. Schiff, G.I. Mackay; in: R. Grisar, G. Schmidke, M. Tacke, G. Restelli (eds.), *Monitoring of Gaseous Pollutants by Tunable Diode Lasers, Proceedings of the International Symposium, Freiburg 1988*, Kluwer Academic Publishers, Dordrecht 1989, pp. 36-39.

67. H.I. Schiff, G.I. Mackay, D.R. Karecki, J.T. Pisano, S.D. Nadler; in: R.Grisar, H. Bottner, M. Tacke, G. Restelli (eds.), *Monitoring of Gaseous Pollutants by Tunable Diode Lasers, Proc. Int. Symp., Freiburg 1991*, Kluwer Academic Publishers, Dordrecht 1992, pp. 21-29.

68. F. Slemr, G.W. Harris, D.R.Hastie, G.I. Mackay, H.I. Schiff; *J. Geophys. Res.* **91D** (1986) 5371-5378.

69. J.A. Silver, A.C. Stanton; *Appl. Opt.* **26** (1987)2558-2566.

70. G. Schmidke, W. Kohn, U. Klocke, M. Knothe, W.J. Riedel, H. Wolf; *Appl. Opt.* **28** (1989) 3665-3670.

71. S.V. Hering *et al.*; *Atmos. Environ.* **22** (1988) 1519-1539.

72. J.E. Sickles *et al.*; *Atmos. Environ.* **24A** (1990) 155-165.

73. D.L. Fox *et al. Atmos. Environ.* **22** (1988) 575-585.

74. K.G. Anlauf, D.C. MacTavish, H.A. Wiebe, H.I. Schiff, G.I. Mackay; *Atmos. Environ.* **22** (1988) 1579-1586.

75. G.L. Gregory *et al.*; *J. Geophys. Res.* **95D** (1990) 10103-10127.

76. F.C. Fehsenfeld *et al.*; *J. Geophys Res.* **95D** (1990) 3579-3597.

77. J. Podolske, M. Loewenstein; *Appl. Opt.* **32**, 5324-5333 (1993)

78. P.S. Connell, R.A. Perry, C. J. Howard; *Geophys. Res. Lett.* **7** (1980) 1093-1096

79. J.R. Podolske, M. Loewenstein, S.E. Strahan, K.R. Chan; *J.Geophys Res.* **94D** (1989) 16767-16772.

80. M. Loewenstein, J.R. Podolske, K.R. Chan, S.E. Strahan; *J. Geophys. Res.* **94D** (1989) 11589-11598.

81. J.A. Ritter, J.D.W. Barrick, G.W. Sachse, G.L. Gregory, M.A. Woerner, C.E. Watson, G.F. Hill, J.E. Collins; *J. Geophys. Res.* **97D** (1992) 16601-16625.

82. S.B. Verma, F.G. Ullman, D Billesbach, R.J. Clement, J. Kim, E.S. Verry; *Boundary-Layer Meteorology* **58** (1992) 289-304

83. F.G. Wienhold, H. Frahm, G.W. Harris; Measurements of N_2O fluxes from fertilised grassland using a fast response tunable diode laser spectrometer, *J. Geophys. Res.* **99D** (1994) 16557.

84. S.M. Anderson, M.S. Zahniser; in: *Measurement of Atmospheric Gases, Proc SPIE* **1433** (1991) 167-178.

85. K.-H. Schlereth, H. Böttner, M. Tacke; "Mushroom" double channel double-heterostructure lead chalcogenide lasers made by chemical etching, *Appl. Phys. Lett.* **56** (1990) 2169-2171.

86. A. Lambrecht, Herres, N., Spanger, B., Kuhn, S., Böttner, H., Evers, J., Tacke, M.; Molecular beam epitaxy of $Pb_{1-x}Sr_xSe$ for the use in IR-devices, *J. Cryst. Growth* **108** (1991) 301-308.

87. K.H. Schlereth, H. Böttner, A. Lambrecht, U. Schießl, R. Grisar, M. Tacke; Buried waveguide and mesa-DH-PbEuSe-lasers grown by MBE, in: P.M. Borrell, P. Borrell, T. Cvitas, W. Seiler (eds.), *Proc. EUROTRAC Symp. '90*, SPB Academic Publishing, The Hague 1991, pp. 361-363.

88. A. Lambrecht, H. Böttner, M. Agne, R. Kurbel, A. Fach, B. Halford, U. Schiessl, M. Tacke; Molecular beam epitaxy of laterally structured lead chalcogenides for the fabrication of buried heterostructure lasers, *Semicond. Sci. Technol.* **8** (1993) 334-336.

89. H. Fischer, H. Wolf, B. Halford, M. Tacke; Low-frequency amplitude-noise characteristics of lead-salt diode lasers fabricated by molecular-beam-epitaxy, *Infrared Phys.* **31** (1991) 381-385.

90. M.A. Fach, H. Böttner, K. H. Schlereth, M. Tacke; Embossed-grating lead chalcogenide buried wave-guide distributed-feedback lasers, *IEEE J. Quantum Electron.* **30** (1994).

91. G. Baldacchini, A. Bizzarri, L. Nencini, G. Buffa, O. Tarrini; New measurements of self broadening and shift of ammonia lines, *J. Quantitative Spectros. Radiative Transfer* **42** (1989) 423-428.

92. G. Baldacchini, A. Bizzari, L. Nencini, V. Sorge, G. Buffa, O. Tarrini; Pressure broadening and shift of ammonia transition lines in the ν_2 vibrational bands, *J. Quant. Spectrosc. Radiat. Transfer* **43** (1990) 371-380.

93. G. Baldacchini, P.K. Chakraborti, F. D'Amato; Infrared diode laser absorption features of N_2O and CO_2 in a laval nozzle, *Appl Phys.* **B55** (1992) 92-101.

94. G. Baldacchini, P.K. Chakraborti, F. D'Amato; Lineshape in a laval molecular beam and pressure broadening of N_2O transition lines, *J. Quant. Spectrosc. Radiat. Transfer* **49** (1993) 439-477.

95. G. Baldacchini, F. D'Amato, E. Righi, F. Cappellani, G. Restelli; Applicazioni di spettroscopia molecolare in fase gassosa con diodi laser accordabili nell'infrarosso, *Il Nuovo Saggiatore* **7** (1991) 5-16.

96. M. Herman, T.R. Huet, Y. Kabbadj, J. Vander Auwera; *l*-type resonance in C_2H_2, *Mol. Phys.* **72** (1991) 75-88.

97. M. Herman, J. Vander Auwera; Spectroscopic analysis of the atmosphere, *Chimie Nouvelle* **10** (1992) 1139-1145.

98. F. Mélen, F. Pokorni, M. Herman; Vibrational band analysis of N_2O_4, *Chem. Phys. Lett.* **194** (1992) 181-186.

99. F. Mélen, M. Herman; Vibrational bands of $H_xN_yO_z$ molecules, *J. Phys. Chem. Ref. Data* **21** (1992) 831-881.

100. F. Mélen, M. Herman; Fourier transform jet spectrum of the ν_9 band of N_2O_4, *Chem. Phys. Lett.* **199** (1992) 124-130.

101. J. Hjorth, F. Cappellani, C.J. Nielsen, G. Restelli; Determination of the $NO_3 + NO_2 \rightarrow NO + O_2 + NO_2$ rate constant by infrared diode laser and Fourier transform spectroscopy, *J. Phys. Chem.* **93** (1989) 5458.

102. R. Grisar, J. Anders, M. Knothe, W.J. Riedel; Application of infrared fibers in diode laser trace gas analysis, in: *Proc. Conf. Infrared Fiber Optics III*, Boston 1992, *Proc. SPIE* **1591** (1992) 201-205.

103. P. Werle, R. Mücke, F. Slemr; *Appl. Phys.* **B57** (1993) 131-139.

104. R. Mücke, J. Dietrich, B. Scheumann, F. Slemr, J. Slemr, P. Werle; Application of tunable diode laser spectroscopy for QA/QC studies: A validation for formaldehyde measurement techniques, in: P.M. Borrell, P. Borrell, T. Cvitas, W. Seiler (eds.), *Proc. EUROTRAC Symp. '94*, SPB Academic Publishing, The Hague 1994, pp. 910-914.

105. P. Werle, R. Mucke, F. Slemr; in: R.Grisar, H. Bottner, M. Tacke, G. Restelli (eds.), *Monitoring of Gaseous Pollutants by Tunable Diode Lasers, Proc. Int. Symp., Freiburg 1991*, Kluwer Academic Publishers, Dordrecht 1992, pp. 169-182.

106. P. Werle; Atmospheric trace gas monitoring using high frequency modulation spectroscopy with semiconductor lasers, in: A.I. Nadezhdinskii, Y.N. Ponomarev, L. N. Sinitsa (eds.), *11th Symposium and School on High-Resolution Molecular Spectroscopy, Proc. SPIE* **2205** (1994) 83-94.

107. P. Werle, F. Slemr, M. Gerhrtz, C. Brauchle; *Appl Phys.* **B49** (1989) 99-108.

108. R. Mücke, F. Slemr, P. Werle; Measurement of atmospheric HO_2 radicals by tunable diode laser absorption spectroscopy: A feasibility study, in: P.M. Borrell, P. Borrell, T. Cvitas, W. Seiler (eds.), *Proc. EUROTRAC Symp. '94*, SPB Academic Publishing, The Hague 1994, pp. 915-919.

109. F.G. Weinhold, H. Fischer, G.W. Harris; *Infrared Phys. Technol.* **37** (1996) 67-74.

110. P. Bergamaschi, M. Schupp, G.W. Harris; Appl. Opt. **33** (1994) 7704-7716.

111. H. Mac Leod, I. Carrasco, H. Poncet, G. Mégie, F. Huard, W. Riedel; Development of an airborne tunable diode laser spectrometer for flux measurements of CO, O_3 and CH_4 by the eddy correlation technique, in: P.M. Borrell, P. Borrell, T. Cvitas, W. Seiler (eds.), *Proc. EUROTRAC Symp. '94*, SPB Academic Publishing, The Hague 1994, pp. 925-929.

112. M.W. Sigrist; Air monitoring by laser photoacoustic spectroscopy, in: M.W. Sigrist (ed.), *Air Monitoring by Spectroscopic Techniques*, J. Wiley, New York 1994, pp. 163-227.

113. H. Sauren, D. Bicanic, H. Jalink, J. Reuss; High sensitivity, interference free Stark tuned carbon dioxide laser photoacoustic sensing of urban ammonia, *J. Appl. Phys.* **66** (1989) 5087.

114. D. Bicanic, H. Sauren, H. Jalink, M. Chirtoc, A. Miklós, K. van Asselt; Progress towards the development and application of laser based photothermal devices for environmental analysis, in: P. Borrell, P.M. Borrell, W. Seiler (eds.), *Proc. EUROTRAC Symp. '90*, SPB Academic Publishing, The Hague 1991, pp. 347-350.

115. D. Bicanic, M. Franco, H. Sauren; The electric resonance tuning and intermodulation for selective and sensitive studies of small polar molecules, in: *Proc. SPIE Euroopto Int. Symp. on Environmental Monitoring*, Berlin 1992.

116. D. Bicanic, M. Franko, H. Sauren; On the potentiality of intermodulated electric resonance, photoacoustic and photothermal spectroscopies for ground based in situ tracking of polar molecules in polluted air, in: H. Schiff, U. Platt (eds.) *Proc. SPIE Conf.* **1715** (1993) 222-234.

117. J. Favier, A. Miklos, D. Bicanic; New and versatile cell for studies of powdered specimens across broad spectral range, *Acta Chim. Slovenica* **40** (1993) 115-122.

118. S.A. Bone, P.B. Davies, S.A. Johnson, N.A. Martin; Diode laser spectroscopy with optoacoustic detection, in: P. Borrell, P.M. Borrell, W. Seiler (eds.), *Proc. EUROTRAC Symp. '90*, SPB Academic Publishing, The Hague 1991, pp. 351-354.

119. V.M. Baev, J. Eschner, M. Paeth, R. Schüler, P.E. Toschek; Intracavity spectroscopy with diode lasers, *Appl. Phys.* **B55** 463-477 (1992).

120. V.M. Baev, J.H. Eschner, J. Sierks, A. Weiler, P.E. Toschek; Dynamics of multi-mode dye lasers, in: *Proc. XVIII Int. Quantum Electronics Conference (IQEC '92)*, Vienna 1992.

Chapter 9

Individual Reports from JETDLAG Contributors

9.1 Development of Infrared Tunable Diode Lasers Specifically Suited for Spectroscopic Applications

A. Lambrecht, H. Böttner and M. Tacke

Fraunhofer-Institut für Physikalische Messtechnik (IPM), Freiburg

Summary

A process for making Buried Heterostructure (BH) lasers was established. Using PbSnSe as active material long wavelength BH-cw-emission at 80 K down to 1026 cm^{-1} was reached. For shorter wavelengths we employed PbSrSe and obtained BH-cw-emission up to 2860 cm^{-1} at 80 K. The general outcome of the analysis of laser data gathered within the project is that BH-lasers fulfill part of the needs of spectroscopists. BH-lasers solve some problems (farfield), but this is paid for by low radiation power. A series of laser linewidth measurements was made on different laser structures. BH-lasers generally show larger linewidths than selected Double Heterostructure (DH) lasers. However, BH-lasers turn out to be less sensitive to optical feedback than DH-lasers. The technical project goals were fulfilled. The BMBF decided not to fund further laser development at present. This will put some risk on a future european supply of lasers.

Introduction

Widespread use of mid-infrared tunable diode lasers is hindered by the rather difficult properties of these devices. Laterally structured lasers promise to have improved farfield emission and modal characteristics. This was the motivation of the present project aimed at a technological improvement of environmental analysis techniques. In this project BH-lasers were made and their properties compared to other lasers. To summarize our results on laser fabrication the maximum operating temperatures (MOT) of the BH-lasers made within the project are compared with existing DH-lasers in Fig. 9.1. The DH-line represents the MOT-data of a large number of lasers with different compositions whereas the

BH-values are obtained within this project. It is clear from the figure that BH-lasers can be made within a broad concentration range and that they reach about the same MOTs as DH-lasers.

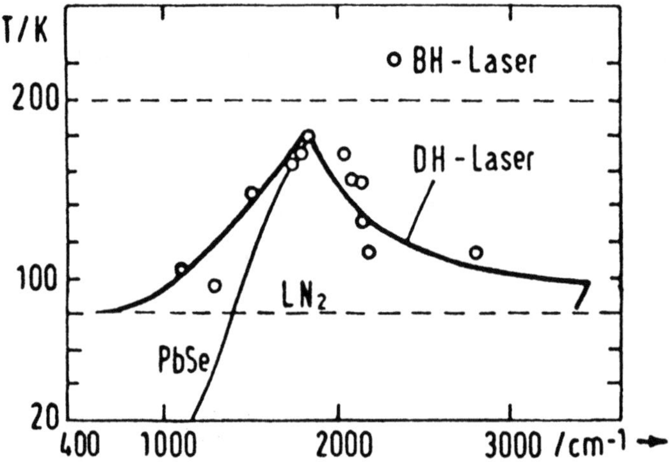

Fig. 9.1: Maximum operating temperatures of BH-lasers(open circles) and DH-lasers (line).

Long wavelength BH-lasers

As an example, the process for BH-laser-fabrication with active layers of PbSnSe will be described (data are given in Table 9.1). Previous work on DH-lasers gives strong evidence, that for an emission at 80 K above 1200 cm^{-1} the use of PbEuSe as confinement material together with active layers of PbSnSe is still reasonable.

But for even longer wavelength we had better results with a PbSe/PbSnSe/PbSe DH- design. We studied both possibilities and reached 1026 cm^{-1} with a PbSe/PbSnSe/PbSe structure. For a PbEuSe/PbSnSe/PbEuSe BH-structure the parallel PbEuSe p/n- junction does not cause much current leakage due to the high difference of the energy gap of the materials. However, for a PbSe/PbSnSe/PbSe-BH-diode current leakage is a severe problem. The comparison of MBE-DH lasers and diffused homojunction lasers for even longer wavelengths shows that in this region gain guiding becomes very efficient, which favours diffused lasers. To go below 1000 cm^{-1} with BH-lasers probably one has to essentially change the basic technological processes.

Short wavelength BH-lasers

The fabrication of DH-lasers for > 3000 cm^{-1} is difficult and needs optimized material and device parameters. We decided to use PbSrSe which seems to be advantageous for p-type material. However, the existing Sr-MBE-source turned out to be quite unstable and difficult to control. Although we managed to shift the BH-limit up to 2860 cm^{-1} (see Table 9.1) we could not fabricate a device operating above 80 K with a cw-emission above this value. Due to rapidly degraded electrical parameters for large gap material, constraints are imposed on the gap of the confinement layer thus reducing the possible difference in gaps between active and confinement layers. For a standard BH-device strong current leakage is expected to occurr. Yield and quality of working BH-lasers mounted from a given wafer decrease rapidly compared to parallel processed DH-lasers. Therefore a new process has to be developed to make BH-lasers for 3000 cm^{-1} and beyond.

Table 9.1: Properties of BH-lasers made in 1992.

MBE run	materials confinement active layer	N	I_{th}(80 K) / mA	T_{max} / K	ν(80 K) / cm^{-1}	$\nu(T_{max})$ / cm^{-1}	P(80 K) / µW	P_{mod}(80 K) / µW	m	f
1415	PbEuSe PbSe (S)	4	101	140	1372	1633	155	116	7	0.7
1432	PbEuSe PbSnSe	11	25	140	1274	1552	140	82	5	0.7
1433	PbSe PbSnSe	4	600	90	1275	1330	484	484	1	0.95
1438	PbSe PbSnSe	7	363	100	1026	1114	402	68	3	0.8
1447	PbEuSe PbSnSe (S)	3	18	140	1335	1632	131	82	8	0.6
1450	PbSrSe PbEuSe	2	149	120	1821	2059	102	68	7	0.7
1453	PbSrSe PbSrSe	3	196	110	2727	2857	255	119	6	0.7
1460	PbSrSe PbSrSe (S)	2	20.3	160	1386	1776.4	139	80	4	0.8
1483	PbEuSe PbSe	1	78	160	1393	1762	133	34	10	0.5

materials:	confinement resp. active layer materials used
(S)	laser made with shadow mask technique [5]
$I_{th}(80)$:	lowest threshold current observed at $T = 80$ K
N:	number of cw-tested diodes (modal characterization)
T_{max}:	maximum cw-operating temperature observed.
ν(80 K):	minimal emission wavenumber at threshold at 80 K
$\nu(T_{max})$:	maximal emission wavenumber at threshold at T_{max}
P(80 K):	total maximum emission power at $T = 80$ K
P_{mod}(80 K):	maximum modal power at $T = 80$ K
m:	number of modes averaged over series of spectra
f:	mode quality factor as explained in the text

The mode numbers m given in Table 9.1 are averages over all sets of mode spectra of all N measured lasers of a given wafer. The mode quality factor f is given by :

$$f = \sqrt{\sum (P_{\mathrm{mode},i} / P_{\mathrm{tot}})^2}$$

f is averaged over the same set of data. It may serve as a measure of the intensity distribution over a series of modes.

Laser parameters

The ***laser farfield analysis*** of different types of DH and BH-lasers shows the main advantage of BH-lasers: The farfields of BH-lasers are well defined, they are stable within the operating parameter range and can be well explained by theory. An improved waveguide model for farfield calculations was developed for this purpose [3]. The usual DH-lasers behave very irregular and unpredictable.

Using our integrated laser characterization system developed within the project ***noise measurements*** were performed with a large number of BH- and DH-lasers. Quite generally peaks in the RIN-curve can be attributed to dips in the laser power and to changes in the mode distribution (Fig. 9.2). Such noise data may now be included into the laser datasheets supplied to EUROTRAC laser users.

A series of ***linewidth-measurements*** were made with different laser types using an automated system. The result is that common diffused lasers operating only below 80 K show the best linewidth data *i.e.* linewidth below 10 MHz for large current ranges (Fig. 9.3). Next to these come some experimental lasers with corrugated structures (DFB/DBR-lasers as proposed as a JETDLAG project). They do not have the high powers obtained with diffused lasers but can be used above 80 K and show a well defined monomode region. The BH-lasers generally have a larger linewidth compared with DH-lasers, possibly due to their low modal power and strong multimode behaviour.

To specify the linewidth quality (*LQ*) of a laser which may be used for spectroscopy the laser linewidth as function of current is recorded. *LQ* may be defined as [4]:

LQ = (current range with linewidth < 50 MHz)/(total current range)

The diffused laser shown in Fig. 9.3 has an average *LQ* of 80 % between 34 and 42 K, which is not reached by the other laser types studied.

The laser linewidth is very sensitive to optical feedback. It can be used to check the feedback sensitivity of different types of lasers. It turned out that DH-lasers are very sensitive to feedback.

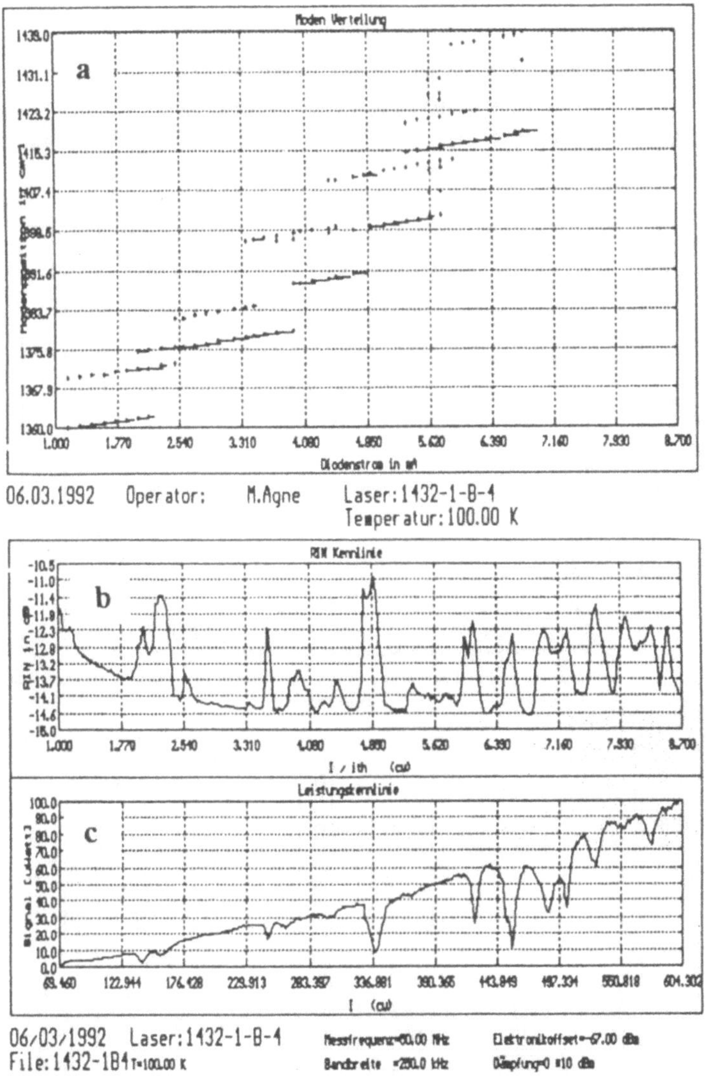

Fig. 9.2: Mode chart (a), relative intensity noise (RIN) and total emission power (b) as function of reduced diode current I/I_{th} for a BH-laser with 10 μm wide active stripe (I_{th} = 70 mA, P_{mod}(450 mA) = 59 μW).

It is hardly possible to adjust DH-laser within a spectrometer system without any influence of feedback to the laser emission. DH-laser-linewidths seem to be mainly determined by external influences. In contrast, BH-lasers are much less sensitive to optical feedback, which is an advantage for spectroscopy applications.

Further remarks

All lasers made within this project were made available to other JETDLAG participants to test these lasers without any cost until the end of our project. From 1993 on the BH-processes can be applied to laser fabrication and may be transferred to an industrial company.

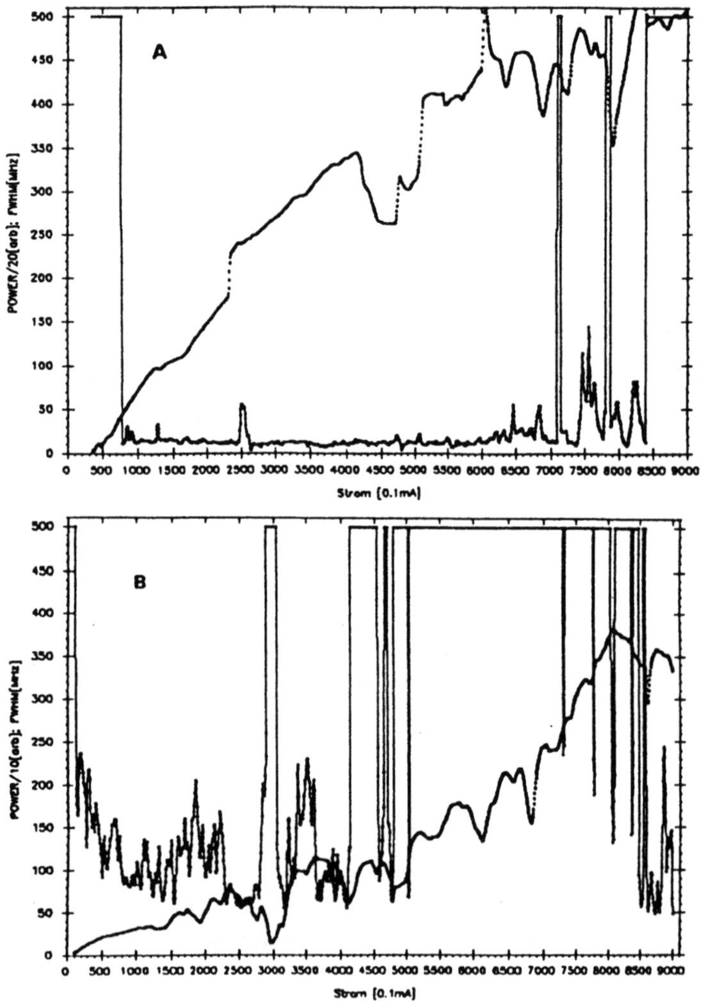

Fig. 9.3: Laser linewidth as function of operating current of the diffused laser 101-4-25(a) and of the 5 μm wide BH-laser 1386-1-3(b) at 40 K The upper curve in (a) is the total emission power, the lower curve is the linewidth measured with a confocal scanning Fabry-Perot interferometer.

Main conclusions

Following the objectives of the project we made a number of laterally structured lasers covering a broad wavelength range. We focused on BH-lasers because they had the most promising design [l]. In the meantime, Laser Photonic Inc. has also reported on good properties of such lasers [2].

A systematic analysis of laser data was made and serves as a data base for laser selection for spectroscopists and for further laser development. The BH-lasers generally do not fulfill all the needs of spectroscopists, as anticipated. They solve some problems (farfield), but this is paid for by low power. Therefore we expect to get a major improvement - feasible for spectroscopists - by incorporation of longitudinal mode selection, as was proposed for EUROTRAC in a DFB/DBR project.

However, the BMBF has decided not to fund this project. Therefore future European development of lead chalcogenide lasers is uncertain. The future existence of a European supplier of such lasers without public support for development will depend solely on the laser market.

Acknowledgements

The authors are indebted to Mrs. M. Korobka, Mrs. O. Prasse, M. Agne, R. Kurbel, B. Halford, G. Spilker, R. Daddato and U. Schiessl of the laser technology group. We are grateful for the project support by the German ministry of research and technology (grant No. 07Eu733.3) and through the GSF.

References

1. A. Lambrecht, H. Boettner, M. Agne, R. Kurbel, A. Fach, B. Halford, U. Schiessl, M. Tacke; Molecular beam epitaxy of laterally structured lead chalcogenides for the fabrication of buried heterostructure lasers, *Semicond. Sci. Technol.* **8** (1993) 334-336.
2. Z. Feit, D. Kostyk, R.J. Woods, P. Mak; Single-mode molecular beam epitaxy grown PbEuSeTe/PbTe buried heterostructure diode lasers for CO_2 high-resolution spectroscopy. *Appl. Phys. Lett.* **58** (1990) 343.
3. M. Agne, A. Lambrecht, U. Schiessl, M. Tacke; Guided modes and far-field patterns of lead chalcogenide buried heterostructure laser diodes, *Infrared Phys. Technol.* **35** (1994) 47-58.
4. G. Spilker; Noise properties of lead chalcogenide diode lasers (in German), *M.Sc. thesis*, University of Freiburg 1992.
5. A. Lambrecht, R. Kurbel, M. Agne; Shadow mask MBE for the fabrication of lead chalcogenide buried heterostructure lasers, *Mat. Sci. Eng.* **B21** (1993) 217-223.

9.2 Development and Application of Multi-laser TDLAS Instruments for Ground-based, Shipboard and Airborne Measurements of Trace Species in the Troposphere

H. Fischer, J. Bonifer, J.P. Burrows[1], D. Klemp[2], U. Parchatka, J. Roths, C. Schiller[3], T. Zenker, R. Zitzelsberger and G.W. Harris[3]

Max Planck Institute for Chemistry, Air Chemistry Dept. Mainz, Germany

Summary

The MPI contribution to JETDLAG spans the period from the subproject's initial development to the present. One of us (GWH) was a coauthor of the original JETDLAG proposal. GWH and JPB were co-PIs of the BMFT project *A four channel TDLAS for trace gas measurements in the troposphere* which was incorporated into JETDLAG in 1987. This project lead to the design of a four laser, high sensitivity TDL system which was used in two major campaigns, aboard FS Meteor in Dec 1987/Jan 1988 and aboard RV Polarstern in Sept/Oct 1988. Following on this successful project a second BMFT funded contribution to JETDLAG, *An airborne, high sensitivity, multi-species TDLAS* (PI GWH) resulted in a more compact, ruggedized version of the instrument suitable for installation and use aboard research aircraft. This Four Laser Airborne InfraRed (FLAIR) spectrometer was first test-flown in June 1990 and took part in the TROPOZ II campaign in Jan 1991. The instrument was used again during the SAFARI 92 campaign in Southern Africa during the Sept/Oct biomass burning season. Since 1992 the FLAIR instrument, under the present PI (HF), has been used in several ground-based measurement programmes within EC projects.

Aims of research

At the start of the JETDLAG programme the TDL technique was perceived as a promising but not yet fully proven method for atmospheric chemistry research and air quality monitoring. Our contribution aimed at the development of special purpose high sensitivity, field instrumentation for atmospheric research purposes and its exploitation for trace gas measurement in the background atmosphere. Our specific goals were to use our TDLAS equipment to measure trace gases related to the production and destruction of ozone on regional scales. By deploying state of the art spectrometers in such field studies we aimed to help bring about the more widespread use of TDL techniques for research purposes in Europe.

Present addresses:
[1] Univ. Bremen, FRG
[2] KFA Jülich, ICH3, FRG
[3] Centre for Atmospheric Chemistry, York Univ., Toronto, Canada

Principal scientific results

Novel features of MPI JETDLAG instrument designs

Both the shipboard and airborne MPI JETDLAG instruments are four laser systems which produce pseudo-simultaneous measurements on the four channels by interrogating each laser in turn within a time comparable to the gas exchange time in the multireflection cell. The first instrument incorporated a 1.5 m baselength single corner cube White cell with a volume of 28 L, while the compact airborne version used a specially designed 1.0 m basepath double corner cube cell with a volume of 6.8 L. The latter cell provided optical pathlengths of 126 m, was highly insensitive to vibration and minimized unwanted etalon effects.

The lasers were mounted in Dewars designed and constructed at the MPI and, in the airborne instrument, multiplexing was achieved using novel, pneumatically driven "pop up" mirrors. The cycle time through the four channels was ~10 s.

Both instruments incorporate corrective optics to focus the astigmatic beam leaving the White cell onto 1 mm HgCdTe detectors.

The spectrometers use beamsplitters to direct part of the laser energy through reference cells to reference detectors and implement accurate linelocking through an optimized combination of laser temperature control and software shifting of spectra on a scan by scan basis.

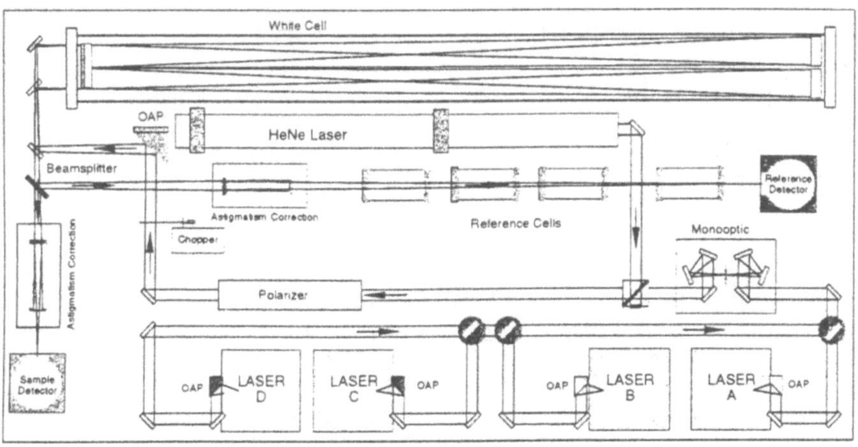

Fig. 9.4. Optical layout of the FLAIR spectrometer.

Calibration of the instruments is carried out under computer control by standard addition at the ambient air inlet, so as to account for potential first order losses of some species in the inlet lines. Stability of the optical subsystem was an extremely important feature of the airborne instrument since the opportunities to re-align optics during TROPOZ and SAFARI were very limited because access to the equipment and electrical power was available only for a short period (~1 h) before flights. During ground and shipboard campaigns the optical alignment could be more thoroughly optimized and slightly better detection limits were usually obtained. A schematic of the optical sub-assembly of the FLAIR instrument is shown in Fig. 9.4.

Shipboard campaigns

The first MPI JETDLAG instrument was used aboard FS Meteor in Dec. 1987 and Jan. 1988 to study the distribution of trace gases in the marine boundary layer off the coast of West Africa. We were able to report the first measurements in this region of H_2O_2, which was typically present at mixing ratios between 2 and 4 ppb, and mapped out latitudinal distributions of this gas together with HCHO, NO_2, CO and ozone over the north tropical and equatorial Atlantic ocean.

Later in 1988, (Sept/Oct) the MPI JETDLAG spectrometer was used aboard RV Polarstern during the multinational Polarstern 88 atmospheric chemistry cruise. Measurements of NO_2 (shown in Fig. 9.5) were made along the line of 30° W longitude throughout the cruise and showed that the elevated levels of O_3 observed between 5° N and 20° S could not be the result of local photochemistry. During the cruise, intercomparison measurements were made between TDLAS and other, less unambiguous methods for HCHO, H_2O_2 and NO_2. In further work an upper limit was placed on the HCl mixing ratio in the remote marine boundary layer near the Azores.

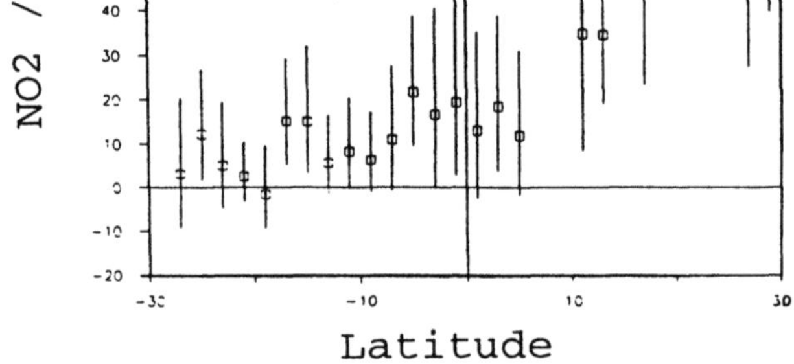

Fig. 9.5: Mean measured NO_2 mixing ratios in 2° latitude bins from 30° N to 30° S at 30° W. From Harris *et al.*, [1].

Airborne campaigns

In January 1991, FLAIR formed part of the instrument package for the French-German campaign TROPOZ II. During this campaign 28 flights, involving 85 flight hours, were carried out. With the TDLAS we obtained vertical profiles of HCHO, H_2O_2, NO_2 and CO between the surface and 11 km altitude and between 50° N and 60° S. Fig. 9.6 provides an overview of the CO measurements. The figure shows data for the northbound flights along the east coast of S. America and the west coast of W. Africa and Europe. A broad region of very homogeneous CO mixing ratios of 50 to 60 ppb can be seen in the south of the measurement region at all altitudes. There is a steep latitude gradient in CO mixing ratios throughout the troposphere, northwards from 30° S in the free troposphere, and starting further south in the lowest 2 km of the atmosphere. A region between 5° N and 25° S can be seen where CO mixing ratios increased with altitude from ca. 80 ppb at 5-6 km to 100-110 ppb at 9-10 km. Back trajectories indicated that these airmasses had three days earlier been over central Amazonia and the measurements suggest that rapid vertical transport from the boundary layer may have taken place.

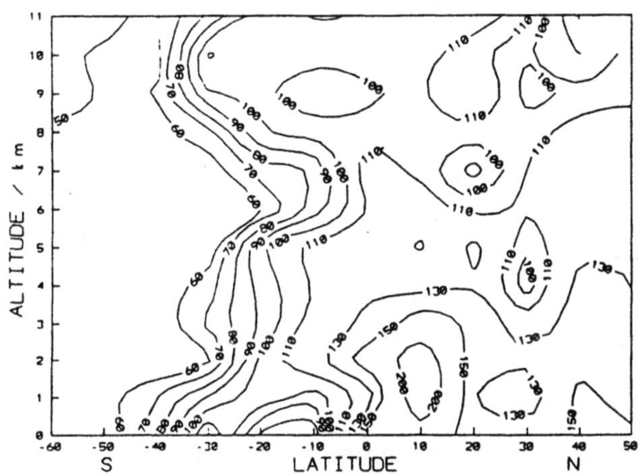

Fig. 9.6: Meridional distribution of CO measured during the northbound flights of TROPOZ II, Jan. 1991. From Roths and Harris [2].

In September/October 1992 the FLAIR-TDLAS was used in a second airborne campaign, SAFARI 92. The main objective of this project was to study biomass burning on a regional scale in southern Africa. Together with other instruments FLAIR was mounted in a DC-3 aircraft to measure NO_2, HCHO, N_2O and CH_4. During 17 measurement flights emission ratios of these trace gases relative to CO_2 - measured by a commercial instrument - were evaluated for controlled Savanna fires in the Krüger National Park as well as for wild fires encountered during survey missions. Additionally the horizontal and vertical distribution of the measured trace gases showed that the troposphere in southern Africa is heavily polluted during the biomass burning season. As a consequence elevated O_3 concentrations were found in the whole region. Fig. 9.7 shows the flight routes during the two survey missions over southern Africa and the frequency distributions of HCHO and CH_4 mixing ratios measured with FLAIR.

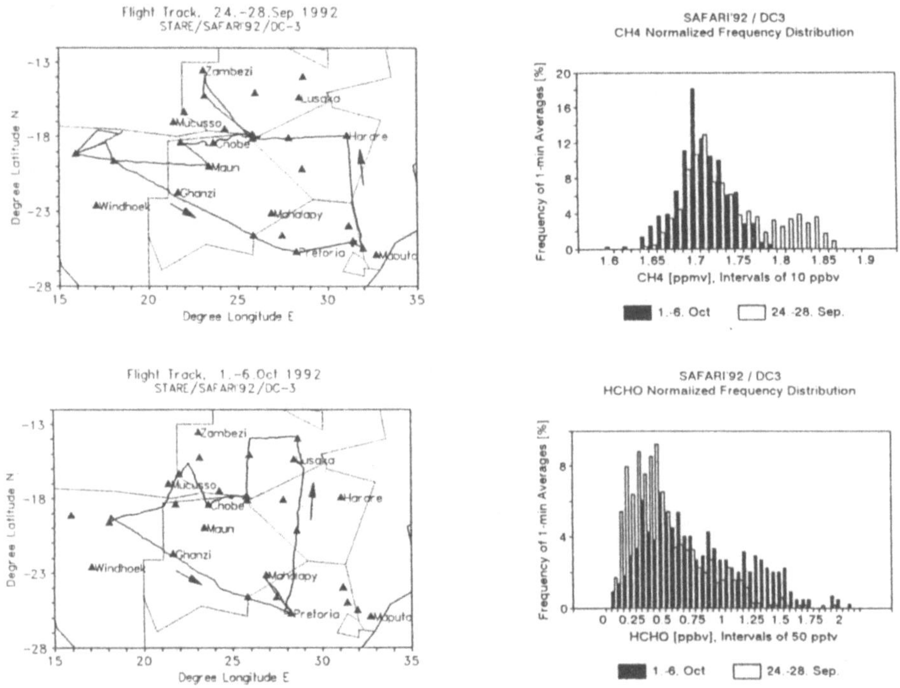

Fig. 9.7: Tracks of survey flights over southern Africa during SAFARI 92 and the frequency distributions of CH_4 and HCHO measured using FLAIR. From Zenker *et al.* [3].

Ground-based campaigns

In 1993 and 1994 the FLAIR instrument was used in a number of groundbased measurement campaigns funded by the CEC. A campaign was carried out at the Pointe de Penmarch (France) in May-June 1993 as part of the CEC-FIELDVOC project, a study of the photooxidation and chemical transformation of DMS under marine boundary conditions. The FLAIR-TDLAS was used for the measurement of NO_2, HCHO, and H_2O_2.

In June and July 1994 a second FIELDVOC campaign was undertaken in Tabua, Portugal. The primary aim was to perform "*in situ*" studies on the photooxidation and chemical transformation of isoprene and other biogenic VOCs at a deciduous forest site (*Eucalyoptus globulus* L.). The FLAIR instrument was used for continuous measurements of H_2O_2, HCHO, CO and NO_2.

In August 1993 a mountain top measurement campaign was carried out within a further EC project, OCTA, at the TOR station, Izaña (Tenerife, 2370 m a.s.l.) to investigate photochemical processes in relatively unpolluted air. A large number of trace gases were measured during the campaign by several groups (HO_2, peroxy radicals, H_2O_2, NO, NO_2, NO_y, PAN, CH_4, NMHC, HCHO, CO, CO_2, O_3). The FLAIR-TDLAS technique was used for the measurement of NO_2, HCHO, and H_2O_2.

Conclusions

In each of our BMFT funded contributions to JETDLAG the aim of the project was the design, construction and testing of special purpose TDLAS equipment, however in each case we were able to go beyond those aims and, after testing, to deploy the equipment in major multi-national atmospheric chemistry research campaigns which have provided valuable data on the distribution of trace gases in the troposphere. We view our contributions to JETDLAG as having been extremely successful and gratefully acknowledge the benefits received from interactions with our colleagues in the sub-project and those elsewhere in EUROTRAC. The instruments we developed within JETDLAG have been used in several campaigns funded by the BMFT and the CEC to study various aspects of tropospheric chemistry, an indication of the success of the subproject in its aim of providing special equipment for atmospheric research. An improved version of the FLAIR instrument with modified electronics for high frequency (MHz range) operation is now available for future European experiments, for example on board the A340-1 aircraft. One of us (DK) has constructed a special TDLAS instrument at KFA-Juelich for use in the TOR subproject at Schauinsland. Within our department, the expertize gained during JETDLAG activities has lead to the construction of two further special purpose TDL instruments, one using Two Tone Frequency Modulation techniques for fast response measurements of N_2O and CO which has been used in eddy correlation studies of the surface fluxes of these gases and of their detailed distribution in the lower stratosphere over the Arctic. The second new instrument is used for studies of the stable isotope ratios ($^{12}C/^{13}C$ and H/D) in sources of atmospheric CH_4. This unique spectrometer allows direct isotope ratio measurements without separation or chemical conversion of the CH_4 and is now in use in studies funded by the BMFT and by the CEC. We believe we have contributed significantly to the JETDLAG goal of widening the acceptance and use of TDLAS for atmospheric chemistry research in Europe.

References

1. G.W. Harris, D. Klemp, T. Zenker, J.P. Burrows, B. Mattieu; Measurements of trace gases using tunable diode lasers during the 1988 Polarstern cruise and intercomparisons with other methods, *J. Atmos. Chem.* **15** (1992) 315-326.
2. J. Roths, G.W. Harris; Vertical and Meridional distribution of CO observed during the TROPOZ II experiment, *J. Atmos. Chem.* in press.
3. T. Zenker, F.G. Wienhold, G.W. Harris, P. LeCanut, M.O. Andreae, R. Koppmann, A.M. Thompson, D.P. McNamara, T.L Kucsera; Regional trace gas distributions and airmass characteristics in the haze layer over southern Africa, presented at the AGU Chapman Conference on Biomass Burning, Williamsburg, Va. 1995.

9.3 High Frequency Modulation Spectroscopy with Tunable Diode Lasers

P. Werle, R. Mücke and F. Slemr

Fraunhofer Institut für Atmosphärische Umweltforschung, Kreuzeckbahnstr. 19, 82467 Garmisch-Partenkirchen, F.R.G.

Summary

The main requirements for *in situ* trace gas analysis, process and air pollution monitoring are specifity, high time resolution and high sensitivity. Tunable diode laser absorption spectroscopy (TDLAS) is increasingly being used to measure trace gas concentrations down to low ppb-levels (volume mixing ratio of 10^{-9}). The application of high frequency modulation (FM) schemes can further improve sensitivity and detection speed of modern instrumentation. The FM technique has been investigated with respect to current limitations of tunable diode laser spectroscopy. Recent research results on FM-spectroscopy and novel findings have been reported: the characterization of system stability using the Allan variance; the influence of optical multipass cells; laser noise and system stability on spectrometer performance; modern signal processing strategies for tunable diode laser spectroscopy; and calibration procedures according to ISO regulations. Applications related to quality assurance and calibration as well as challenges connected with fast and high sensitivity measurements have been addressed.

Original aims of the work

In situ trace gas analysis at sub-ppb levels imposes high demands on analytic instrumentation. Fast, accurate, rugged and operational instruments are needed for environmental and process monitoring, medical diagnostics, plasma analysis and atmospheric research [1-3]. Ultrasensitive instruments free of interference by other atmospheric constituents are required to measure free radicals and other reactive species in the atmosphere [4]. The great number of gaseous pollutants and their generally low variable concentrations with large local differences pose challenging requirements to analytical techniques [5]. Modern research in atmospheric chemistry requires highly sensitive techniques for measurement of concentrations of free radicals which determine the rate of photochemical destruction of most atmospheric pollutants. Since the development of lasers with high spectral brightness and tunability over broad wavelength regions, laser based detection schemes are of growing interest. They are mostly based on absorption measurements. Since the absorption spectrum is characteristic for each molecule, spectroscopic methods allow highly specific detection of many substances. It is obvious, that the mid infrared from 3 to 20 μm, the so-called "fingerprint region", with strong rotational-vibrational absorption bands is the preferable wavelength region. Tunable diode laser absorption spectroscopy (TDLAS) is being frequently used for measurement of trace gas pollutants in the atmosphere [1]. The TDLAS spectrometers usually work with multipass absorption cells to achieve high sensitivity. To alleviate problems by absorption line overlap, these absorption cells are usually operated at low pressure where the linewidth is Doppler limited. In most sensitive instruments, the diode laser is repetitively tuned over an absorption line of a target molecule and the absorption spectra are averaged over a specified time interval. Additional modulation techniques are used to reduce the $1/f$-laser noise. With derivative spectroscopy using lock-in-detection at kHz modulation frequencies, typically, detection limits of the order of 0.1 ppb were achieved for many smaller molecules in the air with spectra averaging times of a few minutes [1-3]. Although these detection limits are sufficient for many applications, still better detection limits are required by modern atmospheric research. Substantial improvements of TDLAS detection speed and detection limits were obtained by introducing the high frequency modulation technique [6]. The FM techniques [6-9] determine the absorption or dispersion of a narrow spectral feature by detecting the heterodyne beat signal that appears when the FM optical spectrum of the probe wave is distorted by the spectral feature of interest [8]. In contrast to conventional derivative spectrometers, where the laser is modulated at several kHz, in a FM spectrometer the laser source is modulated with a rf-current of about 100 MHz for the so-called single-tone technique. However, to achieve the sensitivity improvement using the FM technique and to build instruments for routine high sensitivity measurements under field conditions like trace gas flux measurements [5] (gradient method or eddy correlation technique), many practical problems still have to be solved.

A convenient way to describe the sensitivity at a given signal level is the signal-to-noise ratio (SNR). If the noise in the detection system is the limiting factor for ultimate sensitivity, the detection limit of a spectrometer can be derived from the signal-to-noise ratio referring to the output of the detector preamplifier:

$$SNR = \sqrt{\frac{\langle i_S^2 \rangle}{\sqrt{\langle i_{TN}^2 \rangle + \langle i_{SN}^2 \rangle + \langle i_{1/f}^2 \rangle}}} \quad \xrightarrow{q.l.} \quad \frac{\sqrt{P_{Detector}}}{\sqrt{\Delta f}}$$

The three main noise currents to be considered are the thermal noise of the detector-preamplifier combination (*TN*), the quantum (shot) noise (*SN*) and a $1/f$-type laser excess noise ($1/f$) [10]. While the frequency spectra of thermal noise and quantum noise are generally frequency independent (white noise), the $1/f$-noise contribution is frequency dependent (pink noise). As it can be seen from noise measurements, there are regions at modulation frequencies beyond 100 MHz where the $1/f$-noise contribution can be neglected. If we move in detection frequency range into such a potential *quantum limited* (*q.l.*) regime, $1/f$-noise contributions can be neglected and the total noise can be approximated as the sum of thermal and shot noise. If sufficient laser power is available on the detector, the power independent thermal noise does not contribute significantly to the total noise and shot noise remains the dominating contribution. The SNR under such quantum limited conditions is proportional to the square root of the ratio of the laser power, P_D, available at the detector [11] and the detection bandwidth, Δf. Consequently, there are actually two possible directions towards better detection limits: (i) increasing the laser power on the detector and (ii) reducing the detection bandwidth, which is equivalent to digital signal averaging [12].

Therefore the original aims of the JETDLAG work at the Fraunhofer Institute in Garmisch-Partenkirchen were

- to investigate the new technique of high frequency modulation spectroscopy with lead-salt diode lasers and to develop an FM-spectrometer for sensitive measurements of NO_2
- to develop an instrument for fast (airborne) applications and to investigate the possibility of simultaneous real time measurements of two different species
- and to investigate the feasibility of the measurement of atmospheric HO_2 radicals by FM-TDLAS.

The activities and principal results achieved during the period of the EUROTRAC research within the above mentioned projects will be discussed in the next section.

Activities and principal results achieved during the period of the EUROTRAC research

Technology development : recent findings and novel approaches

System stability and Allan variance

Trace gas measurements near to the detection limit are usually performed by measuring alternatively the spectrum of the ambient air and the spectrum of zero air, *i.e.* air devoid of the target substance. We refer to the latter spectrum as background spectrum. This procedure is based on the inherent assumption that within the time interval needed for the acquisition of both spectra (ambient and background) the fringes do not move. If this assumption is fulfilled, the subtraction of the background spectrum from the ambient spectrum would provide the absorption spectrum of the target species which, to a first approximation, is only subject to random noise. The difference of ambient and background spectra is then fitted to a calibration spectrum. This spectrum is obtained with the absorption cell filled with calibration gas, which for example can be obtained from a permeation based calibration system. From time to time such a spectrum is recorded and stored for signal processing.

Further averaging of these corrected spectra should then improve the detection limit according to a square root relationship [8]. This has been tried but the achieved improvement was always substantially smaller than that one expected on the basis of the square root relationship. The observed deviations are most probably caused by the violation of the above assumption and, consequently, the stability of the TDLAS instrument has to be investigated. It is obvious that every real world system will have an optimum averaging time determined by the drifts in the system such as temperature changes, moving fringes, background changes *etc.* The instrument stability can be described using the Allan variance [12]. In the white noise dominated region the Allan variance is equivalent to the conventional variance and can be used to predict the detection limit of a given system as a function of the integration time [12]. A plot of the Allan variance as a function of the integration time shows a minimum, which corresponds to the optimum integration time, τ_{opt}, typically on the order of 40-100 seconds. This has been confirmed by many other researchers [13]. The optimum integration time is a characteristic property for a given instrument as it reflects the overall system stability and ultimate performance.

Limitations at sub-ppb levels become dominant, when spectrometers are influenced by drift effects and changes in the spectral background structure. In Fig. 9.8 an example of a typical drift is shown. A series of ambient signals as shown in the upper left corner have been recorded. The temporal evolution of the spectral structure is shown in Fig. 9.8a, where the *x*-axis corresponds to the spectral region (analysis window) and the *y*-axis in each quadrant represents a time interval of about one hour. The stable vertical bar in the right half of Fig. 9.8a corresponds to the peak of the absorption line, which shows a slight variation in width. The other non stationary structures are generated by background drifts. The traces in Fig. 9.8b have been recorded when the measurement cell has been flushed with zero air. The sensitivity of a TDLAS instrument should improve by averaging over a long time scale; but at optical densities around 10^{-6} the sensitivity is usually limited by fringes due to unwanted étalons, which we can observe in Fig. 9.8a. In principle, this limitation can be avoided by subtracting the background absorption spectra from the spectrum of ambient air. In Fig. 9.8c a background spectrum, which has been recorded at the beginning of the measurement, has been subtracted from the ambient spectra and it can be seen that the background drift is reduced. However, there are still significant amplitude fluctuations. As shown in Fig. 9.8d, further improvement can be obtained by measuring alternatively background and ambient spectra for the above mentioned subtraction scheme. This subtraction-procedure inherently assumes that the instrument is stable during the time needed for the acquisition of both spectra, *i.e.* that the fringes do not move in the meantime. The exchange time for typical multipass cells varies from less than hundred milliseconds up to several seconds or even much longer for sticky gases like ammonia or nitric acid. Therefore, the exchange time can limit spectrometer performance due to a reduced duty cycle and drift effects. System stability and the detection limits of the instrument can conveniently be described by the Allan variance [12]. If the Allan variance is calculated from a data set as shown in Fig. 9.8c and plotted versus the integration time, we obtain a first Allan plot. The minimum in this Allan plot (Fig. 9.9a) corresponds to the optimum integration

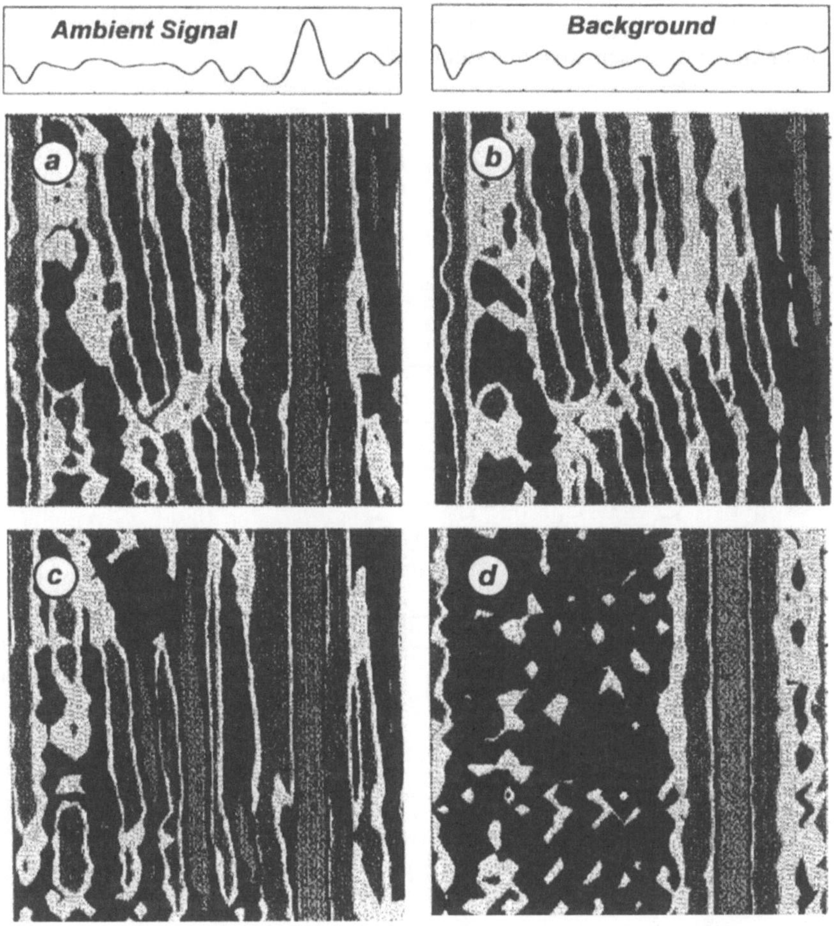

Fig. 9.8: a) FM signal b) background c) signal after subtraction of 1st background d) FM signal after subtraction of actual background (horizontal: wavelength; vertical : time 1 h)

time, τ_{opt}, which is a characteristic property of a given instrument. The experimentally determined optimum integration time defines an upper limit for the time, within which sample and background spectra have to be acquired in order to assure the integrity of data : $\tau_{opt} \geq \tau_{bgr} + \tau_{exch} + \tau_{meas} + \tau_{exch}$, where the time τ_{exch} is required for the complete removal (exchange) of ambient air from the absorption cell. In order to achieve the same signal-to-noise ratio the times for background and sample measurements should be set equal as indicated in Fig. 9.9b. If sample and background spectra are acquired subsequently and the sum of integration and exchange times is choosen to be less or equal to τ_{opt}, as for example with the data from Fig. 9.8d, system drifts will less significantly influence the quality of data. After such a drift correction the data can be stored and analyzed again in terms of the Allan variance, as shown in Fig. 9.9c. This figure also shows that after the background correction made with the optimum integration time, the corrected spectra can be integrated on a much longer period (of the order of 1000 s) allowing much more sensitive measurements.

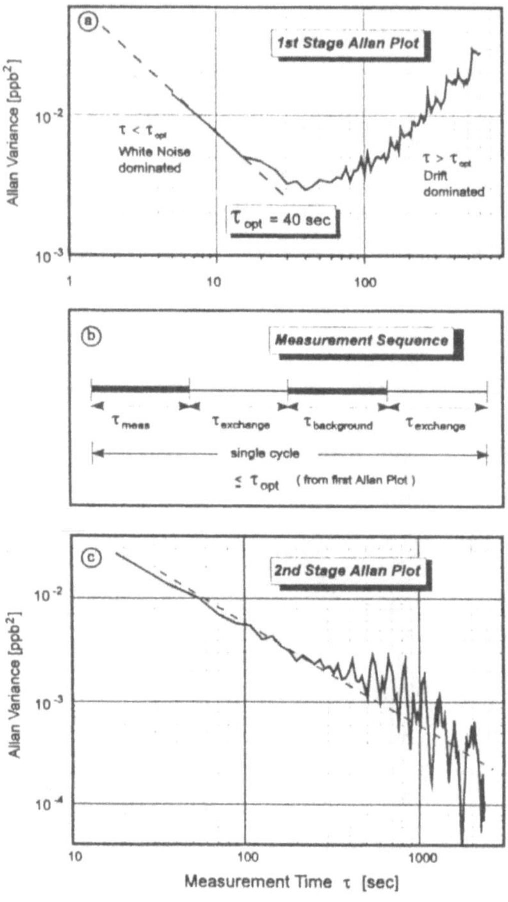

Fig. 9.9: (a) First Allan plot for the determination of the optimum integration time; (b) For an optimized measurement sequence the sum of integration and exchange times should be less than the opt. integration time; (c) Second Allan plot allows signal averaging up to 2000 s.

Signal-to-noise ratio analysis in optical multipass cells for FM-TDLAS and conventional TDLAS

The absorptions that have to be detected are usually small ($< 10^{-5}$), and to achieve sensitivity adequate for environmental monitoring, absorption spectrometers require long optical paths. In instruments with a limited size, long absorption path lengths up to several hundreds of meters have usually been provided by multi-reflection optical systems of which the best known are the systems invented by White and by Herriott. In all these systems the sensitivity gained by lengthening the absorption path is increasingly offset by the attenuation of the radiation power throughput, due to the increasing number of reflections and the imperfect reflectivity of the mirrors. Consequently, each absorption spectrometer has to be operated with an optimal number of reflections in multi-reflection absorption systems to achieve the highest SNR.

In high resolution absorption spectrometers with conventional light sources, the signal-to-noise ratio (SNR) is usually limited by the thermal noise level of the detector-preamplifier combination, which is independent of the light source power. However, the noise in many laser absorption spectrometers is dominated by the excess or shot noise which is dependent on the transmitted laser power, and which in turn is dependent on the number of reflections in a multipass cell [11]. The signal-to-noise ratio in optical multipass cells used for monitoring trace gas concentrations was studied using high frequency (FM) as well as conventional derivative (WM) laser absorption spectrometers. At conditions usually encountered in tunable diode laser spectrometers, the optimal number of passes for the FM spectrometers has been found to be much smaller than the optimal number of passes for the WM spectrometers. This finding implies that if both techniques are operated with an optimal number of passes, the introduction of FM techniques can improve the ultimate SNR in spectrometers using optical multipass cells by only about an order of magnitude. Although still highly desirable, this is substantially lower than the two orders of magnitude potential improvement derived solely from the noise analysis, without considering the use of multipass cells [10]. The smaller number of passes required for optimal SNR in FM spectrometers has other practical consequences. Cells with a smaller number of passes are cheaper to construct and to adjust. As they can be designed with smaller volume, faster measurements can be made. This property is important because the potential high speed of FM measurements can be utilized only in cells with a very small volume. If on the other hand ultrasensitive measurements require long absorption path length, high reflectivity mirrors and high power lasers are a prerequisite to get the full potential of FM techniques.

Signal processing strategies

While many instrumental improvements focus on hardware and optics, much less effort has been put into post detection signal processing [14-16]. Spectroscopic signals vary in frequency due to fluctuations or transients in the laser current and they vary in amplitude due to changes in background radiation and structures such as moving fringes. Therefore, a signal processing concept for TDLAS must provide means for correcting frequency and amplitude fluctuations as well as some capability to cope with changing background structures. Most critical is the impact of drifts of ambient spectra relative to the previously recorded calibration spectrum. These drift effects can lead to a significant underestimation of concentration values, *i.e.* a spectrometer might report extremely low or high concentration values and according to the phase relation of the measurement and background spectra possible variations in calculated data can rather refer to drift effects than to changes in ambient trace gas concentrations. Fluctuations in the laser current due to noise and other interfering signals have a significant influence on the system performance, especially on the confidence range of the measurement and therefore on the detection limit of the system. A useful approach is to correct such drifts and jitter online during the averaging process [16]. Variations in the signal amplitude can be corrected by intensity normalization. A novel idea is based on the measurement of the average modulation signal amplitude instead of measuring detector currents or reference signals [16]. With high frequency modulation techniques, the rf-power level at the detector ideally reflects variations in laser power and system alignment changes due to slow drifts or vibrations. While conventional intensity normalization schemes use a reference channel or the detector bias current, this normalization is independent from changes in background radiation and a detailed investigation of signal stability using the Allan variance analysis [12] indicated improved stability. Background changes can, at least in part, be attributed to the time dependent fringe drifts and in many cases a slope or curvature has been observed. Varying slopes can also be obtained from pressure broadened atmospheric absorption lines (*e.g.* water vapour). Such a situation calls for the introduction of "adaptivity". The multiple linear regression and other adaptive filters [16] use a minimum mean square error algorithm to determine the optimum set of coefficients for a filter. For each ambient spectrum a calculation of the filter coefficients is performed and, therefore, the filter is able to adapt to changing conditions such as signal drift and changing background structures [16]. This filter has the interesting feature that, in principle, besides ambient and calibration spectra no further information is needed. This technique might allow less frequent or even eliminate the background sampling and, therefore, would be well suited for fast continuous measurements.

Laser excess noise and interferometric effects

In addition to system stability, the major prerequisite for ultrasensitive FM-TDLAS spectrometers is the availability of the lead-salt diode laser, which should operate in a single longitudinal mode with low noise and a power output of about 1 mW. The properties of presently available lasers are far from these requirements and have to be improved substantially to exploit the ultimate potential of the TDLAS technique. Single mode operation is required to prevent absorption signals from other modes and to minimize mode competition noise. Wideband noise measurements of lead salt-diode lasers generally show a significant increase in laser noise at frequencies below about 100 - 200 MHz (Fig. 9.10). This excess noise can be attributed to the presence of sometimes spurious side-modes [17] and their mode competition. Consequently, high modulation frequencies are necessary to achieve near shot noise limited performance of the FM spectrometer. On the other hand, a FM spectrometer can be interpreted as a two colour interferometer [17] and, therefore, it is extremely sensitive to thermal drifts in the mechanical setup.

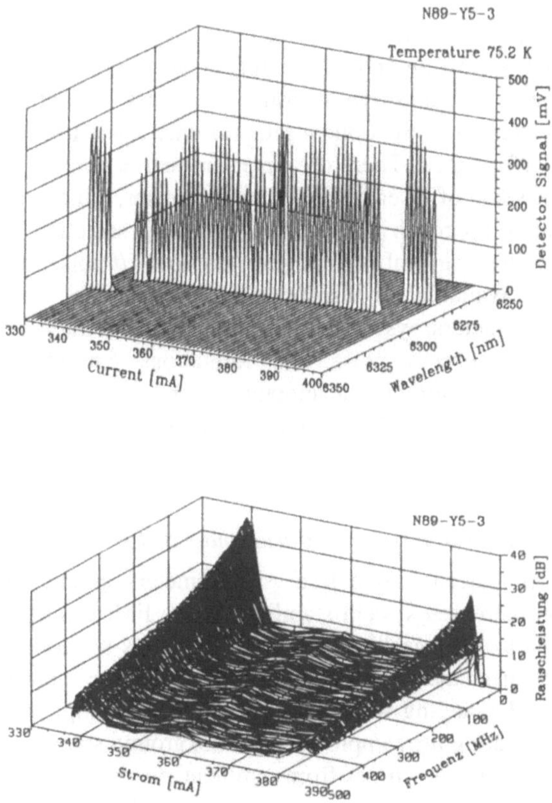

Fig. 9.10: Laser mode map and noise characterstics. Laser excess noise can be observed near the mode jumps at 335/390 mA.

With increasing modulation frequency, *i.e.* higher separation of FM sidebands in the frequency domain, the sensitivity of the FM spectrometer towards fluctuations in the optical path increases. Therefore, low modulation frequencies should be used to achieve a high system stability during long integration times. Signal instabilities at low optical densities caused by interferometric effects might give an explanation for the fact, that the expected improvement in system performance after the application of high frequency modulation schemes to spectroscopic systems designed for conventional derivative spectroscopy has not yet been attained on a routine basis. The interferometric effect at high modulation frequencies is superimposed to the so-called 'étalon-effect'. In most sensitive instruments, the diode laser is repetitively tuned over a molecular absorption line and the spectra are averaged over a specified time interval. When the wavelength of the diode laser is tuned over an absorption line, a periodic fringe structure is superimposed to the desired signal from the absorption of the target gas. If this fringe structure is generated between the reflecting mirrors in the multipass absorption cell, mechanical drifts cause a change of the fringe structure, if the fringe structure cannot be suppressed sufficiently. They usually limit sensitivity due to their variation with time. But even if the losses which are responsible for the 'étalon-effect' could be eliminated, *e.g.* by an optical isolator, the above mentioned 'interferometer-effect' remains and for sensitive measurements, say at optical densities below 10^{-6}, short term changes in the optical setup of only a few μm can significantly influence system performance. Therefore, interferometric stability is required for the opto-mechanical setup of a sensitive FM spectrometer, which can be rather expensive. An alternative approach is the time consuming selection of single mode lead-salt diode lasers with several mW power, with low $1/f$ noise, but unfortunately such lasers are not available for many desired wavelengths. With 'state of the art' FM-TDLAS spectrometers, system optimization is often a trade-off between limitations due to laser excess noise and limitations caused by interferometric effects. Therefore, further development and commercial availability of high power, single mode lead-salt diode lasers with distributed feedback (DFB) structures would help to improve spectrometer performance in the IR for routine applications [17].

Double modulation techniques for the suppression of background drifts

Further progress in sensitivity of current diode laser spectrometers practically depends on improved lasers and increased system stability. As the FM technique originally has been referred to as a zero-baseline technique [6] this is not longer valid for trace gas measurements near the detection limit, because the reference, i.e. the background, is floating and changing with time. Therefore, it is important to measure as fast as possible alternating the sample and the background spectra at low concentration levels. The switching time from ambient to background measurements is usually limited by the exchange time of the gas, which is typically in the order of seconds. For sensitive and precise measurements this might be too long and, therefore, different approaches have to be tried. Design approaches increase the pumping speed and minimize multipass absorption cells. A 36 m Herriott cell with an exchange time as low as 50 ms (1/e) is commercially available (Aerodyne Research, Billerica, USA). A physical approach would be the use of photochemical sample modulation [18]. The Stark- or Zeeman-effect can be used for a fast sample modulation in order to "exchange" the gas in the sample cell very rapidly by shifting the spectral signature of the spectral feature under investigation out of the spectral analysis window in order to record the "background" structure. Fortunately, some molecules of major interest in atmospheric chemistry, as formaldehyde and hydrogen peroxide, exhibit a strong electric dipole moment as indicated in Table 9.2, which shows a list of molecules of interest and the calculated detection limits under the assumption of a minimum detectable absorbance of 10^{-6} within 1 s integration time and an absorption path lenght of about 25 m at a pressure of 30 mbar using a multipass cell [19]. The calculated detection limits scale linear with the minimum detectable absorbance and the absorption path length. However, for the calculations only 25 m absorption path in the multipass cell have been assumed as the power loss due to the multiple reflections on the mirrors overcompensates the sensitivity gain due to a longer absorption path [11]. The optimum path length for a given system can be estimated from system parameters (laser power, mirror reflectivity) and we found 25 m best suited for our experimental setup with a White cell. Higher sensitivities are expected at longer integration times.

Table 9.2: Calculated 1σ-detection limits *DL* for a minimum detectable absorbance of 10^{-6} at 30 mbar in 25 m multi path cell at 296 K, S = linestrength, σ_0 = absorption cross section at line center, γ_V = Voigt-HWHM.

Gas	μ / D	$\tilde{\nu}_0$ / cm^{-1}	λ / µm	S / 10^{-20} cm	γ_V / 10^{-3} cm^{-1}	MHz	σ_0 / 10^{-18} cm^2	c_{min}^{STP} / 10^8 cm^{-3}	DL / ppt	DL^{STP} / µg m^{-3}
HCN	3.00	3337.139	2.997	36.89	5.97	179	23.3	6.3	23	0.028
H_2CO	2.34	2781.035	3.596	11.86	5.17	155	8.41	17	65	0.087
H_2O_2	2.26	1284.205	7.787	4.464	3.52	105	4.27	34	130	0.2
HNO_3	2.17	1694.321	5.902	2.85	3.75	112	2.53	58	220	0.60
CH_3Cl	1.89	3040.220	3.289	0.192	4.12	124	0.174	840	3100	7.1
HF	1.82	7855.643	1.273	7.591	2.09	362	2.73	54	200	0.18
HCOOH	1.70	1113.068	8.984	4.4	3.32	99	4.4	34	130	0.26
SO_2	1.59	1371.695	7.290	4.855	3.86	116	4.11	36	130	0.38
NH_3	1.47	930.757	10.744	52	2.92	88	61.8	2.4	9	0.007
HCl	1.18	2944.914	3.396	50.33	4.30	129	45.1	3.2	12	0.02
H_2S	1.02	1364.601	7.328	0.102	4.89	147	0.0685	2100	8000	12
OCS	0.79	2052.716	7.328	103	3.55	106	100	1.5	5	0.015
HBr	0.79	2649.092	3.775	4.456	2.90	87	5.69	26	96	0.35
O_3	0.53	1052.848	4.872	4.2	2.64	79	5.30	28	100	0.22
HI	0.38	2277.519	4.391	0.208	2.21	66	0.338	430	1600	9.2
NO_2	0.29	1600.413	9.498	21.8	2.78	83	27.7	5.3	20	0.042
N_2O	0.17	2236.234	6.248	100.4	3.55	106	103	1.4	5	0.01
NO	0.16	1875.813	4.472.	3.399	3.26	98	3.91	38	140	0.19
CO	0.10	2169.198	4.610	46.15	3.63	109	48.9	3	11	0.014
C_2H_2	—	3260.427	3.067	22.9	5.43	163	16.5	8.9	33	0.038
CO_2	—	2270.290	4.405	3.358	3.51	105	3.50	42	160	0.31
CH_4	—	3067.300	3.260	21.29	5.77	173	15.3	9.6	36	0.026

Applications of TDLAS to measurements in atmospheric chemistry research

Time series measurements: ambient NO_2 *concentrations*

To test the predictions of the Allan variance analysis ambient measurements were made with a NO_2 doublet line at 1600.413 cm^{-1} since its background was almost free of disturbance from pressure broadened H_2O lines nearby [12]. The optical density of the 100 ppb NO_2 calibration gas in a White cell was calculated to be 5.3 ‰ using a Voigt profile. Consequently, 1 ppb of NO_2 in the following data corresponds to an optical density of about 5.3×10^{-5}. After locking the laser to this line, a series of ambient measurements were made. A characteristic spectrum is shown in Fig. 9.11. The spectra were averaged 256 times within 740 ms to obtain a high SNR. The electronics bandwidth was set to 1.5 kHz. The spectra averaging then resulted in an effective bandwidth of 5.86 Hz. In the subsequent regression analysis the measured spectrum was fitted to the calibration spectrum. The fit reported a concentration of 1.17 ppb with an 1σ error of 31.5 ppt. The data (1.17 ± 0.0315 ppb) of Fig. 9.11, which were well above the detection limit, corresponded to an optical density of $(62 \pm 1.7) \times 10^{-6}$ at 5.86 Hz bandwidth. Under the assumption of the theoretically expected behaviour for further signal averaging - the Allan variance analysis would allow signal averaging up to 60 s - this corresponds to 7×10^{-7} at 1 Hz bandwidth [12]. NO_2 in ambient air was measured continuously in the afternoon near the Fraunhofer Institute in Garmisch-Partenkirchen with a high time resolution of 1.5 s. The observed NO_2 spikes were due to cars coming in and out of the nearby parking lot. The calibration and background data were removed from the plot to avoid confusion. Faster measurements are possible with a properly designed instrument (smaller absorption cell, higher sampling flow rate, more efficient spectra averaging) thus enabling eddy-correlation measurements mentioned later in this section.

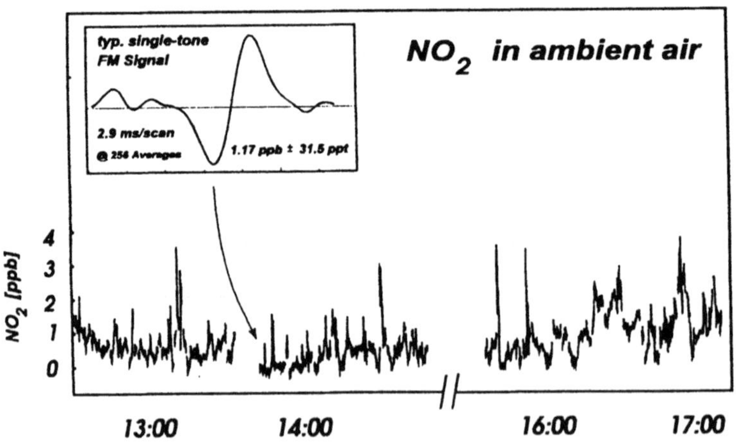

Fig. 9.11: NO_2 monitored in ambient air with a time resolution of 1.5 s and a typical 1.17 ppb NO_2 FM-signal from ambient air.

Intercomparison studies : TDLAS, DOAS and other chemical sensors

Many new instruments has been developed to measure low concentrations of important trace gases in the atmosphere. Among these especially the simple and cheap ones are frequently subject to interferences by other atmospheric species. Due to the complexity of air composition and variability of interferant concentrations, the possibility of interferences cannot be eliminated by simple tests using a few compounds suspected of interference. One of the most efficient tests for interferences is an intercomparison with another more specific measurement technique based on a different physico-chemical detection principle. Due to its universal detection capabilities for all smaller molecules and its specifity, TDLAS technique represents one of the most valuable techniques for verification of cheaper and simpler techniques. Fig. 9.12 shows some typical time series data from an intercomparison of NO_2 measurement techniques between an OPSIS DOAS-instrument, a Luminox chemiluminescence instrument, and a FM-TDLAS system performed near the Fraunhofer Institute. When the air is well mixed due to wind speed above 2 m/s there is a good agreement between the long-path measurement of DOAS and the point measurement of TDLAS, whereas the chemiluminescence instrument measuring on the same point as the TDLAS shows differences. These differences are caused by zero point instabilities of the chemiluminescence instrument, as found in laboratory experiments.

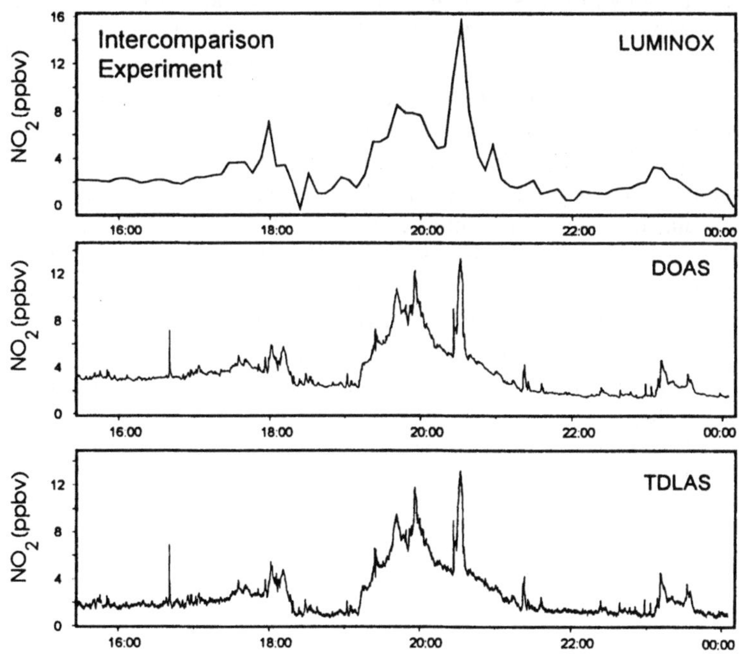

Fig. 9.12: Ambient monitoring of NO_2 on the 15th of August 1993 with different measurement techniques : FM-TDLAS, Luminox and DOAS.

Volatile organic compounds (VOCs) in the atmosphere are of increasing interest as precursors of ozone and other photo-oxidants. Among VOCs, oxygenated compounds such as aldehydes and ketones are of particular importance as one of the major sources of free radicals. Reliable measurements of carbonyl concentrations in the atmosphere are thus required for simulations of photochemical processes. Some chromatographic techniques are used to determine several carbonyl compounds simultaneously. Due to their low temporal resolution these techniques are, however, not suitable for measurements on airplanes, where a high temporal resolution is required to resolve small spatial variations. For this application, continuous measurement techniques for formaldehyde as the most abundant carbonyl compound, supplemented by discontinuous measurements of other carbonyls, may provide data on spatial distributions needed for studies of photochemical processes. Continuous measurements of formaldehyde can presently be carried out using three techniques: a method based on an enzymatic reaction with fluorometric detection, a method based on the Hantzsch-reaction with fluorometric detection (EF), and a direct spectroscopic measurement by tunable diode laser absorption spectroscopy. Formaldehyde measurements with the FM-TDLAS instrument were intercompared with simultaneous measurements using the EF method. In the EF method formaldehyde is stripped continuously from ambient air by 0.01 M HCl. In solution formaldehyde reacts with formaldehyde dehydrogenase (FDH) and nicotinamide adenine dinucleotide (NAD^+) to NADH, which is detected fluorimetrically. The EF instrument is calibrated by standard formaldehyde solutions which in turn are prepared from a formaldehyde stock solution standardized by oxime titration. Prior to simultaneous measurements in ambient air, good agreement was achieved in measurements of gaseous standard mixtures. In order to cover a wide concentration range for measurements in ambient air, polluted indoor air was mixed with relatively clean outdoor air during a part of the intercomparison. The slope of the regression line of 1.0 in the correlation plot in Fig. 9.13 suggests that no major systematic errors are present in the data. The large scatter can partially be ascribed to interferences associated with the less specific EF method and to different time resolutions of the two techniques. Opposite to the mixture of indoor and ambient air, measurements in clean ambient air led to a generally better agreement, both in structure and in mixing ratios [20].

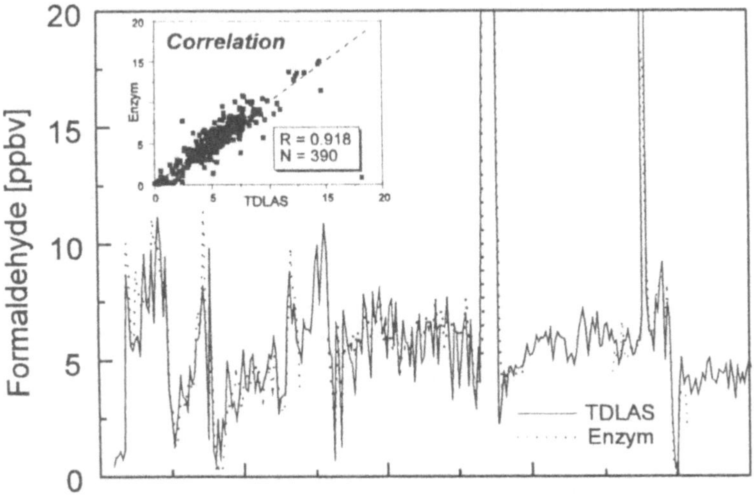

Fig. 9.13: Measurement and correlation between the enzymatic and the TDLAS method.

Assessment of performance characteristics : Calibration of a TDLAS system according to ISO/DIS regulations

The International Organization for Standardization (ISO) has specified in its regulation *Air Quality - Determination of performance characteristics of measurement methods* (ISO/DIS 9169) procedures to quantify calibration function and its linearity, and the instability, lower detection limit, selectivity, and precision of the instrument. To measure these characteristics, calibration gases with mixing ratios in the expected concentration range have to be used. The generation of calibration gases in the low ppb and sub-ppb range, however, is a difficult task. Therefore, part of our work has been focused on a development of a calibration device in compliance with the ISO requirements for an integrated tunable diode laser spectrometer. For this purpose an automated calibration system based on permeation devices with a subsequent flow dilution has been developed [21]. The system can be used as a secondary standard and allows multi-component as well as multi-stage calibrations of TDLAS and similar instruments. The performance of the FM-TDLAS instrument was characterized by two calibration functions: one in the range of 0-10 ppb for determination of the lower detection limit (LDL) and linearity in the lower ppb-level, and the other in the range of 4 - 30 ppb for the assessment of repeatability and linearity in the upper ppb-level (Fig. 9.14). The method was linear within the tested range of 0 to 30 ppb [21].

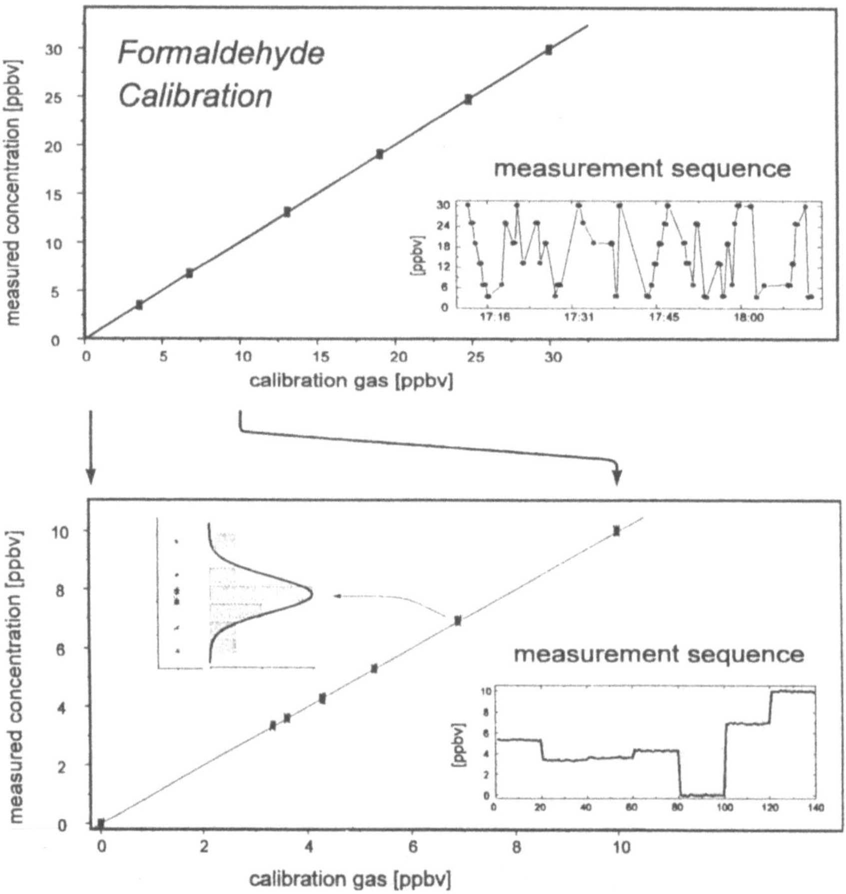

Fig. 9.14: Calibration function to derive system parameters according to ISO/DIS 9169.

Simultaneous multi-component measurements : H_2CO / H_2O_2 *measurements during the Polarstern 1994 cruise*

The importance of hydrogen peroxide, H_2O_2, in atmospheric chemistry arises from its oxidizing potential in the liquid phase and from the fact of being involved in gas- and liquid-phase radical chemistry. Measurements from a ship cruise in 1990 showed a dependence on latitude with generally higher mixing ratio around the equator (*ca.* 1.5 ppb) and lower mixing ratio poleward (down to 0.1 ppb). Formaldehyde is another important photoactive trace component of the atmosphere. Measurements in clean air provide important insight into the removal processes of light hydrocarbons as well as information about the general chemical reactivity of the atmosphere. Only a few measurements of the formaldehyde concentrations in the remote marine troposphere have been reported with differing results. The aim of the measurements during a ship cruise in 1994 was the determination of the background concentrations of H_2O_2 and H_2CO and the expected diurnal variation of H_2CO. Therefore, continuous measurements of hydrogen peroxide and formaldehyde have been done by a tunable diode laser absorption spectrometer (TDLAS) in addition to wet chemical techniques. Two lasers were operated in a liquid nitrogen Dewar at temperatures between 80 and 100 K simultaneously. H_2CO was measured at a wavelength of 1740 cm^{-1} and H_2O_2 around 1280 cm^{-1}. The two laser beams, each modulated individually by high frequency, were combined by a dichroic beam splitter and passed a multipass reflection cell. The signals from the two species were separated in a computer controlled dual channel high frequency (single-tone/two-tone) lock-in type amplifier. It is the first time that a two-component high-frequency modulated TDL system has been used for the detection of trace gases over the Atlantic Ocean on a route from northern to southern hemisphere (50° N - 50° S). For the interpretation of the data all parameters have been recorded (air and water temperature, air pressure, relative humidity, wind direction and speed) using standard meteorological instrumentation. These measurements will give an important contribution for the explanation of the oxidizing potential of the atmosphere.

Trace gas analysis in air samples extracted from ice cores: Measurement of CO_2, N_2O, CH_4, *and* CO *in one sample by FM-TDLAS technique*

Measurements of trace gases in air samples extracted from ice cores drilled in Antarctica or Greenland are the most important way to obtain information about the composition of paleoatmosphere and its change with the changing climate [22]. The present techniques for analysis of trace gases in small air samples on the order of few cm^3 (STP) are currently based mostly on gas chromatographic techniques which in their nature are destructive. Light absorption in IR provide a nondestructive technique enabling the determination of several trace gases in one extracted air sample. In combination with the sensitivity of the FM technique, TDLAS is capable of determination of the most important greenhouse gases such as CO_2, CH_4, N_2O, and CO as well as of isotope ratios such as *e.g.* $^{13}CO_2/^{12}CO_2$. This motivation led us to construct an FM-TDLAS instrument for the analysis of air samples extracted from ice cores. The instrument is designed to measure CO_2, CH_4, N_2O, and CO in a 2 cm^3(STP) large air sample kept in a 70 cm long single pass absorption cell. It consists of two liquid nitrogen cooled laser heads, each with two lasers. Each laser has its own temperature and current control, and is modulated at its optimum frequency and modulation index. The laser beams are collimated and directed consecutively through the absorption cell with the air sample at about 10 mbar. The FM signal is then compared with the calibration signal obtained with the cell filled by calibration gas.

The instrument is capable to detect 5 ppm of CO_2, 20 ppb of CH_4, 1 ppb of N_2O, and 2 ppb of CO in the air sample of 2 cm^3 (STP). These detection limits are sufficient for measurement of preindustrial mixing ratios of about 200 ppm for CO_2, 400 ppb for CH_4, 300 ppb for N_2O, and about 50 ppb for CO and their temporal variations. The sensitivity of the instrument can further be substantially improved by using more efficient signal averaging procedures and by using a multiple regression for spectra evaluation instead of the present peak-to-peak FM-spectra evaluation. Such improvements would open a way to determine even isotope ratios, *e.g.* of $^{13}CO_2/^{12}CO_2$.

Ultrasensitive (ppt) spectroscopy: A feasibility study for measurements of HO_2 *radicals*

As the primary oxidizing agent responsible for the removal of most trace gases, including CH_4 and other hydrocarbons, CO, H_2 and halogenated hydrocarbons, the hydroxyl radical, OH, is central to the chemistry of the troposphere. OH is closely linked to the HO_2 radical through reactions with NO, CO, O_3 and hydrocarbons, and the two are sometimes summarized as HO_x family. HO_2 is present, typically, at concentrations of about 10^8 molecules/cm^3, which is about 100 times that of OH, and has a chemical lifetime of a few minutes. Because of their short lifetime, these radicals are in the stationary state with their precursors and reaction partners. Consequently, a measurement of their concentrations constitutes an excellent test of models of the fast photochemistry of the troposphere. Currently four different approaches to measure HO_2 are partly realized or under discussion [23]: (a) *Cryogenic sampling with electron spin resonance (ESR) detection* is very specific and sensitive, but requires long measurement times; (b) *Chemical amplification (PERCA)* converts one HO_2 or RO_2 molecule to many NO_2 in a chain reaction; the technique is sensitive, but it is not HO_2 specific; (c) *Laser induced fluorescence (LIF)* instruments employ titration of HO_2 to OH *via* added NO, with subsequent OH detection. The technique is not specific to HO_2; (d) *Tunable diode laser absorption spectroscopy (TDLAS)* is very specific but has not yet been demonstrated in the laboratory. We have investigated in detail the feasibility of HO_2 measurements by the FM-TDLAS technique [23]. As the detection of optical densities below 10^{-7} is still a great challenge for TDLAS, the selection of the optimum modulation technique and parameters is necessary for obtaining the required sensitivity. We found that with the suitable frequency modulation index of about 1 the optimum modulation frequency is around twice the absorption linewidth of the measured spectral feature (typically 100-200 MHz) in the single-tone case and at least 5 linewidths (typ. 500 MHz) for the two-tone spectroscopy [7], where the difference of both frequencies should be as small as possible. Using the optimized parameters the two-tone signal amplitude should be about 20 % smaller than for single-tone. Due to high modulation frequencies the two-tone technique requires a broad spectral range free of interferences from other gases. If these results are combined with the previously discussed wideband noise characteristics of lead-salt diode lasers, the main consequence is that high bandwidth detectos should be used in any case. When selecting the optimum spectral region for HO_2 measurements cross sensitivities of many atmospheric trace gases have to be considered. As the strongest absorption lines of HO_2 are covered by other trace gases, the most suitable absorption line of HO_2 is 1410.928 cm^{-1}. A lower pressure and a lower modulation frequency should be used to minimize interferences. Allan plots indicated that a detection limit of 4×10^{-8} might be obtained at 200 m optical path length and a measurement time of 30 minutes, which is sufficient to detect 10 ppt of HO_2. The application of novel signal processing strategies as well as optimized background cycles combined with the known spectroscopic data of HO_2 and new spectral data for interfering substances show that though the measurement of HO_2 radicals is still a great challenge, but it may be feasible using the FM-TDLAS [23].

A fast response chemical FM-TDLAS sensor: micrometeorological trace gas flux measurements by the eddy-correlation technique

The eddy correlation technique directly determines the flux of an atmospheric trace constituent through a plane parallel to the surface. To use the technique several specific criteria have to be fulfilled. Ideally, the meteorological conditions controlling the state of the turbulence should not vary in the course of the measurements. The surface viewed by the sensors should be horizontally uniform, both in its physical and chemical-biological aspects, and should stretch for a distance much greater than the height at which the measurements are made. This height should be much larger than the scale of the surface roughness and the intrinsic scale of the sensors. Because the eddy correlation method may be considered as defining the instantaneous upward or downward transport of the constituent and then averaging contributions to give the net flux, it must take into account the frequency range of the turbulence for vertical transport the constituents in the atmosphere (typ. >1 Hz). The technique requires simultaneous fast and accurate measurements of both the vertical velocity and the trace species in question. Fortunately the technique for the measurement of the turbulence with the necessary resolution is available. Sonic anemometers can readily provide air motion data with the required resolution. Likewise, the ability to handle the air motion and chemical concentration data with modern computer systems is well in hand. The major limitation in using the eddy-correlation technique is presently the lack of appropriate chemical sensors with sufficient time and chemical resolution. The FM-technique is ideally suited for flux measurements by the eddy correlation method and provide an alternative to box chambers. A prototype of a high frequency modulation spectrometer has been developed at our Institute. Originally it has been designed for airborne NO_2/SO_2 measurements [24] and therefore will be modified for high speed (> 5 Hz) micrometeorological measurements.

Conclusions

Tunable diode laser absorption spectroscopy (TDLAS) is a versatile tool for specific, sensitive and fast trace gas analysis. Novel approaches base on the introduction of the high frequency modulation technique, on modern analysis tools (Allan variance) and improved adaptive signal processing concepts. The additional application of sample modulation schemes, as Stark- or Zeeman switching or modulation techniques, for fast ambient/background sampling has been proposed as a promising double modulation technique to obtain improved system performance for trace gas analysis. The TDLAS technique has been applied to validate less specific, simpler techniques, for analysis of trace gases in air extracted from arctic ice cores, and to evaluate system performance characteristics according to ISO regulations. Fast fluctuations in ambient air are well resolved by the FM-TDLAS technique, which is suited for experiments which require high time resolution, such as eddy-correlation flux and aircraft measurements. With improved lasers and systems even reliable measurements in the ppt-range seem to be feasible in the future.

Acknowledgements

This work was funded the German Ministerium für Forschung und Technologie (BMFT) under grant 08431095, 07EU702 and 07EU712 as a contribution to the EUROTRAC subproject JETDLAG and by the Bayerisches Staatsministerium für Wirtschaft und Verkehr (3625-VIII/4c).

References

1. H.I. Schiff, G.I. Mackay, J. Bechara; The use of tunable diode laser absorption spectroscopy for atmospheric measurements, in: M.W. Sigrist (ed.), *Air Monitoring by Spectroscopic Techniques*, John Wiley, New York 1994.
2. D.J. Brassington; Tunable diode laser absorption spectroscopy for the measurement of atmospheric species, in: R.E. Hester, R.J. Clark (eds.), *Advances in Spectroscopy*, Vol. 24: *Spectroscopy in Environmental Science*, John Wiley, New York 1994.
3. D.L. Albritton, F.C. Fehsenfeld, A.F. Tuck; Instrumental requirements for global atmospheric chemistry, *Science* **250** (1990) 75-81.
4. P. Warneck; *Chemistry of the Natural Atmosphere*, Academic Press, London 1988.
5. L. Newman; *Measurement Challenges in Atmospheric Chemistry*, American Chemical Society, Washington, DC 1993.
6. G.C. Bjorklund; Frequency-modulation spectroscopy: a new method for measuring weak absorptions and dispersions, *Opt. Lett.* **5** (1980) 15-17.
7. D.E. Cooper, T.F. Gallagher; Double frequency modulation spectroscopy: high modulation frequency with low-bandwidth detectors, *Appl. Opt.* **24** (1984) 1327-1334.
8. P. Werle, F. Slemr, M. Gehrtz, Chr. Bräuchle; Quantum-limited FM-spectroscopy with a lead-salt diode laser, *Appl. Phys.* **B49** (1989) 99-108.
9. P. Werle; Analytical application of infrared semiconductor lasers in atmospheric trace gas monitoring, *J. Physique IV* **4** (1994) 9-12.
10. P. Werle, F. Slemr, M. Gehrtz, Chr. Bräuchle; Wideband noise characteristics of a lead-salt diode laser: possibility of a quantum noise limited TDLAS performance, *Appl. Opt.* **28** (1989) 1638-1642.
11. P. Werle, F. Slemr; Signal-to-noise ratio analysis in laser absorption spectrometers using optical multipass cells, *Appl. Opt.* **30** (1991) 430-434.
12. P. Werle, R. Mücke, F. Slemr; The limits of signal averaging in tunable diode laser absorption spectroscopy, *Appl. Phys.* **B57** (1993) 131-139.
13. A. Fried, H. Riris, G. Harris, private communication.
14. H. Riris, C.B. Carlisle, R.E. Warren, D.E. Cooper; Signal-to-noise enhancement in frequency modulation spectrometers by digital signal processing, *Opt. Lett.* **19** (1994) 144-146.

15. P. Werle; Signal processing strategies for tunable diode laser spectroscopy, in: H.I. Schiff, A. Fried, D. Killinger (eds.), *Tunable Diode Laser Spectroscopy, Lidar, and DIAL Techniques*, *Proc. SPIE* **2112** (1994) 19-30.

16. P. Werle, B. Scheumann, J. Schandl; Real time signal processing concepts for trace gas analysis by TDLAS, *Opt. Eng.* **33** (1994) 3093-3105.

17. P. Werle; Laser excess noise and interferometric effects in tunable diode laser absorption spectroscopy, *Appl. Phys.* **B60** (1995) 499-506.

18. E.A. Whittaker, H.R. Wendt, H.E. Hunziker, G.C. Bjorklund; Laser FM-Spectroscopy with photochemical modulation, *Appl. Phys.* **B35** (1984) 105-111.

19. P. Werle; Spectroscopic trace gas analysis using semiconductor diode lasers, *Spectrochim. Acta* **A52** (1996) 805-822.

20. R. Mücke, B. Scheumann, J. Slemr, F. Slemr, P. Werle; Measurements of formaldehyde by tunable diode laser spectroscopy and the enzymatic-fluorometric method: An intercomparison study, *Infrared Phys. Technol.* **37** (1996) 29-32.

21. R. Mücke, B. Scheumann, F. Slemr, P. Werle; Calibration procedures for tunable diode laser spectrometers, in *Tunable Diode Laser Spectroscopy, Lidar, and DIAL Techniques*, H.I. Schiff, A. Fried, D. Killinger (eds.), *Proc. SPIE* **2112** (1994) 87-98.

22. H. Oeschger, C.C. Langway, Jr., (eds.); *The Environmental Record in Glaciers and Ice Sheets*, (*Physical, Chemical, and Earth Sciences Research Reports*), J. Wiley, Chichester 1989.

23. R. Mücke, F. Slemr, P. Werle; Measurement of HO_2 radicals by tunable diode laser spectroscopy: A feasibility study, in: P.M. Borrell, P. Borrell, T. Cvitaš, W. Seiler (eds.), *Proc. EUROTRAC Symp. '94*, SPB Academic Publishing, The Hague 1994, pp. 915-919.

24. R. Mücke, P. Werle, F. Slemr; Development of an airborne diode laser absorption spectrometer for fast measurements of two trace gases at lower ppbv down to pptv levels, in: P.M. Borrell, P. Borrell, T. Cvitaš, W. Seiler (eds.), *Proc. EUROTRAC Symp. '92*, SPB Academic Publishing, The Hague 1993, pp. 245-249.

9.4 High Resolution Molecular Spectroscopy: Pressure Broadening and Shift of Ammonia

G. Baldacchini and F. D'Amato

Dipartimento Innovazione, Settore Fisica Applicata,
Centro Ricerche Energia di Frascati, C.P. 65, I-00044 Frascati, Rome, Italy

Summary

The lineshape of the ammonia molecule has been studied for transitions in the ν_2, $2\nu_2$ and ν_4 vibrational bands under various physical conditions by using a tunable diode laser spectrophotometer. In particular self- and foreign-gas (CO_2, N_2, air, H_2 and He) broadenings and shifts have been measured at room temperature, while self-lineshape parameters have been determined as a function of temperature from 200 to 400 K. A comparison with the ATC theory, based on the semiclassical impact approximation, gives a fairly good agreement in general, while at the same time shows the limits of the power law describing the temperature dependence.

Aims of the research

The principal goal of the JETDLAG program was to produce new knowledge and technology leading eventually to the construction of instruments based on tunable diode laser (TDL) absorption spectroscopy for measuring traces of gases (pollutants) in the atmosphere. In order to do that it was also necessary to know high resolution IR spectral data which were missing or disputed. Among them pressure broadenings and shifts of radiative transitions were (and still are) of paramount importance not only for trace gas measurements, but also because they give useful insights into the intermolecular forces and are used for modelling mixtures of gases for laser operation and planetary atmospheres.

For all the previous reasons there have been in the past combined efforts of several laboratories to obtain lineshape measurements, but with unsatisfactory results, mainly for lack of suitable tunable sources. However the situation changed in the last two decades by the availability of laser systems, notably among them TDLs, and techniques appropriate for such endeavour.

As far as the ammonia molecule is concerned, which is the main gas investigated in this line of research, at the beginning of the JETDLAG subproject in 1987 there were several measurements in the microwave and IR region of the spectrum [1]. However these measurements were not systematic, were limited mainly to room temperature, and last but not least the errors on the width and shift parameters were around 10 % or even larger, with few exceptions. Although a comparison with the ATC theory [2, 3], extended to calculate the shift as well as the width [4], gave a fairly good agreement [1], a precise quantitative comparison between theory and experiment was not possible within the still limited experimental knowledge. As a consequence it was decided to decrease the experimental errors of the measurements by stabilizing the frequency of the TDL spectrophotometer and by improving the primitive treatment of the experimental data. Moreover a systematic campaign of measurements was planned together with extensions to foreign gas effects, isotope species and temperature dependence.

Principal experimental results

In the first place while improving the experimental apparatus more measurements were performed so that it was possible to put all together 455 self-broadening, self-shift and foreign gas broadening measurements, performed in part also by other authors, for a close comparison with the theory [5]. The bands and transitions taken into consideration are shown schematically in Fig. 9.15. The conclusions of this great effort was that there were still shortcomings in the accuracy of the experimental data and in the validity of the theoretical approach which prompted us to pursue further the objectives outlined in the previous section.

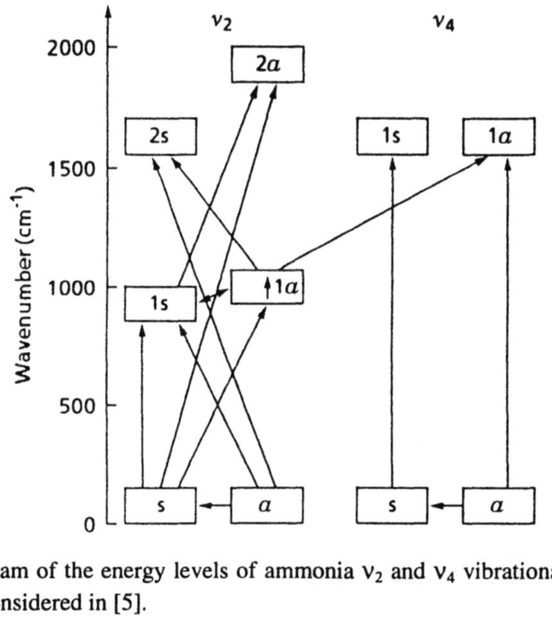

Fig. 9.15: Diagram of the energy levels of ammonia ν_2 and ν_4 vibrational bands showing the transitions considered in [5].

The frequency of the TDL system was stabilized within 20 MHz by acting solely on the electronic instruments (the lack of appropriate funds which was a constant rule throughout the whole project prevented us to build a proper stabilization feedback system). Fortunately the previous limited improvement proved to be satisfactory in the majority of the cases, especially because it was accompanied by the realization of a digital acquisition system [6] which allowed a much easier retrieval and analysis of the experimental data [7].

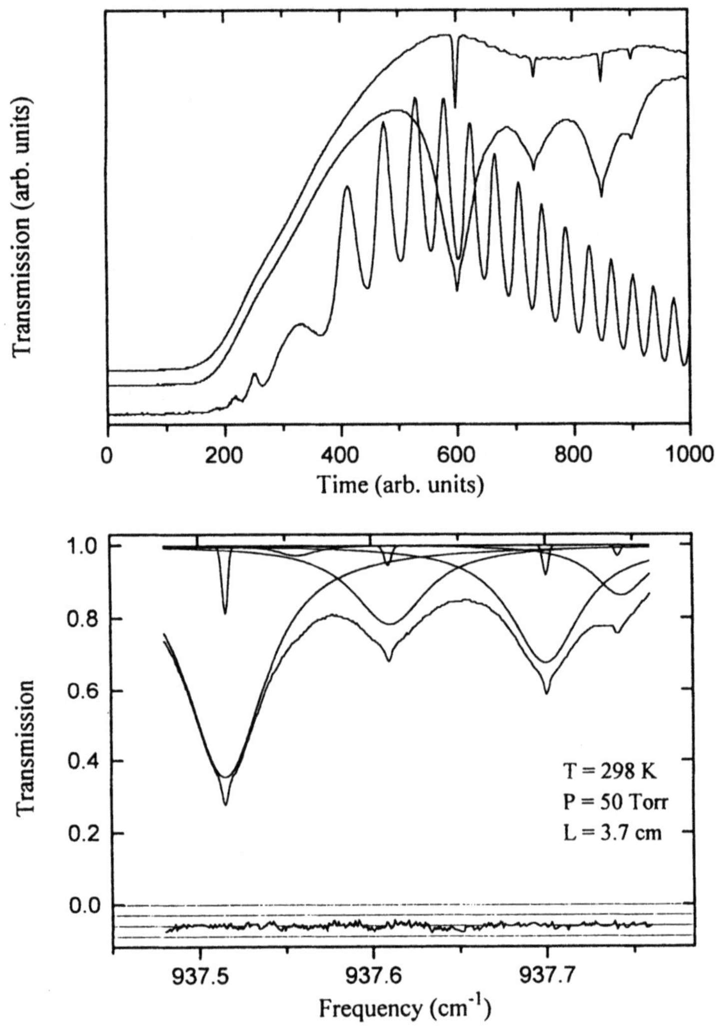

Fig. 9.16: Ammonia absorption as obtained by the TDL spectrophotometer, above, and the same after the treatment with the LINEFIT 2.0 program, below. The residual scale at the very bottom corresponds to 1 % between two lines.

The operative capability of the digital system, whose software is called LINEFIT 2.0, is displayed in Fig. 9.16 where the raw signals from the TDL system are shown above, and the final result worked out by the digital system is shown below. It refers to five transitions of ammonia, from left to right aQ(8,3), 2sQ(11,9), aQ(10,5), aQ(9,4) and aQ(7,1), which mix heavily among them especially at high pressures. The narrow peaks are the absorptions from a low pressure cell, and are used to determine with precision the value of the shifts. Anyway Fig. 9.16, below, shows the single absorption profiles which fit the total experimental absorption signal with a residual given at the bottom.

From such procedure it is possible to obtain both line widths and shifts with errors of 5 % or even less at least for the former. Such measurements have been extended below and above room temperature from 200 to 400 K by using a special cell which can be temperature regulated within 5 K [8]. By repeating the measurements at various pressures and temperatures it has been possible to derive the broadening, *a*, and the shift, *b*, parameters which are reported as a function of temperature in Figs. 9.17 and 9.18 respectively for the line aQ(9,4). As it is observed there is an overall good agreement between experiments and theory which is better for the broadening than for the shift. However this is not so for all the transitions measured especially for the shifts, most of which are very small in absolute value and negative. Moreover the power law which is commonly used to describe the temperature dependence is approximately valid for the broadenings but of difficult application for the shifts.

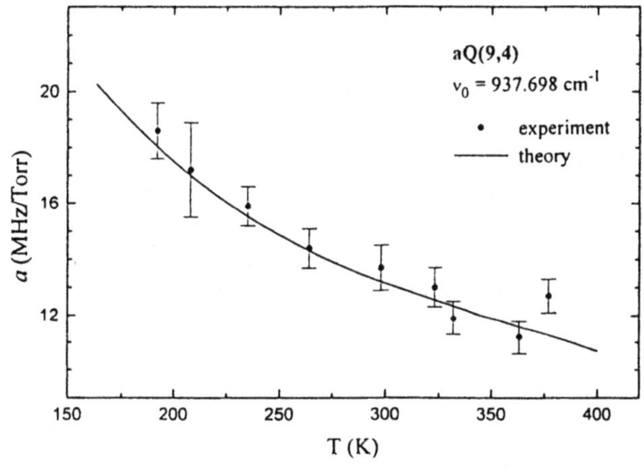

Fig. 9.17: Ammonia broadening parameter, *a*, as a function of temperature.

Achievements of the research

Because of the lacking of a substantial financial support, the results have been obtained later than planned and were not as complete as desired. For instance measurements with isotope species and new and more accurate foreign gas pressure effects have not yet been done. However some remarkable advancements have been accomplished also within these limitations.

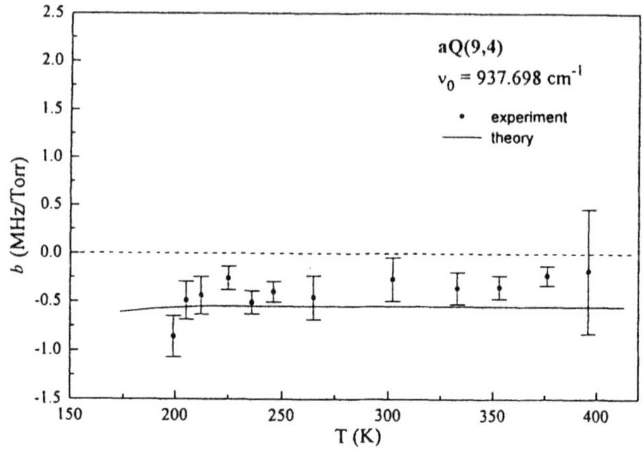

Fig. 9.18: Ammonia shift parameter, *b*, as a function of temperature.

In the field of lineshape studies our confidence in the ATC theory is much more sound now than it was at the beginning of the project. The limits of application of the theory have been well confirmed experimentally. In particular it gives good results only when the molecular dipoles or quadrupoles involved have a large value as it is the case with ammonia, carbon dioxide and nitrogen. In other words when the optical collision happens at a much greater distance than the physical collision, the ATC theory is valid. However self-broadenings are described much better than self-shifts, which are often very small, $b < 1$ MHz/Torr, and so comparable to the minimum measurable value $b \approx 0.2$ MHz/Torr. The ATC theory cannot describe well the collisions between NH_3 and molecules or atoms, which do not posess large dipoles or quadrupoles. We have also found that the broadenings of few transitions do not follow the power law, which is often utilized to describe the temperature dependence. One transition in particular, the aQ(7,1), shows such an evident discrepancy with the power law with a unique exponent, that only an average value of the exponent can describe approximately the data in the range of temperature considered in our work. In conclusion two papers with the complete description of the lineshape parameters of the transition lines shown in Fig. 9.16 are, one finished and sent for publication, and the other in preparation. Moreover experimental work is in progress to measure in detail more transitions, some of them critical with respect to the theoretical expectations.

Apart from the previous principal experimental findings, during the course of the project we have also studied the behaviour of a Laval molecular beam with N_2O and CO_2 molecular gases, the lineshape of the molecular transitions in the molecular beam and the pressure broadening of N_2O. We have also devoted some attention to the revelation of small traces of gases, NH_3, CH_4 and CO in the atmosphere. For carbon monoxide in particular we have started a collaboration with the General Physics Institute of Moscow (Russia) for the development of a simple TDL system for measuring CO gas in open air or in a multipass cell.

However such instrument can also be used to measure other gases if properly equipped.

Acknowledgments

We are indebted to Dr. G. Buffa and Dr. O. Tarrini of the University of Pisa who performed the theoretical calculations on lineshapes, and to Dr. J. Baltussen, Prof. Parimal K. Chakraborti, Dr. A. Ciucci, Dr. M. De Rosa, Dr.E. Righi and M. Sc. F. Pelagalli who contributed at different times to the good outcome of this project, while guests at various titles in our laboratory.

Many thanks are also due to A. Bellatreccia, P. Cardoni, I. Cenciarelli, E. Del Prete, L.Martignon, G. Schina and R. Tiribelli for their valuable technical assistance.

This line of research has greatly benefitted from many discussions during the frequent JETDLAG meetings and from the well managed EUROTRAC project.

Eventually it has been supported by ENEA with contracts and fellowships, by ENEA-INFM with fellowships, by the European Community for conference attendances and CEE-INTAS contracts, and by INFN for financial support.

References

1. G. Baldacchini, S. Marchetti, V. Montelatici, G. Buffa, O. Tarrini; *J. Chem. Phys.* **83** (1985) 4975 and references cited therein.
2. P.W. Anderson; *Phys. Rev.* **76** (1949) 647.
3. C.T. Tsao, I. Curnutte; *J. Quant. Spectrosc. Radiat. Transfer.* **2** (1962) 41.
4. A. Di Giacomo, O. Tarrini; *Nuovo Cimento* **B68** (1970)165.
5. G. Baldacchini, G. Buffa, O. Tarrini; *Nuovo Cimento* **D13** (1991) 719.
6. F. D'Amato, A. Ciucci; *ENEA Report* RT/INN/94/01(1994).
7. G. Baldacchini, A. Ciucci, F. D'Amato; in: P.M. Borrell, P. Borrell, T. Cvitaš, W. Seiler (eds.), *Proc. EUROTRAC Symp. '94*, SPB Academic Publishing, The Hague 1994, p. 906.
8. G. Baldacchini, J. Baltussen, A. Bellatreccia, P. Cardoni, I. Cenciarelli, F. D'Amato; *ENEA Report* RT/INN/90/37(1990).

9.5 IR-Fibre Optical Components for Trace Gas Analysis Equipment

Roland Grisar, J. Anders, M. Knothe and W.J. Riedel

Fraunhofer-Institut für Physikalische Messtechnik (IPM), Heidenhofstrasse 8, 79110 Freiburg, Germany

Summary

Investigations concentrated on infrared optical properties of zirkonium fluoride monomode fibres. As a main outcome, monomode fibres are characterized by nearly Gaussian far fields, which are essentially independent of the launching conditions. They act as spatial filters with suppression of higher laser modes, detrimental in trace gas analysis. Transmission levels for the "useful" modes are up to 60 % for fibre samples of some 10 cm. Under the same condition, optical backward reflection levels as low as 10^{-5} were realized. Direct coupling of a fibre to a diode laser as a prerequisite for pigtailing can result in some 10 % of the total power launched to the fibre. As an issue, monomode infrared fibres potentially could represent highly performing components for trace gas analyzer equipment based on infrared diode lasers.

Aims of the research

The goal of the research was essentially a characterization of zirkonium fluoride infrared monomode fibre components by lead chalcogenide diode laser radiation in view of fibre component application in trace gas analyzers. The main points of interest were optimum launching conditions, far-field behaviour, optical feedback by fibres at extremely low levels, and pigtailing of liquid-nitrogen cooled lead chalcogenide diode lasers.

Principal scientific results

The optical setup specially developed for fibre characterisation within these investigations is shown in Fig. 9.19. It allows for precise control of coupling parameters such as beam aperture, input and output angle. It further enabled investigations of the far-field behaviour as well as of optical feedback which both are decisive parameters in diode-laser based trace gas analyzers. Fibre ends were cut at 10 or 15° in order to suppress radiation from beeing fed back into the aperture of the incoming beam.

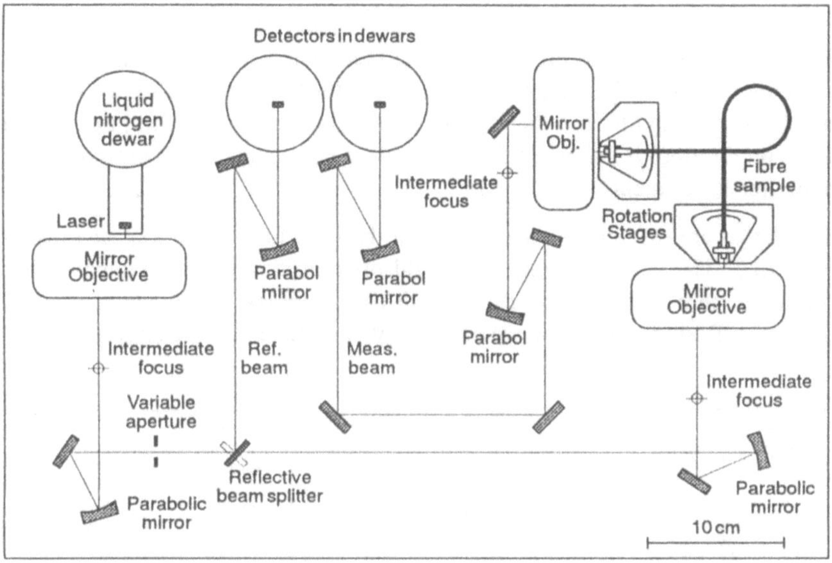

Fig. 9.19: Optical setup for infrared fibre characterisation.

Experimental results in view of optimizing launching beam parameters for a fibre sample are plotted in Fig. 9.20. The parallel beam diameter before aperture transformation and focussing onto the fibre is indicated as a parameter. The highest fibre transmission signal is obtained for a diameter of 8 mm, which corresponds approximately to a 1:4 aperture ratio. For other samples and wavelengths, optimum values between 1:3 and 1:5 were found. The optimum launching angle was found to be 13°, considerably higher than the value of 5° expected from the law of refraction. The optimum angle exhibited considerable scatter for nominally identical samples, probably due to some stress induced by epoxy-gluing the fibre to the ferrule.

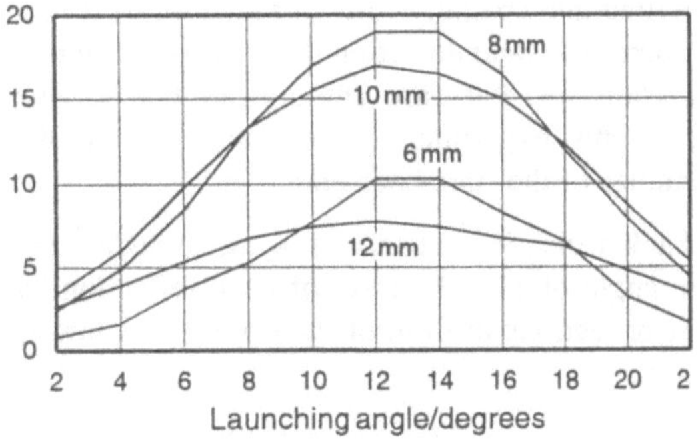

Fig. 9.20: Monomode fibre (6.5 μm core, 10° end faces) transmission at a wavelength of 4.45 μm versus launching angle with the parameter denoting the parallel beam diameter.

Under the same diode laser conditions as encountered for earlier experiments on multimode fibres, monomode type samples consistently exhibited nearly Gaussian far-field distributions as illustrated in Fig. 9.21. The full line gives the normalized measured distribution, the broken line is a Gaussian with the width fitted to the experimental values. The wings of experimental far fields extend somewhat farther than a Gaussian, apart from that the agreement is very good. The laser radiation incident on the fibre is supposed to have a rather erratic geometric emission profile from earlier experiments.

The far-fields are essentially independent of the launching conditions as illustrated in Fig. 9.22. The two normalized far-field distributions were obtained with one 11 mm core fibre (10° end faces), but with the launching angles being 0° and 6°, respectively. Both distributions are identical within the experimental error.

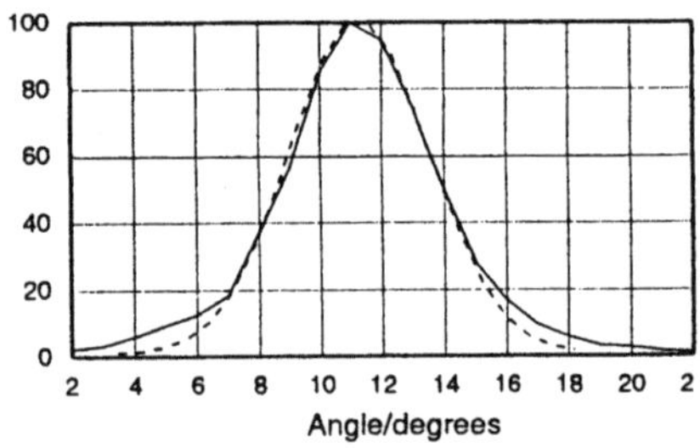

Fig. 9.21: Experimental far-field distribution (full line) of a monomode fibre (6.5 µm core, 15° end faces) at a wavelength of 4.45 µm, and Gaussian fit (broken line) to the experimental data.

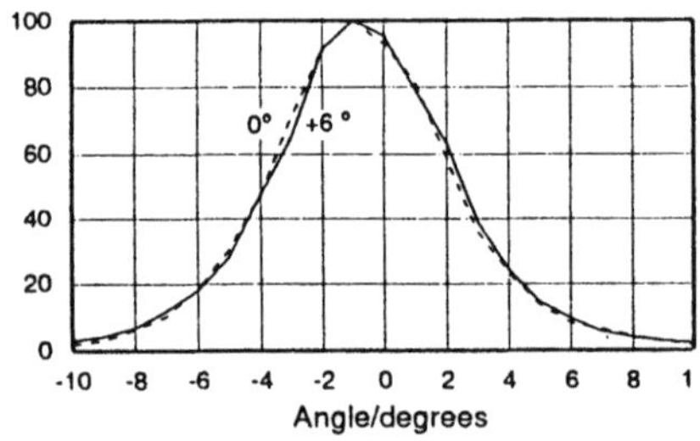

Fig. 9.22: Experimental far-field distributions of a monomode fibre (11 mm core, 10° end faces) at a wavelength of 4.45 mm and two different launching angles.

As can be seen, the monomode fibre acts as a mode stripper with efficient spatial filtering and only transmits a nearly Gaussian mode. For that reason, the signal plotted in Fig. 9.20 does not describe a genuine transmission, since part of the incoming radiation cannot couple to the fibre. A proper transmission experiment requires only an optimum mode being incident on the fibre. Experimentally, this was achieved by coupling the transmitted and spatially filtered radiation by a first passage through a fibre back to the same fibre and detecting the transmitted signal (Fig. 9.23). The maximum transmission for that case amounts to 60 %. The alignment was somewhat delicate in this experiment which may explain the occurence of two peaks at different angles of the fibre back end.

Fig. 9.24 gives transmission and back reflection signals of a monomode fibre versus launching angle. At optimum angle of 11°, the level of back reflection by the fibre sample - corrected for other reflections in front of the fibre - is as low as 10^{-5} of the incoming radiation intensity.

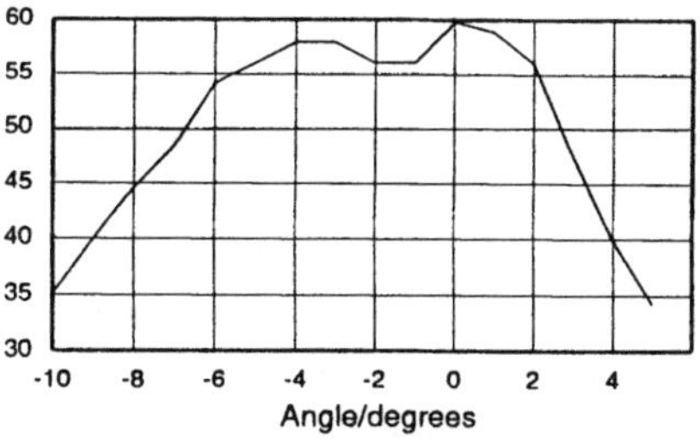

Fig. 9.23: Transmission of a monomode fibre (11 mm core, 15° end faces, 46 cm length) at 4.45 μm wavelength for spatially filtered radiation versus tilting angle of the back end face of the fibre.

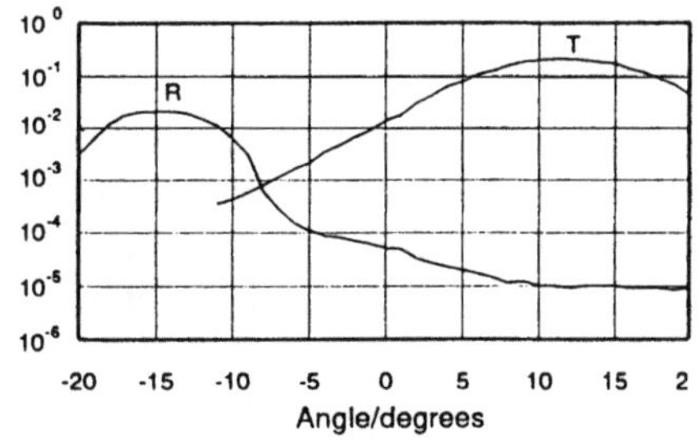

Fig. 9.24: Backward reflection *R* from and transmission *T* of a monomode fibre (11 μm core, 15° end faces, 46 cm length) at 4.45 μm wavelength versus launching angle.

In a further experiment, direct coupling of a 11 μm monomode fibre to a lead chalcogenide diode laser without further optical elements was attempted. The diode laser package was screwed to a brass block and immerged into liquid nitrogen. The monomode fibre end within a ferrule was mounted on a *xyz* translation stage and aligned to the laser emitting spot. The minimum distance possible under these circumstances was 600 to 800 μm.

The transmitted signal was comparable to that obtained under optimum coupling conditions within the standard setup, which means with a number of lossy elements in front of the fibre. The power coupled to the fibre "pigtail" is estimated to amount to 10% of the emitted diode laser power. An increase of this value by a factor of 2 to 4 in the launched power can be expected by reducing the distance between laser chip and fibre to values < 200 μm, which was not possible in the present configuration.

Conclusions

There are some direct issues for insertion of infrared monomode fibres in trace gas analyzer equipment from the present experimental results. So far, only multimode silver halide fibre components have been applied as transmission media to simplified infrared diode laser gas analysers with diode laser source and detector immerged into liquid nitrogen (1,2). The reported sensitivities were, however, far below those required in trace gas monitoring.

As a first result of the present investigations, monomode fibres with some 10 cm in length could be used as transmissive elements of high transparency within trace gas analyzers and enable novel, compact and rugged optical design. At the same time, these devices could be configured to act as highly efficient spatial filters with extremely low optical feedback levels. The resulting nearly Gaussian far-fields could substantially improve and render more reproducible the noise properties of trace gas analyzers, which are known to depend on the laser beam mode quality in a decisive manner (3,4).

A further possibility, which seems now feasible, are pigtailed lead chalcogenide diode lasers with relatively high power levels and excellent far-field properties. It is presumed, that eventual problems in connecting and aligning fibre and diode laser rigidly, and with stable alignment at low temperature, can be solved by existing techniques in a satisfactory manner. The additional profit of such devices would be simple, quick and reproducible exchange and alignment of the narrow-band radiation source to the optics of the trace gas analyzer system.

Acknowledgment

Funding by the German Federal Ministry of Research and Development (BMFT) under No. 07EU732 is gratefully acknowledged.

References

1. A.I. Kuznetsov, A.I. Nadezhdinski, E.V. Stepanov; *Proc. SPIE* **1228** (1990) 262.
2. E.V. Stepanov, A.I. Kuznetsov, K.L. Moskalenko, A.I. Nadezhdinskii; in: R. Grisar, H. Böttner, M. Tacke, G. Restelli (eds.), *Proc. 3rd Int. Symp. on Monitoring of Gaseous Pollutants by Tunable Diode Lasers*, Kluwer Academic Publishers, Dordrecht 1992, p. 203.
3. W.J. Riedel; *Proc. SPIE* **1433** (1991) 179.
4. G. Schmidtke, U. Klocke, M. Knothe, W. Kohn, W.J. Riedel, H. Wolf; in: R. Grisar, G. Schmidtke, M. Tacke, G. Restelli (eds.), *Proc. 2nd Int. Symp. on Monitoring of Gaseous Pollutants by Tunable Diode Lasers*, Kluwer Academic Publishers, Dordrecht 1989, p. 25.

9.6 TDLAS Measurements of HCl, NH_3 and HNO_3 in the Troposphere

D. J. Brassington

Atmospheric Chemistry Research Unit, Imperial College, Silwood Park, Ascot, UK

Summary

The work described in this report originated at the Central Electricity Research Laboratories (CERL) of the UK Central Electricity Generating Board. Initially the emphasis was on measurements relevant to the atmospheric transformation of fossil-fuelled power station emissions. This involved measurements of HCl downwind of a power station stack, and NH_3 measurements at Great Dun Fell during two campaigns aimed at studying cloud chemistry. With the closure of CERL the work transferred to Imperial College where NH_3 measurements have continued. The original single species instrument has now been modified to allow simultaneous measurements of NH_3 and HNO_3

Aims of the research

As a general spectroscopic technique TDLAS allows sensitive measurements of species which are otherwise difficult to measure. HCl, NH_3 and HNO_3 are all examples of species for which no entirely satisfactory alternative measurement technique exists. The aim of this project has thus been to develop TDLAS instrumentation to measure these difficult species whilst at the same time providing a capability to quickly convert the instrument to other species of interest. The main instrument described here is mounted in a lorry and can be used on the move or whilst stationed at a remote site. Work is currently in progress on developing systems for aircraft measurements.

Principal scientific results

Instrument development

The instrument was originally capable of measuring only a single species at any one time [1]. It has since been modified by the addition of an additional laser to allow two-species measurements and it is this configuration which is described here. The optical and inlet systems are shown in Fig. 9.25. One laser is housed in a closed cycle cooler capable of temperatures down to 20 K. The other laser is housed in a liquid nitrogen dewar and this laser must thus be able to operate at above 85 K. The laser outputs are sequentially directed along the optical path through the White cell by means of a stepper motor driven mirror which operates in 0.5 s so allowing rapid alternation of the two lasers. The White cell path-length is typically 64 m and the cell has a volume of 13 L giving an exchange time of 1.2 s when pumped by the 700 L/m rotary pump. This relatively high pumping speed is important in minimising adsorption effects on the inlet system and cell interior. All three species studied here show severe inlet effects of this type. To further reduce such effects the inlet system is of PTFE; a series of computer controlled valves connects the White cell inlet to various sources of zero air and calibration mixtures from permeation sources.

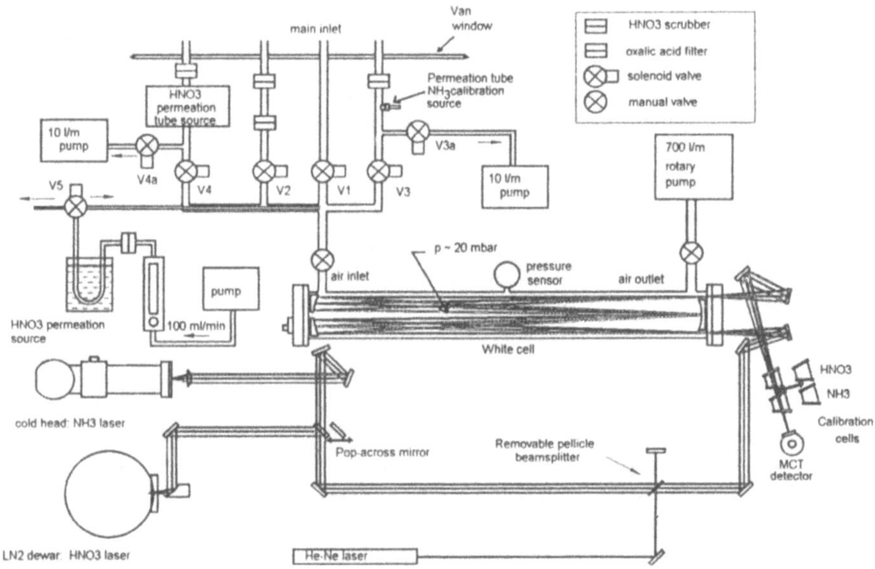

Fig. 9.25: Two species TDLAS system.

The system uses 2nd harmonic wavelength modulation spectroscopy with 50 kHz modulation, coupled with sweep integration of up to 2000 scans at a 100 Hz repetition rate. Ambient and zero-air spectra are acquired alternately with an ambient / zero air pair being acquired in about 40 s. The zero-subtracted spectrum is fitted to a reference spectrum acquired with the reference cell inserted into the beam. Absorption lines at 2775 cm^{-1} for HCl, 967 cm^{-1} for NH_3 and 1694 cm^{-1} for HNO_3 were used. Sensitivities were typically 100 ppt for HCl, 80 ppt for NH_3 and 250 ppt for HNO_3.

In the original instrument configuration part of the beam was split off and directed through a line-lock cell to a second detector, with a PSD monitoring the 1f signal to provide continuous analogue line-locking by feedback to the laser controller. Improvements in laser temperature and current stabilisation have now made this unnecessary and in the current system wavelength stabilisation is achieved in software by monitoring the signal from the reference cell about every 30 s and applying a correction if needed. Even this is sometimes unnecessary and we have on occasion operated for over four hours with no feedback, relying entirely on the inherent stability of the laser controller.

HCl *measurements*

HCl is emitted from coal fired power stations because of the presence of Cl in the coal (typically up to 1% depending on the source of the coal). When monitoring the power station plume a few km from the stack HCl levels are of order a few tens of ppb. This relatively high concentration allowed us to operate the TDLAS in a continuous monitoring mode in which the laser wavelength was not ramped over the line but fixed at line centre. The 2nd harmonic signal is then directly proportional to the HCl concentration and provides a continuous measurement with a time constant limited only by the exchange time (1.2 s) of the White cell. Fig. 9.26 shows an example of such a measurement 3 km downwind of the stack [2]. SO_2 (a product of the sulphur in the coal) was also monitored using a commercial Meloy analyser. The good correlation between the SO_2 and HCl concentrations is expected since both originate from the same source and the variations are due to turbulence in the plume and variations in wind direction., which affect both equally. It can be seen that the variations in the HCl concentration lead those in SO_2 by about 15 s. This is because the HCl inlet was on the upwind side of the van and the SO_2 inlet on the downwind side. The behaviour reversed when the van was subsequently turned round. After 3.5 min the van was driven out of the plume and the concentrations of SO_2 and HCl decrease as expected. The ratio of HCl to SO_2 agrees with that predicted from the chlorine and sulphur content of the coal, assuming complete conversion to HCl and SO_2.

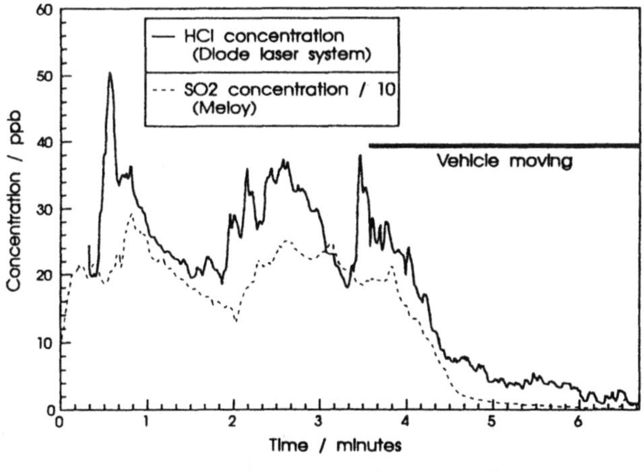

Fig. 9.26: TDLAS measurements of HCl in the plume of a coal fired power station.

NH_3 *measurements*

The summit ridge of Great Dun Fell in Westmoreland provides a useful site for cloud chemistry experiments since it lies approximately at right angles to the prevailing SW winds. When cloud forms on the summit it is thus possible to make measurements upwind of, downwind of, and within the cloud. The TDLAS system has been used to make NH_3 measurements during two such measurement campaigns [3][(3)] . Fig. 9.27 shows measurements from an upwind site on the fell-side at a height of 430 m. For these measurements the weather was damp with occasional rain leading to low NH_3 levels. In dryer conditions levels of up to 15 ppb were seen at this site. The anti-correlation between the NH_3 concentration and external temperature seen after 17:00 is believed to be due to variable mixing of two airstreams, one with high NH_3 and low temperature and the other with lower NH_3 and higher temperature.

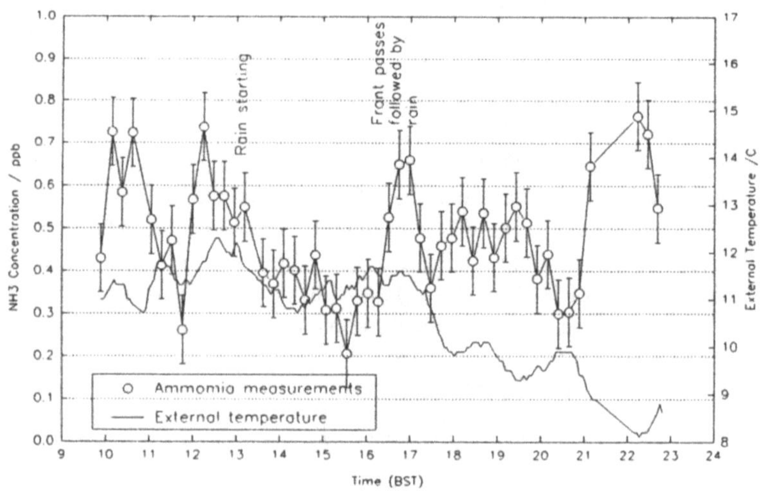

Fig. 9.27: Measurements of NH_3 at Great Dun Fell.

During this campaign NH_3 was measured at other sites by denuder tubes, which needed sampling periods of several hours for one measurement. The ability of the TDLAS instrument to provide time-resolved measurements represents a considerable advance on such alternative methods.

HNO_3 *measurements*

In its current configuration the instrument is capable of quasi-simultaneous measurements of HNO_3 and NH_3. Figs. 9.28 and 9.29 show the data processing for a pair of such measurements. The slope of the regression plot is directly proportional to species concentration. The NH_3 spectrum shows etalon fringes but they are of sufficiently short period that they have little effect on the measurement. The HNO_3 spectrum is dominated by laser noise.

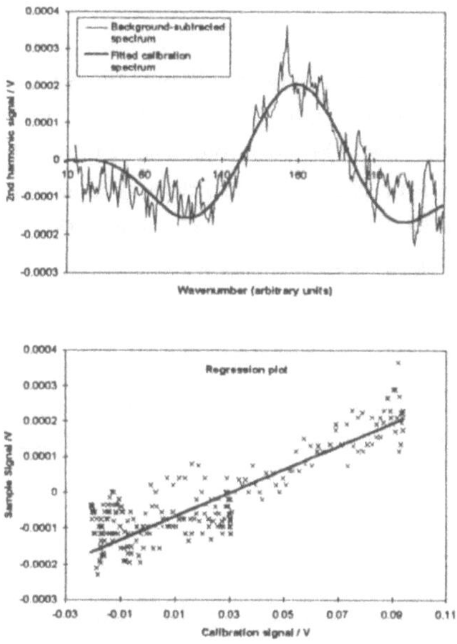

Fig. 9.28: Measurement of 2.6 ppb HNO_3. 18 Aug. 1995.

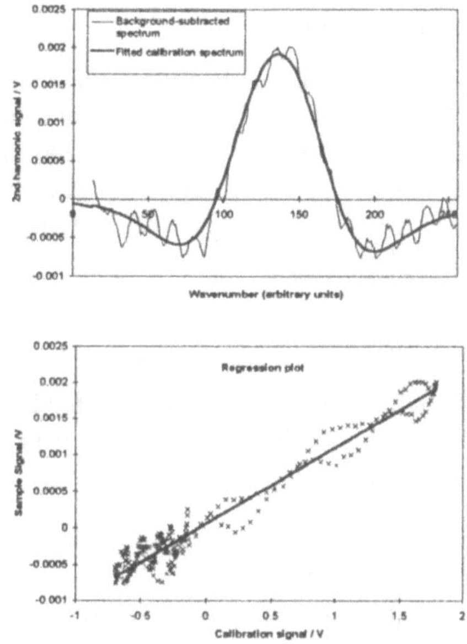

Fig. 9.29: Measurement of 1.9 ppb NH_3. 8 Sept. 1995.

Fig. 9.30 shows measurements of HNO_3 made at Silwood Park, a rural site 40 km SW of London. Concurrent measurements of NO, NO_x and O_3, using commercial chemiluminescent instruments indicate that the high HNO_3 levels on 18 August correspond to polluted air in the last stages of photochemical processing.

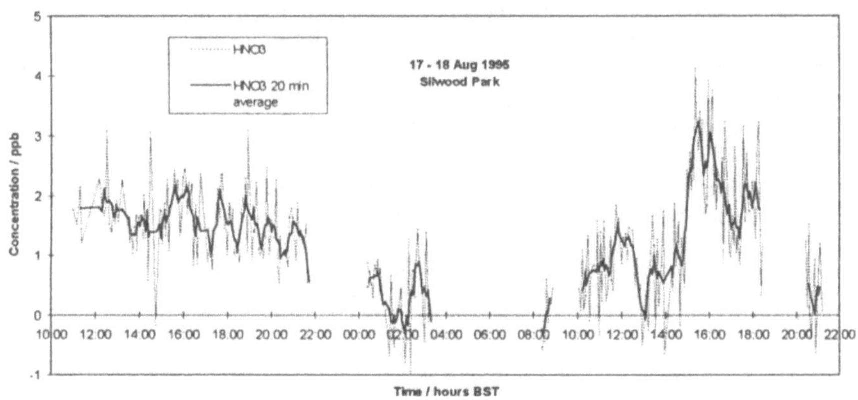

Fig. 9.30: HNO_3 measurements at Silwood Park

NH_3 and HNO_3 in the atmosphere would be expected to be in equilibrium with ammonium nitrate aerosol, provided the concentration product is above a certain minimum value which depends on temperature and humidity. The ability to measure NH_3 and HNO_3 simultaneously will enable this equilibrium to be studied.

Future instruments

Work is in progress in fitting a single channel instrument (built by Unisearch of Canada) to the C130 aircraft of the Met. Research Flight. This will be capable of NO_2 or HCHO measurements and should also be suitable for future flux measurements of CH_4 and N_2O. A compact four channel instrument is also being developed (jointly with the National Physical Laboratory) for use both for tropospheric measurements and on future stratospheric platforms.

Conclusions

TDLAS instrumentation has provided valuable real-time measurements of a range of trace atmospheric species for which no satisfactory alternative measurement techniques exist.

Acknowledgements

The author has benefited considerably from discussions and collaboration with the other members of the JETDLAG community.

The initial stages of this work were performed at the Central Electricity Research Laboratory and supported by the Central Electricity Generating Board. Later work at Imperial College has been funded by the Natural Environment Research Council.

References

1. D.J. Brassington; Tunable diode laser absorption spectroscopy for the measurement of atmospheric species, in: R.J.H. Clark, R.E. Hester (eds.), *Spectroscopy in Environmental Science*, J. Wiley, New York 1995.
2. D.J. Brassington; in: R.Grisar, G. Schmidke, M.Tacke and G.Restelli (eds.), *Monitoring of Gaseous Pollutants by Tunable Diode Lasers, Proc. Int. Symp.*, Freiburg 1988, pp. 16-24
3. D.J. Brassington; in: *EUROTRAC Annual Report '90*, Part 7, JETDLAG, EUROTRAC ISS, Garmisch-Partenkirchen 1991, pp. 25-28.

9.7 Development of an Airborne Tunable Diode Laser Spectrometer for Flux Measurements by the Eddy Correlation Technique

Hélène Mac Leod[1], H. Poncet[1], G. Ancellet[1], I. Carrasco[1], O. Lubin[1], G. Mégie[1], F. Huard[2] and W. Riedel[3]

[1] Service d'Aéronomie du C.N.R.S., U.P.M.C.,75252 Paris cedex 05, France
[2] Société BERTIN, BP 22000, 13791 Aix-en-Provence cedex 3, France
[3] Fraunhofer Institute für Physilalische Meβtechnik, D-7800 Freiburg, Germany

Summary

Ozone and its precursors carbon monoxide and methane play an important role in tropospheric chemistry. Small scale exchange processes between the different layers of the troposphere and, in particular, between the planetary boundary layer and the free troposphere need to be estimated in order to understand the tropospheric ozone budget on a regional scale (mid-latitude industrial regions, tropical areas, *etc.*). The Service d'Aéronomie has developed an airborne tunable diode laser absorption spectrometer during the course of EUROTRAC. This instrument is designed to measure simultaneously the tropospheric concentrations of O_3, CO and CH_4 and to provide fluxes measurements of these gases by eddy correlation technique. The characteristics of the instrument are a short response time (≤ 1 s), a high sensitivity (≤ 1 ppb) and a large dynamical measurement range of concentrations, corresponding to absorptions between 4×10^{-4} and 9×10^{-2}.

The instrumental development was conducted by the Société Bertin under the scientific direction of the Service d'Aéronomie (1991-1994). The diode lasers and associated electronics were provided by the Fraunhofer Institute of Freiburg. A detailed conception of the instrument was conducted in 1991. A model version of the instrument was developed in 1992 for single gas detection. This laboratory instrument was designed for two purposes : to serve as a permanent optical set-up for testing and characterizing infra-red diode lasers; to validate the experimental procedure by measuring standard concentrations of CO and O_3 in short absorption cells (1993-1994). During the same period of time (1994), the airborne instrument was constructed by the Société Bertin and implemented aboard the ARAT. The three first qualifying flights were performed in February 1995 above northern France.

Aims of the research

The major objective of the JETDLAG subproject was to promote the technological development of TDLAS (Tunable Diode Laser Absorption Spectroscopy) for atmospheric chemistry research and air pollution monitoring. This subproject followed four directions : development of ground-based TDLAS instruments, development of special purpose instruments, development of new components and techniques and high - resolution IR-spectroscopy. The Service d'Aéronomie has developed a "special purpose instrument" for fluxes measurements of trace gases by eddy correlation technique. The aim of the research is to study the detailed processes of ozone formation, destruction and transport in the troposphere by estimating the fluxes of ozone and its precursors between tropospheric layers. The TDLAS system was designed to measure simultaneously the tropospheric concentrations of O_3, CO and CH_4 with a sampling frequency superior or equal to 1 Hz and a sensitivity in the order of 0.5 ppb, 1 ppb and 10 ppb for O_3, CO and CH_4 respectively. These minimum concentrations to be detected correspond to one per cent of the average relative concentrations of the three gases in the troposphere. The detection of concentration fluctuations of small magnitude, correlated with vertical wind speed fluctuations, provide fluxes measurements by eddy correlation technique [1]. The JETDLAG instrument developed to reach the above scientific goals is flown aboard a Fokker 27 (ARAT - Avion de Recherche Atmosphérique et de Télédetection) fully equiped for the measurements of atmospheric parameters (temperature, pressure, wind speed and direction, humidity, *etc.*).

Principal scientific results

Laboratory measurements of tropospheric concentrations of carbon monoxide and ozone

Carbon monoxide, methane and ozone are monitored at 4.6 μm, 7.5 μm and 9.5 μm, respectively, using the 2165.6 cm^{-1} or 2169.2 cm^{-1} absorption line for CO, the 1322.0 cm^{-1} or 1327.1 cm^{-1} line for CH_4 and the 1052.8 cm^{-1} line for O_3 [2]. These choices were discussed with W. Riedel (Fraunhofer Institute) with respect to the availability of leadsalt diode lasers tunable in these precise spectral regions. The model version of the JETDLAG instrument (Fig. 9.31) has been designed for successive measurements of O_3, CO and CH_4 in short absorption cells (25 cm) in order to optimize the different parts of the airborne instrument, to quantify the influence of the different noises (optical and electronic) and to validate the data analysis procedure for the restitution of gas concentrations. This instrument will also serve as a permanent set-up for testing and characterizing diode lasers [3]. The detection method used is Wavelength Modulation (WM) at 10 kHz with detection of the signal at its second harmonic (20 kHz). The sampling frequency is 1.25 kHz (125 data points per ramp). Both raw and averaged spectra can be recorded, a sampling frequency of 10 Hz is obtained for a raw spectrum and of 1 Hz for and average of 10 spectra. Two CO diode lasers and three O_3 diode lasers (MUTEK) have been tested and characterized : mode diagram charts, laser power, wavelength and power stability, laser linewidth, laser tuning rate. The two best lasers were selected with respect to single mode behaviour and high emission power.

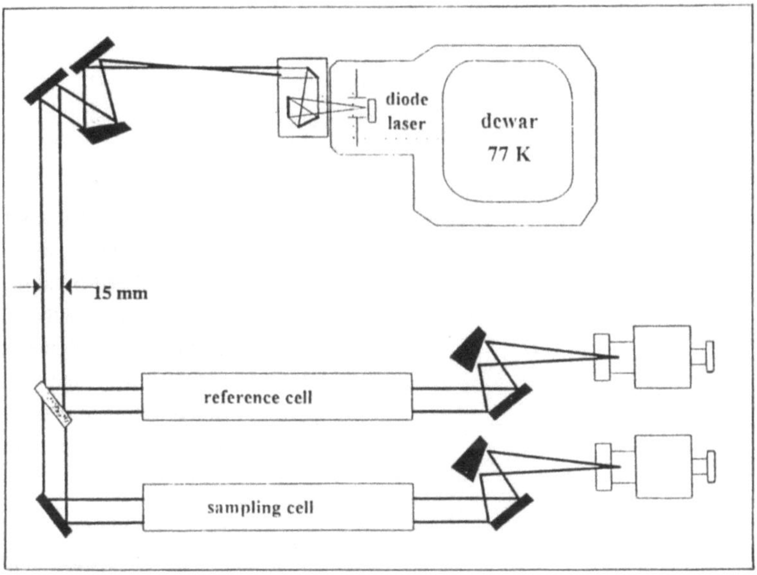

Fig. 9.31: Model version of the JETDLAG TDLAS system

Laboratory measurements of CO concentrations

Standard samples of CO diluted in N_2 (ALPHAGAZ : 52 ppm, 5.9 ppm, and 513 ppb) were chosen in order to produce absorptions covering the dynamical range corresponding to tropospheric concentrations (O_3 : 20-500 ppb; CO : 50-300 ppb; CH_4 : 1.5-1.8 ppm).

A computer program was developed in order to restitute the gas concentration from the second harmonic signal [4]. The algorithm is an inversion procedure based on least square fitting of a simulated second harmonic signal to the experimental signal, assuming a voigt profile.

CO measurements were performed at 20 hPa using the 2169.2 cm^{-1} absorption line. Second harmonic (S2) spectra for the three different CO concentrations are shown in Figs. 9.32a, 9.32b, 9.32c together with the corresponding simulated signals. Table 9.3 presents the restituted CO concentration as well as the error limits on this restitution.

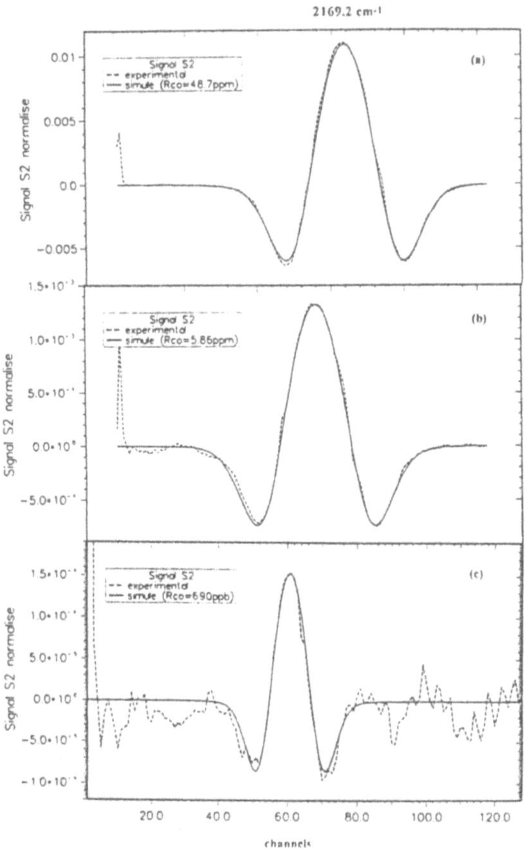

Fig. 9.32: Experimental second harmonic signal (S2) and simulated fitted signal corresponding to the CO absorption line centered at 2169.2 cm^{-1}, obtained at 20 hPa of nitrogen pressure, with a 1 s averaging time (10 spectra).

(a) R_{CO}=52 ppm
- ramp amplitude = 1 mA (6×10^{-2} cm^{-1}).

(b) : R_{CO}=5.9 ppm
- ramp amplitude = 1 mA (6×10^{-2} cm^{-1}).

(c) : R_{CO}=513 ppb
- ramp amplitude = 2 mA (12×10^{-2} cm^{-1}).

Table 9.3: Comparison between standard relative concentrations of CO/N_2 mixtures and restituted concentrations at 20 hPa

R_{CO} / ppm (l = 25 cm)	R'_{CO} / ppb (l = 50 m)	Absorption	$C_{standard}$ / molecule cm^{-3}	$C_{restituted}$ / molecule cm^{-3}
49.8 ± 0.1	249	5.4×10^{-2}	$(2.42 \pm 0.04) \times 10^{13}$	$(2.37 \pm 0.04) \times 10^{13}$
5.90 ± 0.12	29.5	6.1×10^{-3}	$(2.87 \pm 0.04) \times 10^{12}$	$(2.85 \pm 0.02) \times 10^{12}$
0.513 ± 0.051	2.6	5.3×10^{-4}	$(2.50 \pm 0.25) \times 10^{11}$	$(3.3 \pm 0.8) \times 10^{11}$

Laboratory measurements of O_3 *concentrations*

A stainless-steel absorption cell, previously passivated with pure ozone was filled with samples of O_3 diluted in Argon. The O_3 concentration was not precisely known because of slow decomposition on the cell walls.

O_3 measurements were performed at 20 hPa using the 1060.88 cm^{-1} absorption line. An example of the second harmonic spectrum obtained for a relative concentration of ~ 50 ppm is shown in Fig. 9.33 together with the corresponding simulated signals. The simulated signal corresponds to a restituted concentration of:

$R_{O3} \approx 52$ ppm for an optical pathlength of 25 cm

(corresponding to $R_{O3} \approx 260$ ppb for an optical pathlength of 50 m)

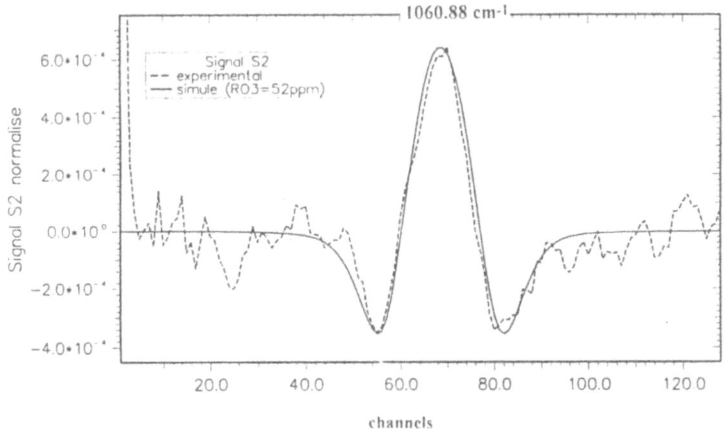

Fig. 9.33: Experimental second harmonic signal (S2) and simulated fitted signal corresponding to the O_3 absorption line centered at 1060.88 cm^{-1} , obtained at 16.8 hPa of argon pressure, with a 1 s averaging time (10 spectra). $R_{O3} \approx 52$ ppm - ramp amplitude = 3 mA (4×10^{-2} cm^{-1}).

Measurements of carbon monoxide and methane concentrations in the troposphere aboard the ARAT

The different parts (emission, absorption, detection, electronics) of the airborne instrument (Fig. 9.34) were designed according to the performances of the laboratory spectrometer. The three diode lasers are contained in a single dewar with three independent temperature control systems. The laser beams are optically multiplexed before entering a multipass White cell and unmultiplexed at the exit using dichroic plates. The beams are focused onto 3 detectors (InSb for CO and HgCdTe for O_3 and CH_4). The output signals are preamplified and fed into a data acquisition/storage system driven by a personal computer. The temperature of the lasers are controlled between 80 and 130 K. Reference cells are filled with concentrated gases for precisely tuning the diode lasers to the chosen gas absorption lines. Tropospheric air is pumped through the multipass cell using a ~16 l/s rotary pump. The pressure inside the multipass cell is controlled between 20 and 50 hPa independently of outside pressure. The cell volume is ~8 l insuring full renewal of the air sample at least once per second. The absorption pathlength is 50 m corresponding to 80 passes. The sensitivity of the instrument, for this 50 m pathlength, is well in the expected range : 0.5 ppb, 1 ppb and 10 ppb for O_3, CO and CH_4 respectively. The instrument was delivered to the Service d'Aéronomie in October 1994. It was tested aboard the ARAT on ground in November and three qualifying flights were performed in February 1995 above northern France, from Creil (100 km north of Paris). Examples of CO and CH_4 spectra obtained during these flights are presented below.

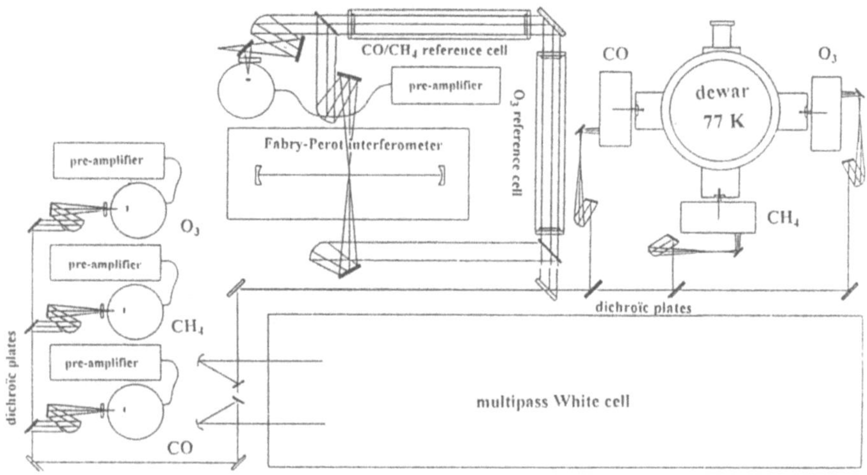

Fig. 9.34: Airborne JETDLAG TDLAS system

Flight of February 14, 1995

The aim of this first flight was to assess the general behaviour of the instrument in clear weather at two different altitudes (2000 m and 4000 m) for investigating two different levels of the CO and CH_4 tropospheric profiles.

Flight of February 16, 1995

The plan of this flight was to follow a lower altitude flight-track (1000 m), then to climb in spiral for exploring the CO and CH_4 profiles as a function of altitude. This flight was interrupted by a failure of the liquid nitrogen cooling system of the CO diode laser.

Flight of February 17, 1995

The aim of this flight was to detect O_3 by flying at 600 m above a polluted area, i.e. the oil reffineries near Le Havre. The sensitivity of the instrument was not sufficient for detecting O_3 using the 1066.88 cm^{-1} absorption line. This line is too

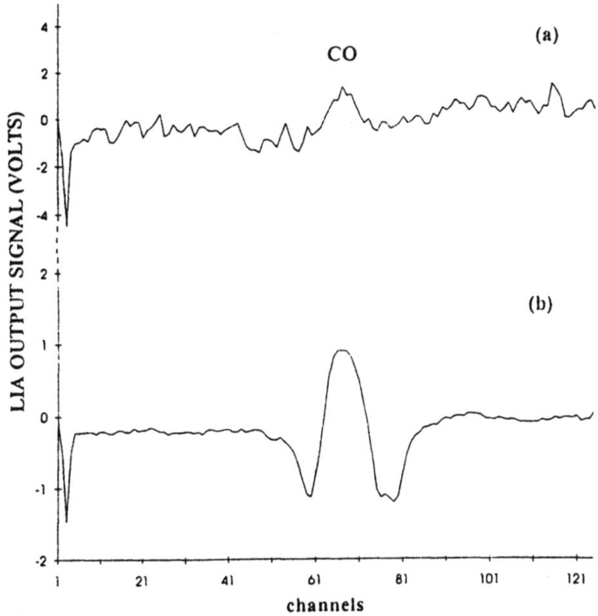

Fig. 9.35: Example of second harmonic signal (S2) obtained for CO (2169.2 cm^{-1}) during the flight of February 17, 1995, obtained at 50 hPa White cell pressure, without averaging (scanning period = 100 ms). (a) CO measured S2 signal ; optical pathlength : 50 m; (b) CO reference S2 signal ; optical pathlength : 20 cm ; relative concentration : 49.8 ppm.

close to a water vapor line and its strength is ~ 10 times weaker than the one of the selected 1052.8 cm^{-1} line. Unfortunately none of the available ozone diode lasers emitted at this wavenumber with sufficient power in a single mode. It will be necessary to obtain a new ozone diode laser from the Fraunhofer Institute of Freiburg for the next flight campaign.

During this flight, profiles of CO and CH_4 were continuously recorded. An example of the spectra obtained for atmospheric CO and CH_4 (multipass White cell) and for CO and CH_4 samples in the reference cell are shown in Figs. 9.35 and 9.36.

Conclusions

During the course of EUROTRAC, the Service d'Aéronomie and the Société BERTIN have developed an airborne tunable diode laser absorption spectrometer for simultaneous measurements of ozone, carbon monoxide and methane with a sampling frequency superior or equal to 1 Hz and a sensitivity in the 1 ppb range. Both the experiments with the model version of the instrument and the experiments conducted during the qualifying fligths have shown that the

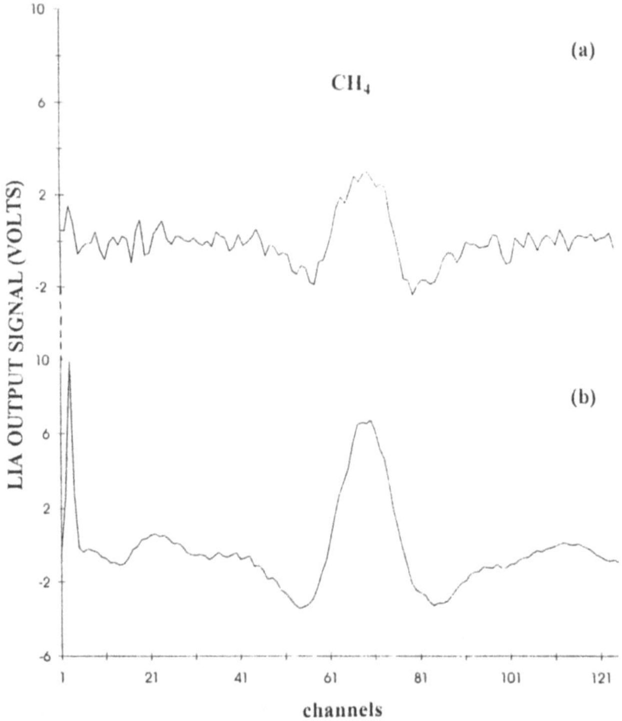

Fig. 9.36: Example of second harmonic signal (S2) obtained for CH_4 (1322.0 cm^{-1}) during the flight of February 14, 1995, obtained at 50 hPa White cell pressure, without averaging (scanning period = 100 ms). (a) CH_4 measured S2 signal ; optical pathlength : 50 m; (b) CH_4 reference S2 signal ; optical pathlength : 20 cm ; relative concentration : 200 ppm.

expected performances were met and allow fluxes measurements of trace gases by eddy correlation technique. However, due to the delivery of the airborne instrument in late 1994, the second flight campaign aiming at validating the fluxes measurements will only take place in October 1995.

Acknowledgments

This work has been supported by the French "Ministère de l'Enseignement Supérieur et de la Recherche" through the "Secrétariat Français d'EUREKA".

References

1. J.A. Ritter, D.H. Lenschow, J.D.W. Barrick, G.L. Gregory, G.W. Sachse, G.F. Hill, M.A. Woerner; Airborne flux measurements and budget estimates of trace species over the Amazon basin during the GTE/ABLE 2B expedition, *J. Geophys. Res.* **95** (1990) 16875.

2. I. Carrasco; Etude et réalisation d'un spectromètre à diodes laser pour la mesure de constituants minoritaires dans la troposphère, *Thèse de doctorat,* Université Pierre-et-Marie-Curie, Paris 1994.

3. H. Mac Leod, Poncet H., Carrasco I., Mégie G., Huard F., Riedel W., Development of an Airborne Tunable Diode Laser Spectrometer for Fluxes Measurements of CO, O_3 and CH_4 by Eddy Correlation Technique, in: P.M. Borrell, P. Borrell, T. Cvitaš, W. Seiler (eds.), *Proc. EUROTRAC Symp. '94*, SPB Academic Publishing, The Hague 1994, p. 925.

4. O. Lubin, Spectrométrie par diodes laser accordables appliquée à la mesure de constituants atmosphériques minoritaires, *Diplome d'Etudes Approfondies* Université Pierre-et-Marie-Curie, Paris 1994.

Chapter 10

JETDLAG Publications 1989–1995

1989

Baldacchini, G., G. Buffa, O. Tarrini; Diode laser application for measurements of foreign-gas broadening of ν_2 band in NH_3, in: R. Grisar, G. Schmidtke, M. Tacke, G. Restelli (eds), *Monitoring of Gaseous Pollutants by Tunable Diode Lasers*, Kluwer Academic Publ., Dordrecht 1989, pp. 240-245.

Baldacchini, G., A. Bellatreccia, L. Nencini; Applicationi e prospettive della spettroscopia molecolare con diodi laser accordabili, *Report ENEA* RT/TIB/89/29, 1989.

Baldacchini, G., A. Bizzarri, L. Nencini, G. Buffa, O. Tarrini; New measurements of self broadening and shift of ammonia lines, *J. Quantitative Spectros. Radiative Transfer* **42** (1989) 423-428.

Bicanic, D., F. Harren, E. Woltering, J. Reuss; Trace detection in agriculture and biology, in: P. Hess (ed), *Photoacoustic Photothermal and Photochemical Processes in Gases, Topics Curr. Phys.* **46**, Chapter VII, Springer Verlag, Berlin 1989, p. 213.

Bicanic, D., H. Jalink, H. Sauren, F. Harren, J. Reuss, B. Zuidberg, E. Woltering; Photoacoustic investigation of pesticides in: R. Grisar, G. Schmidtke, M. Tacke, G. Restelli (eds), *Monitoring of Gaseous Pollutants by Tunable Diode Lasers*, Kluwer Academic Publ., Dordrecht, 1989, pp. 280-288.

Bicanic, D., H. Jalink, H. Sauren, K. van Asselt, A. Miklos, F. Harren, J. Reuss, B. Zuidberg; Recent developments in photoacoustic sensing of the atmospheric pollution by infrared lasers, in: L. Braser, W. Mulders (eds), *Man and His Ecosystem*, Volume 3, Elsevier Sci. Publ., Amsterdam 1989, pp. 676.

Buffa, G., O. Tarrini; Ammonia spectrum as a test for two different pressure shift theories *Appl. Opt.* **28** (1989) 1800.

Harris, G.W., J.P. Burrows, D. Klemp, T. Zenker; Measurements of trace species in the remote maritime boundary layer using TDLAS in: R. Grisar, G. Schmidtke, M. Tacke, G. Restelli (eds), *Monitoring of Gaseous Pollutants by Tunable Diode Lasers*, Kluwer Academic Publ., Dordrecht 1989, pp. 68-74.

Haurie, Y., C. Pagny, F. Cappellani, and G. Restelli; Formulae, algorithms, procedures, and program for application of infrared spectroscopic measurements, *Eur-Report* 12527 EN, CEC, Brussels 1989.

Hjorth, J., F. Cappellani, C.J. Nielsen, and G. Restelli; Determination of the $NO_3 + NO_2 \rightarrow NO + O_2 + NO_2$ rate constant by infrared diode laser and Fourier transform spectroscopy, *J. Phys. Chem.* **93** (1989) 5458.

Jalink, H., D. Bicanic; The concept, design and the use of the photoacoustic heat-pipe cell, *Appl. Phys. Lett.* **35** (1989) 1507.

Jalink, H., D. Bicanic; A novel wide temperature range photoacoustic cell for molecules in the vapour phase, in: R. Grisar, G. Schmidtke, M. Tacke, G. Restelli (eds), *Monitoring of Gaseous Pollutants by Tunable Diode Lasers*, Kluwer Academic Publ., Dordrecht 1989, pp. 274-279.

Sauren, H., D. Bicanic, H. Jalink, J. Reuss; High sensitivity, interference free Stark tuned carbon dioxide laser photoacoustic sensing of urban ammonia, *J. Appl. Phys.* **66** (1989) 5087.

Sauren, H., L. van Hove, D. Bicanic, H. Tonk, H. Jalink, J. Quist; On the absorption properties of ammonia to various surfaces, in: R. Grisar, G. Schmidtke, M. Tacke, G. Restelli (eds), *Monitoring of Gaseous Pollutants by Tunable Diode Lasers*, Kluwer Academic Publ., Dordrecht 1989, pp. 196-200.

Van Roozendael, M., M. Herman; Optoacoustic experiments in the gas phase, *Opt. Commun.* **72** (1989) 71-75.

Van Roozendael, M., M. Herman; Mechanism of the response of the $N_2O_4 = 2\ NO_2$ chemical equilibrium to an optical perturbation, *Chem. Phys.* **139** (1989) 391-400.

Werle, P., F. Slemr; Entwicklung eines frequenz-modulierten, quantenbegrenzten Absorption-spektrometers mit Bleisalz-Dioden-Laser zur Messung atmosphärischer Spurengase, *Abschlußbericht zum BMFT-Vorhaben* Nr. 07431095, April 1989.

Werle, P., F. Slemr, M. Gehrtz, C. Bräuchle; Quantenbegrenzte FM-Spektroskopie mit Bleisalz Diodenlasern, in: *Verhandlungen der Deutschen Physikalischen Gesellschaft*, Essen 1989, p. 50.

Werle, P., F. Slemr, M. Gehrtz, C. Bräuchle; Wide-band noise characteristics of a lead salt diode laser: possibility of quantum limited TDLAS performance, *Appl. Opt.* **28** (1989) 1638.

Werle, P., F. Slemr, M. Gehrtz, C. Bräuchle; Quantum limited FM-spectroscopy with a lead salt diode laser, *Appl. Phys.* B **49** (1989) 99.

1990

Baldacchini, G., A. Bizzari, L. Nencini, V. Sorge, G. Buffa, O. Tarrini; Pressure broadening and shift of ammonia transition lines in the ν_2 vibrational bands, *J. Quant. Spectrosc. Radiat. Transfer* **43** (1990) 371-380.

Baldacchini, G., J. Baltussen; Linefit, a digital acquisition and fitting program for molecular spectroscopy with tunable diode lasers, *Report ENEA* RT/TIB/90.47, 1990.

Baldacchini, G., J. Baltussen, A. Bellatreccia, P. Cardoni, I. Cenciarelli, F. D'Amato; Realizzazione di una cella per misure di spettroscopia molecolare a temperatura variabile, *Report ENEA* RT/INN/90.52, 1990.

Bicanic, D., H. Jalink, B. van Veldhuyzen; The compact resonant heat-pipe photoacoustic cell used for high temperature studies and as a gas chromatographic detector, *Measurement Sci. Technol.* **1** (1990) 247-249.

Fischer, H.; Dynamische Eigenschaften von Bleisalz-Diodenlasern, *IPM-Research Report* No. 21, 1990.

Fusina, L., G. Baldacchini; The spectroscopic parameters of the $\nu_2 = 2$ state for $^{14}NH_3$ and $^{15}NH_3$, *J. Mol. Spectrosc.* **141** (1990) 23-28.

Guilmot, J.M., M. Carleer, M. Godefroid, M. Herman; The ν_1 fundamental band of *trans*-HNO_2, *J. Mol. Spectrosc.* **143** (1990) 81-90.

Harren, F., J. Reuss, E. Woltering, D. Bicanic; Laser photoacoustic investigation of gases of agricultural interest, *Appl. Spectrosc.* **44** (1990) 1360-1368.

Harris, G.W., D. Klemp, T. Zenker, J.P. Burrows, B. Mattieu, P. Jacob; Polarstern 1988: Measurements of trace gases using tunable diode lasers and intercomparisons with other methods, in: G. Restelli, G. Angeletti (eds), *Proc. 5th Europ. Sympos. on Physico-chemical Behaviour of Atmospheric Pollutants*, Kluwer Academic Publ., Dordrecht 1990, pp. 644-650.

Jalink, H., D. Bicanic; Resonant photoacoustic heat-pipe for spectral investigation of low volatility liquids and solids, in: J. Murphy, L. Aamodt, B. Royce (eds), *Photoacoustic and Photothermal Phenomena II, Springer Series in Optical Sciences* **62**, Springer Verlag, Berlin 1990, pp. 530-532.

Kleiner, I., M. Godefroid, M. Herman, A.R.W. Mc Kellar; The fundamental torsion band in acetaldehyde, *J. Mol. Spectrosc.* **142** (1990) 238-253.

Schlereth, K.-H., H. Böttner, M. Tacke; "Mushroom" double channel double-heterostructure lead chalcogenide lasers made by chemical etching, *Appl. Phys. Lett.* **56** (1990) 2169-2171.

Van Roozendael, M., M. Herman; Optoacoustic study of the $N_2O_4 = 2\ NO_2$ chemical equilibrium, *Chem. Phys. Lett.* **166** (1990) 233-239.

1991

Angeli, G., A. Miklos, A. Solyom, A. Lorincz, A. Thony, M. Sigrist; Automatic frequency tracking system for resonant photoacoustic cell, *Rev. Sci. Instrum.* **62** (1991) 810-813.

Baldacchini, G., A. Bellatreccia, P.K. Chacraborti, F. D'Amato, E. Righi; Rivelazione di trace di metano nell'atmosfera per scopi medici, *Report ENEA* RT/INN/91.18, 1991.

Baldacchini, G., F. D'Amato, E. Righi, F. Cappellani, G. Restelli; Applicazioni di spettroscopia molecolare in fase gassosa con diodi laser accordabili nell'infrarosso, *Il Nuovo Saggiatore* **7** (1991) 5-16.

Baldacchini, G., G. Buffa, O Tarrini.; A review of experiments and theory for collisional line shape effects in the rotovibrational ammonia spectrum, *Il Nuovo Cimento* **13**D (1991) 719-733.

Bicanic, D., H. Sauren, H. Jalink, M. Chirtoc, A. Miklós, K. van Asselt; Progress towards the development and application of laser based photothermal devices for environmental analysis, in: P. Borrell, P.M. Borrell, W. Seiler (eds.), *Proc. EUROTRAC Symp. '90*, SPB Academic Publishing bv, The Hague 1991, pp. 347-350.

Bone, S.A., P.B. Davies, S.A. Johnson, N.A. Martin; Diode laser spectroscopy with optoacoustic detection, in: P. Borrell, P.M. Borrell, W. Seiler (eds.), *Proc. EUROTRAC Symp. '90*, SPB Academic Publishing bv, The Hague 1991, pp. 351-354.

Bone, S.A., P.B. Davies, S.A. Johnson, N.A. Martin; Diode laser spectroscopy with optoacoustic detection, in: P. Borrell, P.M. Borrell, W. Seiler (eds.), *Proc. EUROTRAC Symp. '90*, SPB Academic Publishing bv, The Hague 1991, pp. 351-354.

Chirtoc, M., D. Bicanic, V. Tosa; Versatile IPPE method and instrument for real time observation of water vapour condensation in the atmosphere, *Rev. Sci. Instrum.* **62** (1991) 2257-2261.

Fischer, H., H. Wolf, B. Halford, M. Tacke; Low-frequency amplitude-noise characteristics of lead-salt diode lasers fabricated by molecular-beam-epitaxy, *Infrared Phys.* **31** (1991) 381-385.

Fischer, H., M. Tacke; High-frequency intensity noise of lead-salt diode lasers, *J. Opt. Soc. Amer.* **88** (1991) 1824-1830.

Herman, M., T.R. Huet, Y. Kabbadj, J. Vander Auwera; *l*-type resonance in C_2H_2, *Mol. Phys.* **72** (1991) 75-88.

Herman, M., Y. Kabbadj, M. Carleer, J. Vander Auwera, J.M. Guilmot, T. Huet, A. Vandaele, F.Vanhorenbeke; High resolution IR-spectroscopy, in: P. Borrell, P.M. Borrell, W. Seiler (eds.), *Proc. EUROTRAC Symp. '90*, SPB Academic Publishing bv, The Hague 1991, p. 355.

Huet, T.R., M. Herman, J.W.C. Johns; The bending vibrational levels in C_2D_2, *J. Chem. Phys.* **94** (1991) 3407-3414.

Josek, K., P. Werle, F. Slemr, C. Bräuchle; Stabilitätsuntersuchungen an Frequenzmodulierten Diodenlaser-Spektrometern, in: *Verhandlungen der Deutschen Physikalischen Gesellschaft*, Freiburg 1991, p. 850.

Kabbadj, Y., M. Herman, L. Fusina, G. Di Lonardo, J.W.C. Johns; The bending vibrations in C_2H_2, *J. Mol. Spectrosc.* **150** (1991) 535-565.

Klann, R., R. Buhleier, Th. Elsaesser,, A. Lambrecht; Picosecond optical nonlinearities in lead chalcogenide semiconductors, *Appl. Phys. Lett.* **59** (1991) 885.

Kleiner, I., J.M. Guilmot, M. Carleer, M. Herman; The v_4 fundamental bands of *trans*- and *cis*-HNO_2, *J. Mol. Spectrosc.* **149** (1991) 391-397.

Lambrecht, A., Herres, N., Spanger, B., Kuhn, S., Böttner, H., Evers, J., Tacke, M.; Molecular beam epitaxy of $Pb_{1-x}Sr_xSe$ for the use in IR-devices, *J. Cryst. Growth* **108** (1991) 301-308.

Miklos, A., H. Sauren, D. Bicanic; An experimental methodology for characterizing the responsivity of the photoacoustic cell for gases at reduced pressure by means of the vibrating strip as the calibrating sound source, *Measurement Sci. Technol.* **2** (1991) 957-962.

Mücke, R., P. Werle, F. Slemr; Entwicklung eines flugzeuggetragenen Dioden-Laser-Absorptionsspektrometers für schnelle Messung von zwei Spurengasen im unteren ppbv- und oberen pptv-Bereich, *Abschlußbericht des Vorhabens* 07EU702, Garmisch-Partenkirchen, 1991.

Mücke, R., P. Werle, F. Slemr and W. Prettl, "Comparison of time and frequency multiplexing techniques in multicomponent FM spectroscopy, *Proc. SPIE* **1433** (1991) 136-144.

Mücke, R., P. Werle, F. Slemr, W. Prettl; Mehrkomponenten-Diodenlaser-Spektrometer für die Spurengasanalytik: Vergleich von Zeit- und Frequenzmultiplex-Techniken, in: *Verhandlungen der Deutschen Physikalischen Gesellschaft*, Freiburg 1991, p. 851.

Notholt, J., F.Cappellani, H. Roesdahl, G. Restelli; Spectral Data of Formic Acid: Absolute Band Intensities and Air Broadening Coefficient, in: P. Borrell, P.M. Borrell, W. Seiler (eds.), *Proc. EUROTRAC Symp. '90*, SPB Academic Publishing bv, The Hague 1991, pp. 357-359.

Riedel, W.J.; Optics for tunable diode lasers, *Proc. SPIE Measurement of Atmospheric Gases*, **1422** (1991) 179-189.

Sauren, H., D. Bicanic, K. van Asselt; Photoacoustic detection of ammonia at the sum and difference sidebands of the modulating laser and Stark electric fields, *Infrared Phys.* **31** (1991) 475-484.

Sauren, H., T. Regts, K. van Asselt, D Bicanic.; Simplifying laser photoacoustic trace detection of ammonia by effective suppression of water vapour and of carbon dioxide as the major absorbing atmospheric constituents, *Environ. Technol.* **12** (1991) 719-724.

Schlereth, K.H., H. Böttner, A. Lambrecht, U. Schießl, R. Grisar, M. Tacke; Buried waveguide and mesa-DH-PbEuSe-lasers grown by MBE, in: P. Borrell, P.M. Borrell, W. Seiler (eds.), *Proc. EUROTRAC Symp. '90*, SPB Academic Publishing bv, The Hague 1991, pp. 361-363.

Slemr, F.; Report on JETDLAG, in: P. Borrell, P.M. Borrell, W. Seiler (eds.), *Proc. EUROTRAC Symp. '90*, SPB Academic Publishing bv, The Hague 1991, p. 345.

Springholz, G., A. Holzinger, H. Clemens, G. Bauer, H. Böttner, M. Maier, P. Norton; Eu-interdiffusion in $Pb_{1-x}Eu_xSe$/PbSe multiquantum well structures, *J. Cryst. Growth* **113** (1991) 593-598.

Stocker, W., S.N. Magonov, H.-J. Cantow, H. Böttner, S. Schelb, M. Meihofer, M. Tacke; Scanning tunneling microscopy of lead selenide monocrystals and epitaxial layers on barium fluoride, *Fresenius J. Anal. Chem.* **341** (1991) 184-188.

Strauss, E., J.P. Favier, D. Bicanic, C. van Asselt, M. Lubbers; Sensitive colorimetric analysis of ammonium ion in water by laser photothermal detection, *The Analyst,* **116** (1991) 75-78.

Vander Auwera, J., J.W.C. Johns, O.L. Polyansky; The far infrared spectrum of C_3O_2, *J. Chem. Phys.* **95** (1991) 2299-2316.

Werle, P., F. Slemr; High frequency modulation spectroscopy, in: P. Borrell, P.M. Borrell, W. Seiler (eds.), *Proc. EUROTRAC Symp. '90*, SPB Academic Publishing bv, The Hague 1991, p. 365.

Werle, P., F. Slemr; Signal-to-noise ratio analysis in laser absorption spectroscopy using optical multipass cells, *Appl. Opt.* **30** (1991) 430-434.

Werle, P., K. Josek and F. Slemr; Application of FM spectroscopy in atmospheric trace gas monitoring: A study of some factors influencing the instrument design, *Proc. SPIE* **1433** (1991) 128-135.

1992

Baev, V.M.; Dynamik von Vielmoden-Lasern, in: *Verhandlungen der Deutschen Physikalischen Gesellschaft*, Hannover 1992, p. 1394.

Baev, V.M., J. Eschner, M. Paeth, R. Schüler, P.E. Toschek; Intracavity spectroscopy with diode lasers, *Appl. Phys.* B **55** (1992) 463-477.

Baev, V.M., J.H. Eschner, J. Sierks, A. Weiler, P.E. Toschek; Dynamics of multi-mode dye lasers, in: *Proc. XVIII Int. Quantum Electronics Conference (IQEC '92)*, Vienna 1992, WeA1.

Baev, V.M., P.E. Toschek; Sensitivity limits of intracavity laser spectroscopy, in: *Proc. Int. Symp. on Environmental Sensing*, Berlin 1992, p. 31 (paper 1715-74).

Baldacchini, G., J. Baltussen; Pressure broadening and line shifts in the spectrum of ammonia, *EUROTRAC Newslett.* **9** (1992) 23-25.

Baldacchini, G., P.K. Chakraborti, F. D'Amato; Infrared diode laser absorption features of N_2O and CO_2 in a laval nozzle, *Appl Phys.* B **55** (1992) 92-101.

Baldacchini, G., P.K. Chakraborti, F. D'Amato; Lineshape in a laval molecular beam and pressure broadening of N_2O transition lines, *Report ENEA* RT/INN/91/56, 1992.

Bicanic, D., H. Jalink, M. Chirtoc, H. Sauren, M. Lubbers, J. Quist, E. Gerkema, K. van Asselt, A. Miklos, A. Solyom, G. Angeli, P. Helander, H. Vargas; Interfacing photoacoustic and photothermal techniques for new hyphenated methodologies and instrumentation suitable for agricultural, biological, and medical applications, in: D. Bicanic (ed), *Photoacoustic and Photothermal Phenomena III, Springer Series in Optical Sciences* **69** (1992) 20-28.

Bicanic, D., M. Franco, H. Sauren; The electric resonance tuning and intermodulation for selective and sensitive studies of small polar molecules, in: *Proc. SPIE Euroopto Int. Symp. on Environmental Monitoring*, Berlin 1992.

Bicanic, D., A. Miklos, H. Sauren; Laser photoacoustic and thermo-optical sensing of the atmosphere: From the laboratory tool towards a field instrument, *EUROTRAC Newsletter,* **9** (1992) 8-13.

Bicanic, D., P. Torfs, M. Lubbers, A. Tam; Horticultural sensing by photoacoustics and thermal lensing, *Acta Horticulturae* **304** (1992) 28-42.

Eschner, J.H., R. Böhm, V.M. Baev, P.E. Toschek; Intra-cavity spectroscopy with diode lasers, in: *Proc. XVIII Int. Quantum Electronics Conference (IQEC '92)*, Vienna 1992, FrO2.

Grisar, R., J. Anders, M. Knothe, W.J. Riedel; Application of infrared fibers in diode laser trace gas analysis, in: *Proc. Conf. Infrared Fiber Optics III*, Boston 1992, SPIE vol. 1591, pp. 201-205.

Harris, G.W., D. Klemp, T. Zenker, J.P. Burrows, B. Mattieu; Measurements of trace gases using tunable diode lasers during the 1988 Polarstern cruise and intercomparisons with other methods, *J. Atmos. Chem.* **15** (1992) 315-326.

Harris, G.W., D. Klemp, T. Zenker; An upper limit on the HCl near-surface mixing ratio at 28° N, 30° W measured using TDLAS, *J. Atmos. Chem.* **15** (1992) 327-332.

Hasan, K.H., J.M. Hollas, F. Vanhorenbeke, M. Herman; Hybrid character of the 0-0 band of the S_1-S_0 absorption spectrum of 1,2,4-trifluorobenzene by rotational band contour analysis, *J. Mol. Spectrosc.* **151** (1992) 522-528.

Herman, M.; Chemistry at the microscopic level with high resolution FTIR, in: E.F. Vansant (ed), *Proc. Second Int. Workshop on FTIR*, Antwerp 1992, pp. 83-100.

Herman, M., J. Vander Auwera; Spectroscopic analysis of the atmosphere, *Chimie Nouvelle* **10** (1992) 1139-1145.

Herrmann, K.H., K.-P. Möllmann, J.W. Tomm, H. Böttner, A. Lambrecht, M. Tacke; Some band structure related optical and photoelectrical properties of $Pb_{1-x}Eu_xSe$ ($0 < x < 0.2$), *J. Appl. Phys.* **72** (1992) 1399.

Holland, J.K., M. Carleer, R. Pétrisse, M. Herman; Supersonic molecular jet Fourier transform interferometry applied to nitric oxides, *Chem. Phys. Lett.* **194** (1992) 175-180.

Josek, K., P. Werle, F. Slemr, C. Bräuchle; Untersuchungen zur Empfindlichkeitssteigerung an Frequenzmodulierten Diodenlaser-Spektrometern (FM-TDLAS), in: *Verhandlungen der Deutschen Physikalischen Gesellschaft*, Hannover 1992, p. 1409.

Mélen, F., F. Pokorni, M. Herman; Vibrational band analysis of N_2O_4, *Chem. Phys. Lett.* **194** (1992) 181-186.

Mélen, F., M. Herman; Vibrational bands of $H_xN_yO_z$ molecules, *J. Phys. Chem. Ref. Data* **21** (1992) 831-881.

Mélen, F., M. Herman; Fourier transform jet spectrum of the m_9 band of N_2O_4, *Chem. Phys. Lett.* **199** (1992) 124-130.

Mürtz, M., M. Schaefer, M. Schneider, J.S. Wells, W. Urban, U. Schießl, M. Tacke; Stabilization of 3.3 and 5.1 µm lead-salt diode lasers by optical feedback, *Opt. Commun.* **94** (1992) 551.

Sauren, H., D. Bicanic; Novel characterization techniques for agricultural and environmental applications: Photoacoustic detection of ammonia in a simulated atmosphere of varying water content, *Anal. Instrum.* **20** (1992) 63-79.

Seemann, J., R. Böhm, J.H. Eschner, V.M. Baev, P.E. Toschek; Halbleiterdioden-Ringlaser, in: *Verhandlungen der Deutschen Physikalischen Gesellschaft*, Hannover 1992, p. 1407.

Solyom, A., G. Angeli, A. Miklos, D. Bicanic; Calibration of a windowless photoacoustic cell for detection of trace gases, *Anal. Chem.* **64** (1992) 155-158.

Solyom, A., G. Angeli, D. Bicanic, M. Lubbers; Determination of ammonia - CO_2 photoacoustic spectroscopy compared to conventional spectrophotometry, *The Analyst* **117** (1992) 379-382.

Werle, P., R. Mücke, F. Slemr; Development of a prototype IR-FM absorption spectrometer: design criteria and system performance, in: R. Grisar et al. (eds), *Monitoring of Gaseous Pollutants by Tunable Diode Lasers*, Kluwer Academic Publ., Dordrecht 1992, pp. 169-182.

Wietzorrek, J., P. Werle, R. Mücke, F. Slemr, C. Bräuchle; Untersuchung der Moden- und Strahlstruktur von Bleisalz-Laserdioden, in: *Verhandlungen der Deutschen Physikalischen Gesellschaft*, Hannover 1992, p. 1408.

1993

Baldacchini, G., A. Bellatreccia, F. D'Amato; Assorbimento di Ammoniaca in Celle per Spettroscopia Molecolare, *Report ENEA* RT/INN/93/07, 1993.

Baldacchini, G., P.K. Chakraborti, F. D'Amato; On the spectral linewidth and lineshape of ammonia in free jets, *Report ENEA* RT/INN/92/47, 1993.

Baldacchini, G., P.K. Chakraborti, F. D'Amato; Lineshape in a laval molecular beam and pressure broadening of N_2O transition lines, *J. Quant. Spectrosc. Radiat. Transfer* **49** (1993) 439-477.

Baldacchini, G., P.K. Chakraborti, F. D'Amato; On the spectral linewidth and lineshape of ammonia in free jets, in: R. Stamm, B. Talin (eds), *Spectral Line Shapes*, Volume 7, Nova Science Publishers, Inc. 1993, pp. 455-456.

Bicanic, D., M. Chirtoc, M. Lubbers, H. Jalink; A photoacoustic detector of the total carbon content in soil water solutions, *Measurement Sci. Technol.* **4** (1993) 1016-1020.

Bicanic, D., M. Chirtoc, K. van Asselt, H. Jalink, K. Haupt, P. Torfs, H. Sauren, E. Gerkema; New trends and perspectives of photoacoustic and photothermal spectroscopies in agricultural and ervironmental sciences, *Acta Chim. Slovenica* **40** (1993) 175-202.

Bicanic, D., M. Franko, H. Sauren; On the potentiality of intermodulated electric resonance, photoacoustic and photothermal spectroscopies for ground based in situ tracking of polar molecules in polluted air, in: H. Schiff, U. Platt (eds.) *Proc. SPIE Conf. 1715,* Bellingham, Washington 1993, pp. 222-234.

Bone, S.A., P.G. Cummins, P.B. Davies, S.A. Johnson; Measurement of water vapour pressure and activity using infrared diode laser absorption spectroscopy, *Appl. Spectrosc.* **47** (1993) 834.

Carrasco, I., G. Ancellet, P. Chazette, H. Mac Leod, G. Mégie, F. Huard, P. J. Crepin; Tunable diode laser spectrometer for flux measurements by eddy correlation technique, in: P.M. Borrell, P. Borrell, T. Cvitaš, W. Seiler (eds.), *Proc. EUROTRAC Symp. '92*, SPB Academic Publishing bv, The Hague 1993, pp. 732-735.

Favier, J., A. Miklos, D. Bicanic; New and versatile cell for studies of powdered specimens across broad spectral range; *Acta Chim. Slovenica* **40** (1993) 115-122.

Josek, K., R. Mücke, F. Slemr, P. Werle, J. Wietzorrek, P. M. Borrell, P. Borrell; A feasibility study of the direct measurement of tropospheric HO_2 radicals by tunable diode laser spectroscopy, in: P.M. Borrell, P. Borrell, T. Cvitaš, W. Seiler (eds.), *Proc. EUROTRAC Symp. '92*, SPB Academic Publishing bv, The Hague 1993, pp. 240-244.

Klann, R., T. Hoefer, R. Buhleier, T. Elsässer, A. Lambrecht; Ultrafast recombination processes in lead chalcogenide semiconductors studied via picosecond optical nonlinearities, *Semicond. Sci. Technol.* **8** (1993) 305-308.

Lambrecht, A., H. Böttner, M. Agne, R. Kurbel, A. Fach, B. Halford, U. Schiessl, M. Tacke; Molecular beam epitaxy of laterally structured lead chalcogenides for the fabrication of buried heterostructure lasers, *Semicond. Sci. Technol.* **8** (1993) 334-336.

Lambrecht, A., R. Kurbel, M. Agne; Shadow mask MBE for the fabrication of lead chalcogenide buried heterostructure lasers, *Materials Sci. Eng.* **B 21** (1993) 217-223.

Mélen, F., M. Carleer, J. M. Guilmont, J. K. Holland, D. Hurtmans, J. Vander Auwera, M. Herman; High resolution laboratory FTIR spectroscopy, in: P.M. Borrell, P. Borrell, T. Cvitaš, W. Seiler (eds.), *Proc. EUROTRAC Symp. '92*, SPB Academic Publishing bv, The Hague 1993, p. 418.

Miklos, A., D. Bicanic, J. de Bont; On the CO laser photoacoustic absorption spectrum of isoprene - a reactive biogenic forest emittant, *Infrared Phys.* **34** (1993) 635-640.

Mücke, R., P. Werle, F. Slemr; Development of an airborne diode laser absorption spectrometer for fast measurements of two trace gases at lower ppbv down to pptv levels, in: P.M. Borrell, P. Borrell, T. Cvitaš, W. Seiler (eds.), *Proc. EUROTRAC Symp. '92*, SPB Academic Publishing bv, The Hague 1993, pp. 245-249.

Renard, E.; Mid-infrared optical investigations on PbEuSe eptaxial layers, University of Karlsruhe and IPM Freiburg, 1993

Sauren, H., E. Gerkema, D. Bicanic, H. Jalink; Real time and in situ determination of ammonia concentrations in the atmosphere by means of intermodulated stark resonant CO_2 laser photoacoustic spectroscopy, *Atmos. Spectrosc.* **27** (1993) 114-120.

Scheumann, B., Ch. Schrötter, P. Werle; Rauschverhalten von Bleisalz Lasern in Abhangigkeit von der Modenstruktur, in: *Verhandlungen der Deutschen Physikalischen Gesellschaft*, Berlin 1993.

Tacke, M.; Mid IR diode lasers for diode laser trace gas analysis, in: J. Peeters (ed), *Chemical Mechanisms Describing Tropospheric Processes, CEC Air Pollution Research Report* **45**, Guyot, Bruxelles 1993.

Werle, P., R. Mücke, F. Slemr; The limits of signal averaging in atmospheric trace gas monitoring by tunable diode laser absorption spectroscopy, *Appl. Phys.* **B57** (1993) 131-139.

Xu, J., B. Halford, M. Tacke; The electrical characteristics of Pb_{1-x},Eu_x,Se homojunctions, *Semicond. Sci. Technol.* **8** (1993) 360-363.

1994

Agne, M., A. Lambrecht, U. Schießl, M. Tacke; Guided modes and far-field patterns of lead chalcogenide buried heterostructure laser diodes, *Infrared Phys. Technol.* **35** (1994) 47-58.

Baldacchini, G., A. Bellatreccia, A. Ciucci, F. D'Amato, A.I. Nadezhdinskii, N. Lemechov, N. Sobolev; Misura della concentrazione del CO_2 nell'atmosfera con diodi laser sintonizzabili, in: *Atti del 3. Convegno Nazionale "Intrumentazione e Metodi di Misura Elettroottici"*, Maggio, Italy 1994, pp. 196-200.

Baldacchini, G., A. Ciucci, F. D'Amato, G. Buffa, O. Tarrini; Temperature dependence of pressure broadening in the spectrum of ammonia, *EUROTRAC Newsletter* **13** (1994) 22-25.

Baldacchini, G., A. Ciucci, F. D'Amato; Ammonia lineshape and LINEFIT 2.0 program, in: P.M. Borrell, P. Borrell, T. Cvitaš, W. Seiler (eds.), *Proc. EUROTRAC Symp. '94*, SPB Academic Publishing bv, The Hague 1994, pp. 906-909.

Bicanic, D., A. Solyom, G. Angeli, H. Wegh, M. Postumus, H. Jalink; On the extent of unwanted infrared photoacoustic signals from polymer sampling tubings exposed to ultraviolet radiation, *Infrared Phys. and Technol.* **35** (1994) 637-633.

Bozoki,Z., A. Miklos, D. Bicanic; Photothermal elastic matrix, *Appl. Phys. Lett.* **64** (1994) 1362-1366.

Fach, M.A., H. Böttner, K.-H. Schlereth, M. Tacke; Embossed-grating lead chalcogenide buried wave-guide distributed-feedback lasers, *IEEE J. Quantum Electron.* **30**(2) (1994).

Favier, J.P., D. Bicanic, C. van Asselt, A. van Wijk; Organic compounds measured with infrared (3.39 μm) photoacoustics, *J. de Physique* **IV C7** (1994) 495-497.

Groot, T., D. Bicanic, M. van Montfort, P. Torfs, R. Braun, C. van Asselt; The analysis of error in trace gas concentration measurements by intermodulated photoacoustic stark spectroscopy (IMPASS), *J. de Physique* **IV C7** (1994) 519-522.

Herrmann, K.H., J.W. Tomm, S. Haertel, W. Hoerstel,, K.-P. Möllmann, W. Heimbrodt, M. Tacke, H. Böttner; Luminescence of a quantum-confined dense electron-hole plasma from PbSe nanostructures, *J. Luminesc.* **60&61** (1994) 371-375.

Herrmann, K.H., M. Happ, K.-P. Möllmann, F. Peuker, J.W. Tomm, H. Böttner, M. Tacke; Evidence of hybridization from a direct comparison of semimagnetic and nonsemimagnetic ternary lead chalcogenides, *Japan J. Appl. Phys.* **33** (1994) 1928-1930.

Mac Leod, H., I. Carrasco, H. Poncet, G. Mégie, F. Huard, W. Riedel; Development of an airborne tunable diode laser spectrometer for flux measurements of CO, O_3 and CH_4 by the eddy correlation technique, in: P.M. Borrell, P. Borrell, T. Cvitaš, W. Seiler (eds.), *Proc. EUROTRAC Symp. '94*, SPB Academic Publishing bv, The Hague 1994, pp. 925-929.

Mücke, R., B. Scheumann, F. Slemr, P. Werle; Calibration procedures for tunable diode laser spectrometers, in: H.I. Schiff, A. Fried, D. Killinger (eds), *Tunable Diode Laser Spectroscopy, Lidar, and DIAL Techniques, Proc. SPIE* **2112** (1994) 87-98.

Mücke, R., F. Slemr, P. Werle; Grenzen der Signalmittelung in der Diodenlaserspektroskopie, in: *Verhandlungen der Deutschen Physikalischen Gesellschaft*, Hamburg 1994, p. 712.

Mücke, R., F. Slemr, P. Werle; Weiterentwicklung eines frequenz-modulierten Absorptionsspektrometers mit Dioden-Lasern und sein Einsatz zur Messung der HO_2-Radikale in der Luft, *Abschlußbericht des BMFT-Vorhabens* 07EU712, Wissenschafts-Verlag Dr. Wigbert Maraun, Frankfurt am Main / *Schriftenreihe des Fraunhofer-Instituts für Atmosphärische Umweltforschung*, Garmisch-Partenkirchen, Band 32 (1994).

Mücke, R., F. Slemr, P. Werle; Measurement of atmospheric HO_2 radicals by tunable diode laser absorption spectroscopy: A feasibility study, in: P.M. Borrell, P. Borrell, T. Cvitaš, W. Seiler (eds.), *Proc. EUROTRAC Symp. '94*, SPB Academic Publishing bv, The Hague 1994, pp. 915-919.

Mücke, R., J. Dietrich, B.Scheumann, F. Slemr, J. Slemr, P. Werle; Application of tunable diode laser spectroscopy for QA/QC studies: A validation for formaldehyde measurement techniques, in: P.M. Borrell, P. Borrell, T. Cvitaš, W. Seiler (eds.), *Proc. EUROTRAC Symp. '94*, SPB Academic Publishing bv, The Hague 1994, pp. 910-914.

Shi, Z., A. Lambrecht, M. Tacke; Determination of band-edge offset by weak field hall measurement on MBE Pb/Se/PbEuSe multi-quantum well structures on KCl, *Solid-State Electronics* **37** (1994) 1113-1116.

Tomm, J.W., K.-P. Möllmann, F. Peuker, K.H. Herrmann, H. Böttner, M. Tacke; Valence band hybridizing in europium-alloyed lead selenide, *Semicond. Sci. Technol.* **9** (1994) 1033-1041.

Werle, P.; Signal processing strategies for sub-ppbv trace gas analysis by tunable diode laser spectroscopy, in: P.M. Borrell, P. Borrell, T. Cvitaš, W. Seiler (eds.), *Proc. EUROTRAC Symp. '94*, SPB Academic Publishing bv, The Hague 1994, pp. 920-924.

Werle, P.; Atmospheric trace gas monitoring using high frequency modulation spectroscopy with semiconductor lasers, in: A.I. Nadezhdinskii, Y.N. Ponomarev, L. N. Sinitsa (eds.), *11th Symp. and School on High-Resolution Molecular Spectroscopy, Proc. SPIE* **2205** (1994) 83-94.

Werle, P.; High frequency modulation spectroscopy : A sensitive detection technique for atmospheric pollutants, in: G. Harding, R.C. Lanza, L.J. Myers, P.A. Young (eds.), *Substance Detection Systems, Proc. SPIE* **2092** (1994) 4-15 .

Werle, P.; Signal processing strategies for tunable diode laser spectroscopy, in: H.I. Schiff, A. Fried, D. Killinger (eds.), *Tunable Diode Laser Spectroscopy, Lidar, and DIAL Techniques, Proc. SPIE* **2112** (1994) 19-30.

Werle, P.; Analytical application of infrared semiconductor lasers in atmospheric trace gas monitoring, *J. Physique IV* **4** (1994) 9-12.

Werle, P., B. Scheumann, J. Schandl; Real time signal processing concepts for trace gas analysis by TDLAS, *Opt. Eng.* **33** (1994) 3093-3105.

1995 + in press

Baldacchini, G., A. Ciucci, F. D'Amato, G. Buffa, O. Tarrini; Temperature dependence of self-broadening of ammonia transitions in the ν_2 band, *J. Quant. Spectrosc. Radiat. Transfer* **53** (1995) 671.

Baldacchini, G., F. D'Amato, M. De Rosa, A.I. Nadezhdinskii, N. Lemekhov, N. Sobolev; Measurement of atmospheric concentration with tunable diode lasers, in: *4th Internat. Symp. on Monitoring Gaseous Pollutants by Tunable Diode Lasers,* Freiburg 1994.

Bicanic, D., M. Chirtoc, I. Chirtoc, P. Helander, J.P. Favier; New technique for measuring absorption coefficients of strongly absorbing liquids: Optothermal study of sunflower oil, oleic acid and its chloroform solutions at 3.39 microns, *Spectros. Lett.* **28** (1995) 101-110.

Bicanic, D., M. Franko, J. Gibkes, E. Gerkema, J.P. Favier, H. Jalink; Application of photoacoustic and photothermal non-contact methods in selected areas of environmental and agricultural sciences, in: A. Mandelis (ed.) *Progress in Photoacoustic and Photothermal Science,* American Institute of Physics, New York, in press.

Chirtoc, M., D. Dadarlat, D. Bicanic, J.S. Antoniow, M. Egee; Application of photothermal calorimetry in agriculture, medicine and environmental sciences, in: A. Mandelis (ed.) *Progress in Photoacoustic and Photothermal Science,* American Institute of Physics, New York, in press.

Chirtoc, M., D. Bicanic, I. Chirtoc, M. Lubbers, B. Arnscheidt, J. Pelzl; Laser photoacoustic spectroscopy of liquids using a large dynamic range optothermal window cell: Application to ammonium in water, fatty acids and lubricating oils, *J. Molec. Sci:* **348** (1995) 469-472.

Harris, G.W., F.G. Wienhold, T. Zenker; Airborne observations of strong biogenic NO_x emissions from the Namibian savanna at the end of the dry season, *J. Geophys. Res.* in press.

Mücke, R., B. Scheumann, J. Slemr, F. Slemr, P. Werle; Measurements of formaldehyde by tunable diode laser spectroscopy and the enzymatic-fluorometric method: An intercomparison study, *Infrared Phys. Technol.* **37** (1996) 29-32.

Roths, J., G.W. Harris; The tropospheric distribution of carbon monoxide as observed during the TROPOZ II experiment, *J. Atmos. Chem.* **24** (1996) 1577-188.

Tacke, M.; New developments and applications of tunable IR lead salt lasers, *Infrared Phys. Technol.* **36** (1995) 447-463.

Werle, P.; Laser excess noise and interferometric effects in frequency modulated diode laser spectrometers, *Appl. Phys.* **B60** (1995) 499-506.

Werle, P.; High frequency modulation spectroscopy : Progress in techniques and components, *Infrared Phys. Technol.* **37** (1996) 59-66.

Werle, P.; Spectroscopic trace gas analysis using semiconductor diode-lasers, *Spectrochim. Acta* **A52** (1996) 805-822.

Zenker, T., A.M. Thompson, D.P. McNamara, T.L Kucsera, G.W. Harris, F.G. Wienhold, P. LeCanut, M.O. Andreae, R. Koppmann, Regional trace gas distributions and airmass characteristics in the haze layer over southern Africa during the biomass burning season (Sept/Oct 92): Observations and modelling for the STARE/SAFARI'92 DC3, in: J.S. Levine (ed.), *Biomass Burning and Global Change*, MIT Press, Cambridge, Ma. 1966, in press.

Theses

M. Sc. / Diploma

Agne, M. Nah- und Fernfeldverteilungen von Bleisalz-Diodenlasern, Fakultät für Physik, Albert-Ludwig University of Freiburg, 1992.

John, J. Mid-infrared diode lasers with distributed periodic feedback, University of Freiburg and IPM Freiburg, 1993.

Kurbel, R. Entwicklung und Charakterisierung von Schattenmasken-PbSe-BH-Lasern, Fakultät für Physik, Albert-Ludwig University of Freiburg, 1992.

Lambrecht, U. Abstimmung und Linienbreite von Diodenlasern in der Infrarotspektroskopie, Albert-Ludwig University of Freiburg, 1990.

Pelagalli, F.; Spettroscopia molecolare con diodi laser accordabili per la rivelazione di tracce di gas, University of Rome, 1994.

Spilker, G. Rauscheigenschaften von Bleichalkogenid-Diodenlasern, Fakultät für Physik, Albert-Ludwig University of Freiburg, 1992.

Wietzorrek, J. Untersuchung der Moden- und Strahlstruktur von Bleisalz-Diodenlasern, Fakultät für Chemie und Pharmazie, Ludwig-Maximilian University of Munich, 1992.

Ph. D.

Carrasco, I.; Etude et réalisation d'un spectromètre à diode pour la mesure de constituants minoritaires dans la troposphère, University Pierre-et-Marie-Curie (Paris VI), 1994.

Josek, K.; Untersuchungen zur Machbarkeit der direkten Messung von HO_2-Radikalen in der Atmosphäre mit frequenzmodulierter Dioden-Laser Absorptionsspektroskopie, Ludwig Maximilian University of Munich, 1993.

Mücke, R. Frequenzmodulations-Spektroskopie mit Bleisalz-Diodenlasern zur Messung von atmosphärischen Spurengasen, University of Regensburg, 1992.

Xu, J. Electronic properties of the interfaces of PbSe / PbEuSe diodes and related material properties, University of Freiburg and IPM Freiburg, 1993.

Transport and Chemical Transformation
of Pollutants in the Troposphere

Volume 8, Part III

Tropospheric Optical Absorption Spectroscopy

Development of High Performance Differential Optical Absorption Spectroscopy

Paul C. Simon (Brussels)
Coordinator

TOPAS Steering Group

Reginald Colin	Brussels
Bo Galle	Gothenburg
John Plane	Norwich
Ulrich Platt	Heidelberg
Jean-Pierre Pommereau	Verrières le Buisson

Chapter 11

General Scientific Report

Paul C. Simon

Institut d'Aéronomie Spatiale de Belgique, Bruxelles, Belgium

11.1 Summary

The aim of the Tropospheric OPtical Absorption Spectroscopy (TOPAS) subproject of EUROTRAC was the development of high performance instruments based on Differential Optical Absorption Spectroscopy (DOAS) in order to measure minor tropospheric constituents using their absorption properties in the ultraviolet, visible and near infrared.

The target molecules detectable by this method are ozone (O_3), sulphur dioxide (SO_2), nitrogen dioxide (NO_2) and trioxide (NO_3). Other tropospheric constituents like HONO, HCHO and CS_2 as well as several hydrocarbons (toluene, benzene, xylene and naphthalene) are also measurable by DOAS.

Several DOAS systems were developed in the framework of this subproject. There are generally based on grating spectrometers with different detection systems. The Belgian groups used Fourier Transform Spectrometers (FTS), adapted for the first time to DOAS measurements.

New absorption long path systems were designed and extensively used, with the new concept of the combined transmitting/receiving telescope offering several advantages with respect to the classical double-ended system. Furthermore, a multipass cell was built and tested during campaigns.

The instruments were improved by using a variety of new type of detectors. Careful characterisation was carried out during the time-frame of this subproject. Several performances were improved *e.g.* in terms of measurement frequency. Dedicated softwares for instrument automation and data analysis were developed and improved, taking into account the specific problems imposed by each type of instrument and its associated detector.

Two intercomparison campaigns for DOAS instruments were organised, the first in 1992 in Brussels (Belgium) and the second in 1994 in Weybourne (UK).

The measurements made during the first campaign by eight different instruments were carefully analysed and yielded an absolute accuracy limit for NO_2, O_3 and SO_2 measurements made by the DOAS technique using an absorption path of a few hundred meters in a relatively non-polluted urban troposphere.

The dispersion ($\pm 1\sigma$) of the DOAS measurements performed during the coordinated comparison periods for the three target molecules is:

NO_2: $\pm 5 \times 10^{10}$ molec/cm^3

O_3: $\pm 6 \times 10^{10}$ molec/cm^3

SO_2: $\pm 1 \times 10^{10}$ molec/cm^3

The second campaign compared seven different instruments through ten days of continuous measurement under field conditions. The comparisons concerned SO_2, NO_2, O_3, NO_3, HCHO and HONO. The DOAS were compared with commercial point monitors for NO_2, O_3 and SO_2 and a home-built monitor for HONO. The preliminary results are very encouraging although the full analysis of the data has not yet been completed. Two other comparisons were also made during the campaign. Firstly, the DOAS were tested using cells containing known amounts of NO_2 and SO_2. The agreement was to within about 7 %. Secondly, the group's software was tested using synthetic spectra including NO_2, O_3 and HCHO. Agreement was found within about a 5 % error. This represents a considerable improvement on the results of the first comparison exercise.

These two campaigns demonstrated the necessity for quality assurance and quality control procedures in order to obtain reliable tropospheric data relevant for high quality scientific studies of the troposphere.

On the other hand, intensive laboratory measurements were conducted to support the instrument developments and the aforementioned retrieval softwares.

Finally, scientific studies were performed with several of the TOPAS instruments during other field campaigns and are briefly presented.

One instrument has been fully commercialised, namely the "Système d'Analyse par Observations Actives" (SANOA) developed by the CNRS French group. For the first time, a Fourier transform spectrometer was used in the UV-visible range for DOAS and its performances demonstrated. This instrument is commercialised by BRUKER (Germany). Furthermore, HOFFMANN MESSTECHNIK (Germany) has also developed an instrument for DOAS in the troposphere.

11.2 Aims of the TOPAS subproject

The Tropospheric OPtical Absorption Spectroscopy (TOPAS) subproject of EUROTRAC is one of the three subprojects concerned with instrument development and validation. It was officially accepted by the EUROTRAC SSC in 1988. Its objective was to develop high performance instruments based on Differential Optical Absorption Spectroscopy (DOAS) in order to measure minor tropospheric constituents using their absorption properties in the ultraviolet, visible and near infrared (wavelengths smaller than 1000 nm).

The target molecules detectable by this method are ozone (O_3), sulphur dioxide (SO_2), nitrogen dioxide (NO_2) and trioxide (NO_3). Other tropospheric constituents like HONO, formaldehyde (HCHO) and CS_2 have absorption bands in the UV range as well as several hydrocarbons (toluene, benzene, xylene and naphthalene) and are also measurable by DOAS.

The hydroxyl radical has also absorption lines around 308 nm but its detection requires very high resolution and very long optical path because of its weak tropospheric abundance.

In addition, new laboratory measurements of spectroscopic parameters used for the retrieval of trace species in the troposphere were needed to improve the performance of DOAS and were included in the TOPAS objectives.

The kick-off meeting was held at the University of Brussels in December 1988. Only two institutions respectively from France and Sweden were funded by their national agencies at that time. In April 1990, three Belgian institutions were funded. This chaotic start was a very serious drawback detrimental to the sub-project during its first phase. Nevertheless new instrument developments were successfully performed as well as the first measurements in the field.

Thanks to the availability of several instruments for tropospheric DOAS including, for the first time, a Fourier Transform Spectrometer its was decided in agreement with the SSC of EUROTRAC to focus the TOPAS objectives during the last years on intercomparison campaigns in order to compare the performances of the available instruments and to fully characterise and validate this method.

The priority was given first to O_3, NO_2 and SO_2 measurements and second to NO_3, HONO and HCHO, without excluding the hydrocarbons. Two new institutions from UK and Germany joined the sub-project in 1992. The Swedish group did not get further funding after 1994 and could not participate in the planned activities of the last two years of TOPAS.

11.3 Principal results

Because of the nature of the TOPAS subproject objectives defined in Section 1.2, the principal results are directly related to the instrument developments, the various optical path devices used with the instruments and the software developments in relation with the retrieval of the concentration of the measured tropospheric constituents.

Two field campaigns were organised for DOAS instruments for intercomparison purposes. In addition, some contributors participated in other field campaigns, not co-ordinated by TOPAS, but fully relevant for this subproject.

On the other hand, intensive laboratory measurements were conducted to support the instrument developments and the aforementioned retrieval softwares.

Finally, scientific studies were performed with several of the TOPAS instruments during other field campaigns and will be briefly presented.

11.3.1 Instrument developments

Several DOAS systems were developed in the framework of this subproject. There are generally based on grating spectrometers with different detection systems. The Belgian groups used Fourier Transform Spectrometers (FTS), adapted for the first time to DOAS measurements. They began first by using a fixed high resolution FTS BRUKER 120 HR FTS located at the University of Brussels and therefore used in urban conditions. A mobile high resolution FTS (BRUKER 120 M) was acquired in 1993 and adapted to field atmospheric measurements in the UV-visible range in 1994.

The main advantages of Fourier transform spectrometry compared to grating instruments are (i) the large spectral coverage achieved in one sampling, allowing to study many molecules simultaneously, (ii) an internal high precision wavenumber calibration device (He-Ne laser line), (iii) a constant, linear dispersion wavenumber scale, (iv) a simple, well-known instrumental line shape function when the instrument is perfectly tuned and (v) an available high resolution (0.01 cm^{-1} in the UV-Visible) when required by sharp structured absorption features.

The grating spectrometers, usually commercially available, were adapted for tropospheric DOAS measurements by all groups. They have generally small focal lengths, from 125 to 500 mm, with large aperture (*f*-number from 2.9 to 6.9). The bandpass varies between 0.4 and 1.2 nm. The detectors are either photodiode arrays (PDA), NMOS type detectors, charged coupled devices (CCD), or photomultiplier tubes associated with a slotted disc system.

Table 11.1 summarises the principal characteristics of the various DOAS instruments available, at the end of the TOPAS subproject.

Table 11.1: DOAS instruments available in the framework of TOPAS in 1995.

Institution	Acronym	Instrument and detector	Measured molecules
Swedish Environmental Research Institute, Sweden	SERI1	Retro/grating PMT + slotted disc	NO_2, NO_3, O_3, SO_2,
	SERI2	Mirror/grating PDA(1024)	NO_2, O_3, SO_2, HONO
	SERI3	Source/grating PMT + slotted disc	NO, NH_3, SO_2
Université Libre de Bruxelles, Belgique	ULB	Mirror/Fourier Transform Spectrometer (mobile) Solar blind UV and visible photodiodes	NO_2, O_3, SO_2
Service d'Aéronomie, CNRS, France	CNRS	Source/grating PDA(512)	NO_2, O_3, SO_2, HCHO HONO, hydrocarbons
Universität Heidelberg, Germany	UH1	Retro/grating PDA(1024)	NO_2, SO_2, O_3, HONO
	UH2	White cell/grating PDA(1024)	NO_2, NO_3
University of East Anglia, UK	UEA	Retro/grating PDA(1024)	NO_2, NO_3, O_3, SO_2, HONO
Institut d'Aéronomie Spatiale de Belgique	IASB	Mirror or source/grating PDA(1024), grating CCD(512)	NO_2, O_3, SO_2

Source: single path with source at a given distance from the spectrometer
Retro: double path with retroreflector
Mirror: double path with mirror
Grating: grating spectrometer
PMT: photomultiplier tube
PDA: photodiode array
CCD: charge coupled device

11.3.2 Absorption device developments and improvements

DOAS measurements were originally performed using a double-ended system with a transmitting and a receiving telescope positioned some kilometres apart. In a number of applications however it is more convenient, and sometimes even necessary, to use single-ended systems. For this purpose a combined transmitting/receiving system was designed [1, 2]. The system uses different parts of the parabolic main mirror for transmitting and receiving, and uses retroreflectors at the remote end of the path. The design is simple, easy to align, has good light efficiency and is compact. With this system a number of advantages can be gained. As the lamp is located close to the spectrometer, lamp reference spectra can easily be obtained and the lamp can be intermittently blocked out by an automatic shutter to record background noise and light levels. Multiple path measurements can easily be performed by realignment of the main mirror towards different retroreflectors, opening up interesting measurement strategies. Finally, measurements can be made also on remote locations with the need for electric power and a stable platform at one end only.

This single-ended system was used by several groups. The Swedish group has implemented such a device for background trace monitoring at Draget, close to the EUROTRAC TOR Station at Rörvik. It is also used with a second instrument developed by the same group and optimised for urban air monitoring. A coaxial transmitting and receiving telescopes was also implemented by the British group at the Weybourne Atmospheric Observatory (WAO) in North Norfolk, providing an overall path of 5.5 km. The new long path DOAS system built by the German group is also based on this new coaxial design.

On the other hand, a long path system was set up on the campus of the University of Brussels. The light from xenon or tungsten lamps is collimated onto a 30 cm Cassegrain-type telescope, which projects the beam onto a slightly parabolic mirror, situated 394 m away. This mirror sends the beam back into a second 30 cm Cassegrain telescope, which collimates the light onto the entrance aperture or slit of the spectrometer.

Single path mobile system was designed by CNRS and used by both with the "Système d'ANalyse par Observations Actives" (SANOA) instrument developed by the French group [3, 4] and with the mobile FTS spectrometer acquired by the University of Brussels. On both cases, absorption measurements by several tropospheric constituents are performed over distances ranging from 150 to 400 m.

A totally different system, namely a multipass cell, was designed and tested in the framework of this sub-project by the German group. This multireflection cell is based on the original design of White [5] and described by Ritz *et al.* [6]. The length of the cell was chosen to be 5 m allowing, with 144 reflections in the cell, light paths as long as 2 km. This high number of reflections could only be achieved with dielectric coated mirrors, which limits the spectral region that can be observed. With the chosen mirror coating, only NO_2 and NO_3 were measured with the cell. The mechanical stability of the mirror mounting proved to be a major problem, as changes in the adjustment reduced the light intensity. The mechanical set-up was therefore rebuilt for maximum stability and an computer controlled adjustment routine was developed. The time between two manual readjustments was prolonged to 3 days, thus enabling the cell to perform automatic long term measurements. Another problem with multireflection cells is the stray light in the cell. Two different sources could be identified: solar light scattered in the cell that can easily be measured by taking spectra without the light of the lamp and the lamp light itself which produces stray light in the cell. The latter is difficult to measure as the stray light intensity depends on the path length illuminated by the lamp and the path length observed by the spectrograph. By measuring the stray light obtained by interrupting the light path in the middle and than applying a mathematical correction, this stray light contribution could be removed from the spectra. The detection limits determined in field measurements in Heidelberg and Norwich are 1 ppb and 10 ppt for NO_2 and NO_3, respectively when using a 2160 m light path.

11.3.3 Software developments

The DOAS technique known since the 1930's for stratospheric total ozone [7] and since the 1970's for total NO_2 measurements [8-10] was pioneered by Platt and Perner [11] for tropospheric composition monitoring. It consists in looking at the narrow spectral features of the absorbents after removing the broad band absorptions and attenuations due to Rayleigh molecular and Mie aerosol scattering. The DOAS technique can provide simultaneously concentrations of several atmospheric species. Detailed descriptions of the method are given by Platt [12] and Plane and Smith [13].

In contrast to absorption spectroscopy in the laboratory, the true intensity $I_0(\lambda)$, as would be received from the light source in the absence of any absorption, is usually difficult to determine for measurement in the atmosphere (*e.g.* DOAS in the troposphere and for ground-based stratospheric monitoring). It would involve in principle removing the air from the absorption path. The solution to this apparent dilemma lies in measuring the so called "differential" absorption. This quantity can be defined as the part of the total absorption which is due to a given molecular species and which varies "rapidly" with wavelength, in contrast to the background absorption which varies "slowly" with wavelength. Accordingly the total intensity can be split into two parts $I(\lambda) = I_B(\lambda) + \delta I(\lambda)$, where $I_B(\lambda)$ is due to the background absorption varying slowly with wavelength (λ) and describing a general "shape" and $\delta I(\lambda)$ shows rapid variations due to absorption lines or bands. The meaning of "rapid" and "slow" variation of the intensity as a function of wavelength is, of course, related to the observed wavelength interval and to the width of the absorption bands or lines to be detected. The extinction due to Rayleigh and Mie scattering are smoothly varying with wavelength. Similarly, the cross-section is written as $\sigma(\lambda) = \sigma_B(\lambda) + \delta\sigma(\lambda)$, where $\sigma_B(\lambda)$ is the broad band cross-section and $\delta\sigma(\lambda)$ the differential cross section. This slow variation with wavelength is removed by the application of various filtering techniques, in order to leave the differential absorption peaks of the atmospheric trace gases. This filtering process is an important part of DOAS method: there are several mathematical methods used to derive $\delta I_0(\lambda)$ from experimental spectra. It is important that $\delta\sigma(\lambda)$ be derived with the same technique as the one used to derive $\delta I_0(\lambda)$. The methods developed by each group are adapted according to the types of spectrograph and detector used. Several methods such as polynomial interpolation, Fourier transform filtering and filtering after removal of residual emission peaks from the source were carefully compared in order to optimise the procedure. The overall routine to derive concentrations of atmospheric trace gases is a combination of linear least square procedure for the fitting of the calculated differential optical thickness with the observed one, and a non-linear least square procedure for alignment correction between atmospheric and reference spectra. No unique method is adaptable to both Fourier and grating instruments and this called for an overall intercomparison of instruments and algorithms used by the TOPAS participants which were carried out during the two campaigns described below.

Special attention was paid in the estimation of the error on the concentration derived from instruments using photodiode array detector. To calculate the correct error on the concentrations, the errors on the optical densities of the absorption bands and the error on the alignment have to be considered. A new method to calculate this total error was included in the analysis routine [14]. Monte-Carlo tests showed, that the error can now be calculated within 10 % of the statistical fluctuations due to noise in the spectra.

11.3.4 DOAS intercomparison campaigns

Two intercomparison campaigns for DOAS instruments were organised in the framework of TOPAS, the first in 1992 in Brussels (Belgium) and the second in 1994 in Weybourne (UK)

The Brussels campaign

The campaign was organised under the leadership of the Laboratoire de Chimie Physique Moléculaire (CPM) of the Université Libre de Bruxelles (ULB).

Twenty two persons were present representing eight groups and five nations (Belgium, France, Germany, Sweden and UK). Four commercial companies, selling or developing DOAS instruments were invited to participate provided they brought one of their instruments. HOFFMANN MESSTECHNIK (Germany) participated with a prototype instrument. ATMOS EQUIPEMENT (France) was represented by the CNRS French group who used one of the Atmos instruments. BRUKER (Germany), who is presently selling a high resolution portable Fourier transform spectrometer to be used in the UV-Visible region, accepted, but was unfortunately not ready with its instrument in September 1992 and did not take part in the campaign. OPSIS (Sweden) declined the invitation.

The following list gives the names of the participants and their affiliation:

Referee: Claude Camy-Peyret,
Laboratoire de Physique Moléculaire et Atmosphérique, Université Pierre et Marie Curie, Paris, France.

DOAS instrumentation:

ULB Reginald Colin, Michel Carleer, Cathy Clerbaux, Jean-Pierre Walgraeve,
Laboratoire de Chimie Physique Moléculaire, Université Libre de Bruxelles, Bruxelles, Belgium.

IASB Paul C. Simon, Ann-Carine Vandaele, Christian Hermans, Willy Decuyper,
Institut d'Aéronomie Spatiale de Belgique, Bruxelles, Belgium.

UMH Caroline Fayt,
Université Mons-Hainaut, Mons, Belgium.

UH Frank Heintz, Thomas Rudolph, Irene Pundt, Martin Hausmann,
Institut for Umweltphysik, Universität Heidelberg, Heidelberg, Germany.

SERI Bo Galle, Bjorn Bergqvist,
Swedish Environmental Research Institute, Göteborg, Sweden.

CNRS Jean-Pierre Pommereau, Florence Goutail, Manuel Nunes-Pinharanda, Pierre Hulin,
Service d'Aéronomie CNRS, Verrières-le-Buisson, France.

UEA John Plane, Nicola Smith,
School of Environmental Sciences, University of EastAnglia, Norwich, UK.

HOFF Alexander Hoffmann, Bernd Fromer,
Hoffmann Messtechnik, Rauenberg, Germany.

The campaign took place from the 7th to the 18th of September 1992 on the campus of the Université Libre de Bruxelles, which is situated south of the city. Most instruments began their measurements on the morning of the 10th. Seven instruments described in Table 11.2 were installed in a laboratory situated on the 5th floor of a 10 floor building at an altitude of ~20 m from ground level and ~120 m above sea level. Through open windows all instruments pointed in the NE direction towards a set of retroreflectors placed at various distances. In addition to these set-ups, the 15 m open White absorption cell and spectrograph (UH2 in Table 11.2) was installed on the roof of the laboratory building (altitude 50 m).

In order to reach a maximum efficiency in the comparisons of results, three molecules (NO_2, O_3 and SO_2) were chosen as imposed targets in spite of the fact

Table 11.2: Description of the participating instruments during the 1992 intercomparison campaign.

Institution	Acronym	Instrument and detector	Optical path	Measured molecules	Recording time
Swedisch Environmental Research Institute	SERI1	grating spectrom. PMT + slotted disc	A, D	NO_2, O_3, SO_2, H_2CO	20 min/molec
	SERI2	grating spectrom. CCD (1024)	C	NO_2, O_3, SO_2	20 min
Université Libre de Bruxelles	ULB	Fourier Transform Spectrom. Solar blind UV photodiode	C	NO_2, O_3, SO_2	45 min
Service d'Aéronomie CNRS	CNRS	grating spectrom. PDA (512)	B, C	NO_2, O_3, SO_2, H_2CO, HNO_2, hydrocarbons	5 min
Universität Heidelberg	UH1	grating spectrom. PMT + slotted disc	A, D	NO_2, SO_2	20 min/molec
	UH2	grating spectrom. PDA (1024)	White cell	NO_2, NO_3	10 min
University of East Anglia	UEA	grating spectrom. PDA (1024)	C	NO_2, O_3	15 min
Institut d'Aéronomie Spatiale de Belgique	IAS	grating spectrom. PDA (1024)	C	NO_2, SO_2	6 min

A : Retroreflector at 230 m   B : Source at 230 m   C : Mirror at 390 m   D : Retroreflector at 1010 m that some instruments could not cover all three species. The spectral region to be investigated was left to the appreciation of each group according to its instrument characteristics. A common set of absorption cross sections for these three molecules was used.

Four optical path lengths were available from mirror, retroreflectors and sources placed on several buildings within and outside the ULB campus. Path lengths of 230 (B), 460 (A), 780 (C) and 2020 m (D) were used. In several instances, the light from one optical path (780 m) was distributed simultaneously to several spectrographs.

Co-ordinated observation periods of 12 (night or day) and 24 consecutive hours were devoted to simultaneous measurements of the imposed species. Additional species (*e.g.* HCHO, HONO, hydrocarbons, *etc.*) were measured by some groups.

An overall view of the variations in concentration of the NO_2, O_3 and SO_2 species during the full ten days of measurements in presented in Fig. 11.1. This figure is only an illustration of the behaviour of the three measured constituents during the campaign. As expected in an urban site, these variations were quite large and allowed to test the instruments in both high and low concentration conditions. The average concentrations for NO_2 and for O_3 were around 5×10^{11} molecules cm^{-3}, with the classical anticorrelation in their diurnal variability. For SO_2, the average concentration was $\sim 10^{11}$ molec. cm^{-3} with a few spurious maxima values reaching 5 to 6×10^{11} molec. cm^{-3}. For an urban troposphere, the level of pollution was reasonably low during the campaign period which was dominated by south western winds. These winds did not pass over the more industrial northern part of Brussels.

The four coordinated periods are shown on Fig. 11.1 and designated as I, II, III and IV. Periods I and II were of 8 hour duration, from 10.00 am to 6.00 pm and took place on the 10th and 11th of September. The most important period was period III which lasted 24 hours non stop, starting at 4.00 pm on the 14th of September. Period IV lasted two days beginning at midnight on the 16th.

Two instruments (CNRS and SERI1) were fully automated and measured continuously, night and day, during the whole period of the campaign. The other instruments, which required human surveillance, were confined to measure during the co-ordinated periods. Some instruments did not measure either O_3 or SO_2 (see Table 11.1) and unfortunately, technical reasons prevented some instruments from taking part in some of the co-ordinated periods.

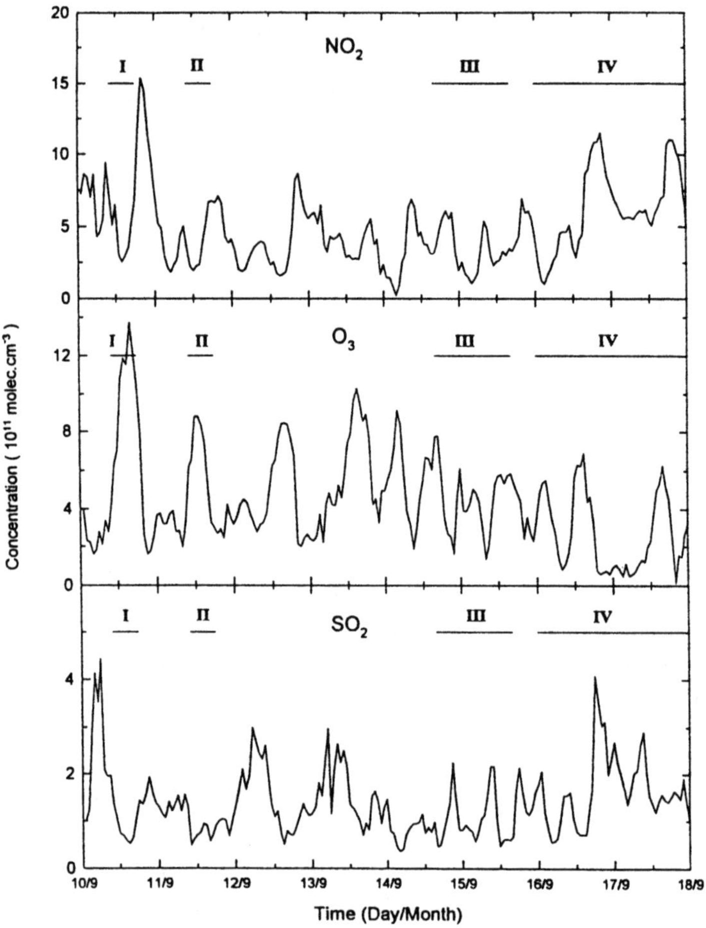

Fig. 11.1: Temporal variations of the concentrations of NO_2, O_3, and SO_2 during the campaign with the four coordinated periods.

The results obtained during the 24 hour period (Period III) will be examined first. Fig. 11.2 presents these measurements: all eight instruments performed NO_2 measurements (top panel), five obtained O_3 measurements (middle panel) and six SO_2 data (bottom panel). A similar scatter of the points is observed for all three species and is of the same order of magnitude for high and low concentrations. Using exclusively the results of the fast instruments (all but ULB and SERI2), relative standard deviations are readily calculated for time intervals of 6 min. As

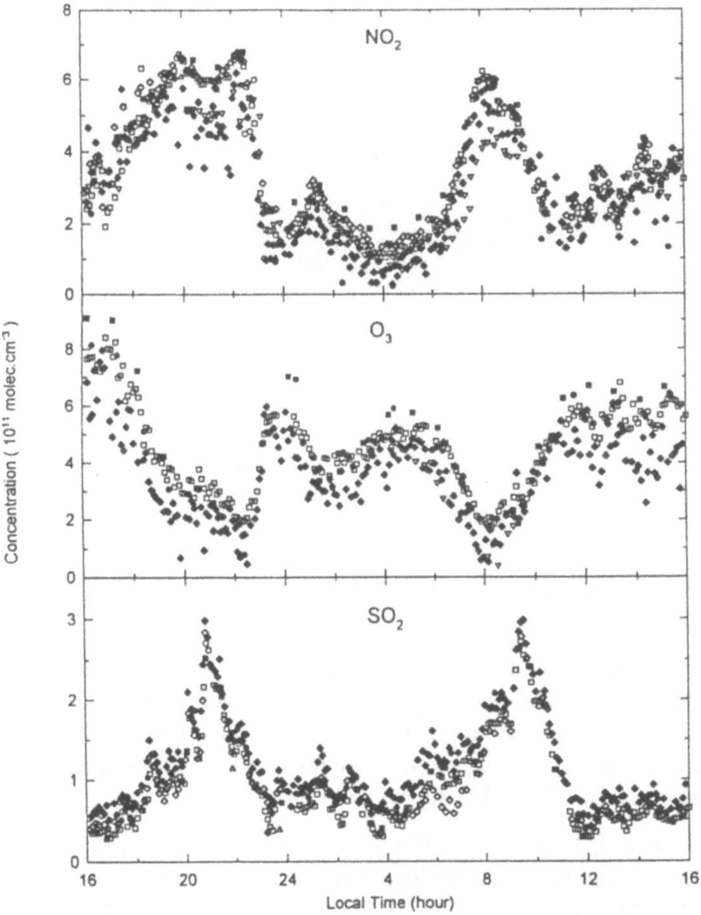

Fig. 11.2: Concentration evolution of NO_2, O_3, and SO_2 during the 24 hour period starting at 4.00 pm on the 14th of September (● ULB, ◊ IAS, and ■ SERI, ◆ CNRS, Δ and ❍ UH, ∇ UEA).

an example these are presented for O_3 in Fig. 11.3 which also shows the average O_3 concentration throughout this period. The relative standard deviations are small for high concentrations (±5 %) and large for small concentrations (±10 %); they are roughly proportional to the inverse of the concentration. A similar situation is observed for NO_2 and SO_2 but in the latter case the deviations are smaller by a factor of 2. A similar situation to that described above is observed during periods I, II and IV.

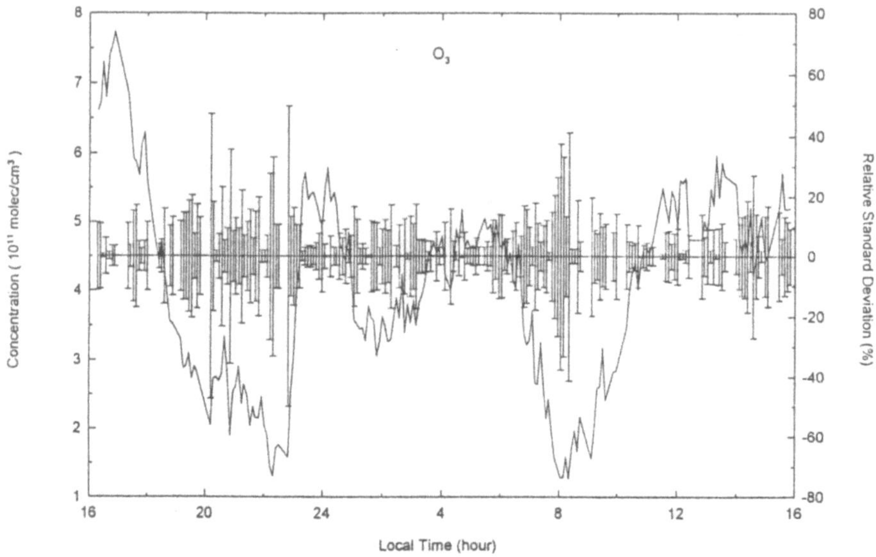

Fig. 11.3: Relative standard deviation of the O measurements made by the fast instruments during Period III with the average concentration (solid line).

The short paths (A, B, and C) measurements integrated over a period of 45 min, are presented in Fig. 11.4, where the difference of each integrated value with the average value is shown as a function of time. Only those instruments which performed at least one measurement every 60 min throughout most of period III were retained: five for NO_2, four for O_3 and four for SO_2.

The comparison between the different data shows that rapid fluctuations in concentrations are not responsible for the discrepancies between instruments using almost the same paths (A, B, and C), because the overall accuracy for each molecule has not improved with respect to the comparison of the fast (non-integrated) measurements presented above. The absolute accuracy does not vary in spite of a fairly large concentration variation of the three species investigated during period III (by a factor of ~ 7, ~ 8 and ~ 6 for NO_2, O_3 and SO_2 respectively). It may therefore be stated that the dispersion of the results are mainly due to the instruments themselves including the use of individual algorithms for the retrieval of the data. The dispersion ($\pm 1\sigma$) of the DOAS measurements on NO_2, O_3 and SO_2 performed by this large variety of UV-Visible instruments using a path length of a few hundred meters is:

NO_2: $\pm 5 \times 10^{10}$ molec/cm^3

O_3: $\pm 6 \times 10^{10}$ molec/cm^3

SO_2: $\pm 1 \times 10^{10}$ molec/cm^3

The lower value for SO_2 was expected because the signature (cross section) of this molecule is stronger than that of the two others.

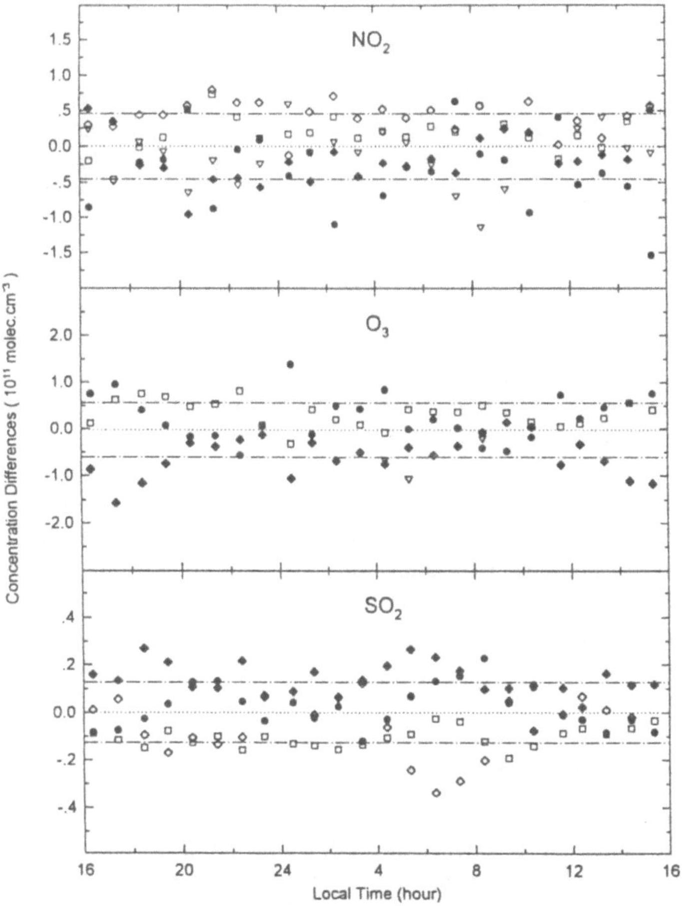

Fig. 11.4: Difference of the concentrations integrated on 45 min with the average value for NO_2, O_3, and SO_2 during Period III (● ULB, ◊ IAS, and ■ SERI, ◆ CNRS, Δ and ○ UH, ∇ UEA)

From the various comparisons made with the campaign data, it is clear that the discrepancies between the instruments are not random and are closely linked to each instruments. It is difficult in the present state to attribute these differences to a particular cause. They are definitely connected to the different algorithmic treatments used be each group in the DOAS technique. In spite of many efforts on behalf of the participants to the campaign, it was impossible in a short time to conceive a uniform retrieval method (same algorithm) on account of the difference in conception of the various instruments (grating or Fourier, slotted disk, CCD or PDA). The technique used to adapt the resolution of the cross sections to the resolution of the instruments, the fact that several molecules are measured simultaneously or not, the type of regression procedure used, the fact that a dark current need to be measured or not, the filtering procedure used to eliminate the broad band component of the recorded spectrum, are a few examples of the necessity of different approaches to the treatment of the raw data.

Significant progress was made by all the participants due to the synergy of this type of campaign. After a first comparison of results made without any constraints, it was obvious that a common set of absorption cross sections had to be used by each group and also triggered improved measurements of these [15]. The campaign also served to exchange various instrumental and software improvements between the groups.

In order to improve the accuracy of the presently available type of instruments, further work was required. In particular it would be interesting to compare results obtained by this technique to those obtained by other techniques, in particular chemical analysis. In order to do so, it would be useful to conduct a campaign in a more stable environment, where concentrations vary less than in an urban site. This would allow to reduce the difficulties encountered here related to the various integration times used by the different instruments. Also, cell measurements should be performed by all instruments, although such measurements do but poorly represent the difficulties associated with real atmospheric measurements, they would probably reveal more clearly differences in results due to the different retrieval methods used. Finally, the atmospheric pollution conditions encountered in this urban campaign did not allow to study the detection limits of each experiment set-up.

The Weybourne campaign

The second TOPAS comparison campaign was held at the Weybourne Atmospheric Observatory between the 26th September and 7th October 1994. The laboratory is situated on the north coast of Norfolk, UK (latitude 57° N, longitude 7° E) and is part of the TOR network (TOR station 11). This site was chosen because, as well as being the base for the UEA group's measurements it offered a location that was well equipped for the measurement of trace gases and was exposed to a wide range of conditions, from polluted European to very clean Arctic air.

The second exercise sought to extend the findings the first campaign held in Brussels in several ways: firstly, the seven DOAS instruments involved in the campaign were run continuously throughout the campaign to observe their performance in varied conditions and to test their reliability; secondly, the DOAS were compared with commercial point monitors for NO_2, SO_2, and O_3; thirdly, a quantitative comparison of cell measurements of both NO_2 and SO_2 was made; fourthly, trials of the participants analysis software were conducted; and lastly, as well as observing atmospheric concentrations of NO_2, O_3 and SO_2 the 1994 study also included measurements HONO.

Twenty four scientists, from seven different research institutions, participated in the Weybourne comparison exercise. The following list gives the names of the participants and their affiliation:

Referees: Hugh Coe (UEA, UK) and Rod Jones (University of Cambridge, UK)

DOAS instrumentation:

CNRS Jean-Pierre Pommereau, Florence Goutail, Miguel Nunes-Pinharanda, Service d'Aéronomie CNRS, Verrières-le-Buisson, France.

IASB Paul C. Simon, Ann-Carine Vandaele and Christian Hermans, Institut d'Aéronomie Spatiale de Belgique, Bruxelles, Belgium.

UEA John Plane, Beverley Allan, Nicola Smith, School of Environmental Sciences, University of EastAnglia, Norwich, UK.

UH Ulrich Platt, Thomas Etzkorn, Matthias Helmling, Christoff Janssen, Jochen Stutz, Thomas Wagner Institut for Umweltphysik, Universität Heidelberg, Heidelberg, Germany.

ULB Reginald Colin, Michel Carleer, Jean-Michel Guilmot Laboratoire de Chimie Physique Moléculaire, Université Libre de Bruxelles, Bruxelles, Belgium.

Other Measurements:

UEA Stuart Penkett, Rachel Burgess, Kevin Clemitshaw, School of Environmental Sciences, University of EastAnglia, Norwich, UK.

UB Roy Harrison, John Peak, University of Birmingham, UK

Table 11.3 gives a list of the DOAS instruments compared during the comparison exercise together with some detail.

Currently the data are undergoing analysis within the TOPAS sub-project and so no association has been made between an individual instrument and the responsible institute. The DOAS are labelled by the number in the first column of Table 11.3, these references are retained throughout.

Table 11.3: Instruments used during the second intercomparison campaign

			Fitting Region / nm		
	Path	System/Detector	NO_2	O_3	SO_2
1	155 m	Grating/Diode Array	330-370	254-275	280-310
2	155 m	Grating/CCD	262-268		295-305
3	155 m	Fourier	263-364	263-364	263-364
4	292 m	Plane mirror/Grating/ Diode Array	330-370	254-275	280-310
5	1600 m	White Cell/Grating/ Diode Array	430-443		
6	5000 m	Retro/Grating/Diode Array	350-380	285-300	285-300
7	5000 m	Retro/Grating/Diode Array	345-380	300-335	300-335

Other measurements were also made during the campaign using commercially available instrumentation and these data were compared with the DOAS measurements. Specifically, these were: an Ecophysics NO analyser and NO_2 photolytic converter; a Thermo-Electron SO_2 analyser and a Thermo-Electron O_3 analyser. Also, a home built HONO instrument was run by the University of Birmingham.

As mentioned above, the experiment was split into three comparison exercises:

the first used synthetic spectra to test the participants analysis software; the second used a cell made up of known concentrations of NO_2 and SO_2; and in the third exercise the instruments were run alongside the commercial point monitors for 10 consecutive days in the field.

Five test spectra were made up to give typical atmospheric absorptions for a 5 km path length. Each contained NO_2, O_3 and HCHO in varying concentrations and increasing amounts of noise. Fig. 11.5 shows the results from the fifth, and most difficult, of these tests. Here again, no association is made between the test results and the institution responsible because the participants have not yet fully reviewed these data. Each institute is identified by a letter. Institutes B and D had to operate their software in a non-ideal way to interpret these test spectra and so their results do not wholly reflect the performance of their software when analysing their own spectra. However, looking at the data as a whole the NO_2 and O_3 values were retrieved to within < 5 % in the majority of cases and the HCHO values to within 10 %.

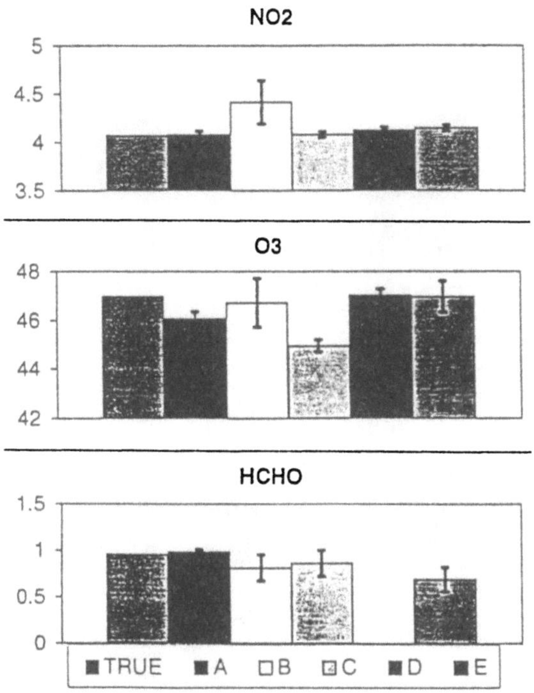

Fig. 11.5: Concentrations in ppb assuming a 5 km path of NO_2, O_3 and HCHO derived from Test V for the five groups.

A cell was made up with known amounts of NO_2 and SO_2 that gave typical atmospheric absorptions for a 5 km path. Fig. 11.6 shows the results from the these measurements with the cell make-up in the left hand column. The values are expressed as ppb having measured the local temperature and after assuming a path length of 5 km. As the figure shows the results for the comparison are very encouraging. All the instruments, with the exception of number 4, agree with the cell make-up to within 7% for NO_2 and all except 1 and 4 agree to within 7 % for SO_2.

Ambient atmospheric measurements were made by all seven instruments over the 10 day period between 26th September and the 7th October 1994. The gases considered during this part of the exercise were NO_2, SO_2, O_3 and HONO. NO_3 was also measured by instruments 6 and 7 on three nights during this period and so there are gaps in these data. NO_3 was not strictly part of this comparison exercise and will be considered separately at a later date.

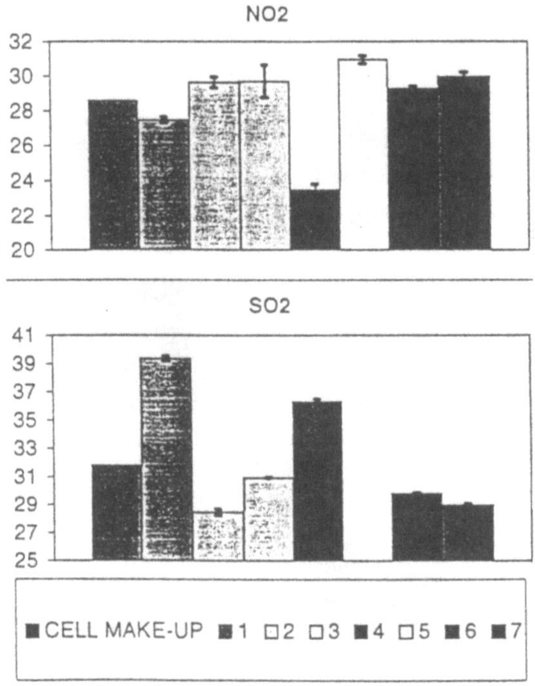

Fig. 11.6: Concentrations in ppb assuming a 5 km path of NO_2 and SO_2 in the test cell measured by the seven instruments.

Fig. 11.7 shows measurements of NO_2 from 6 of the DOAS instruments (instrument 2 did not measure NO_2) and also from the Ecophysics instrument. Overall the agreement is very good, however, there are from time to time some significant differences between the instruments. These are currently being considered by the TOPAS community in a series of workshops to discuss these data. The amount of data gathered here is enormous and data evaluation and reduction is difficult, however, some example periods are presented here:

Fig. 11.8 shows the O_3 time series for two short path length instruments and a long path instrument for the daytime of 29th September (Julian day 272). All three instruments agree well with the Teco O_3 instrument throughout except for instrument 2 after 16:00. The reasons for this are still being explored. The long path instrument has considerably lower errors than the other two instruments as should be expected from the greater absorptions it observes.

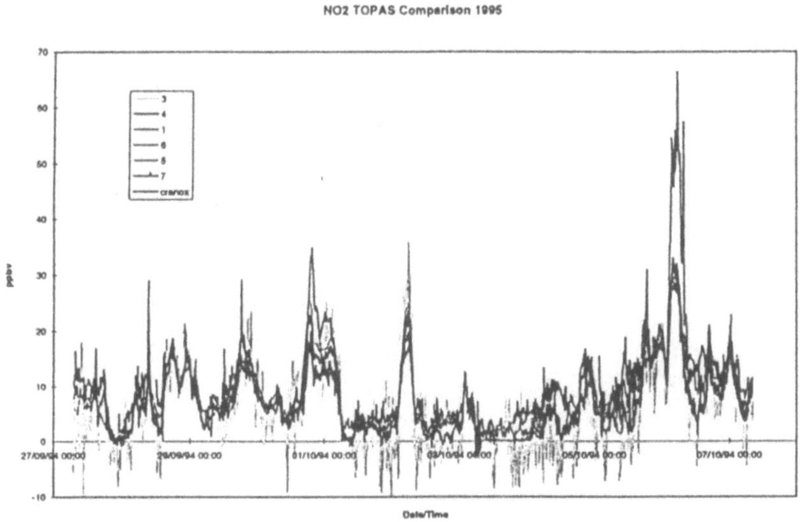

Fig. 11.7: Measurement of NO_2 from six of the DOAS instruments and the Ecophysics instrument.

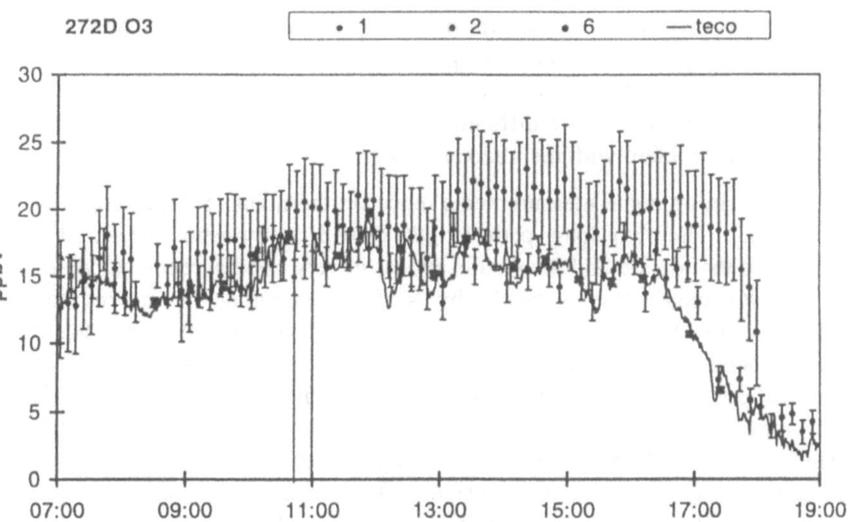

Fig. 11.8: O_3 time series for three DOAS instrument and the TECO O_3 instrument for the daytime of 29th September.

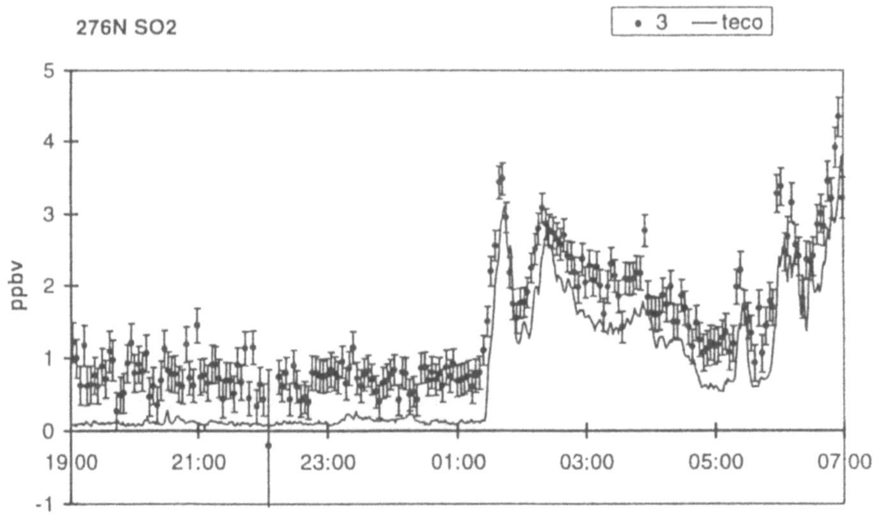

Fig. 11.9: Variation of SO_2 during the night the 3rd-4th October measured by the Fourier transform spectrometer and the TECO analyser.

Fig. 11.9 shows the variation SO_2 during the night of the 3rd-4th October (276-277) as measured by the Fourier transform instrument and the Teco analyser. Both these instruments are in excellent agreement and show similar variations in the concentrations of SO_2 throughout this period. However, the absolute concentrations measured are different by about 0.5 ppb. This is almost certainly due to a zero problem with the Teco analyser.

Fig. 11.10 shows some measurements made on the night of 27th-28th September (270-271). The two long path length DOAS instruments measured HONO throughout that period and the results appear in very good agreement. Also shown are HONO measurements made by a point sampling instrument from the University of Birmingham, these also are in agreement with the DOAS measurements. The differences between all three instruments are generally less than 100 ppt. These data are plotted together with NO and NO_2 data from the University of East Anglia Ecophysics instrument. At approximately 04:00 an increase of HONO from 100 ppt to 300 ppt was observed coinciding with an increase in NO_2. During this period NO remained below the detection limit (60 ppt). After sunrise at 06:00, when NO first appeared, the HONO concentration began to decrease, presumably because of photolysis. These results strongly indicate that night-time production of HONO is occurring in the absence of NO.

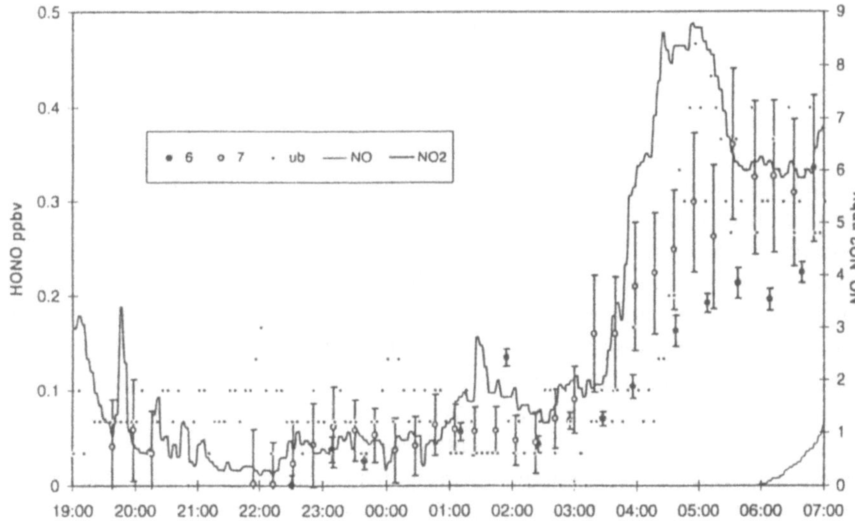

Fig. 11.10: Measurement of HONO, NO and NO_2 on the night of 27th-28th September.

The TOPAS community has amassed extensive comparison data during the Weybourne campaign. Analysis and interpretation of that data set is currently underway both within the respective groups and also in a series of discussion meetings. Although this process is some way from completion, it can clearly be seen that the results are both useful and encouraging both as a DOAS comparison exercise and as an atmospheric chemistry field measurement campaign.

11.3.5 Other field campaigns and related scientific studies

In 1991, the latest version of the French instrument "Système d'Analyse par Observations Actives" (SANOA) was tested in urban (Paris) and rural (Vosges) sites in order to improve its performances. In Paris, the SANOA provided measurements of NO_2, SO_2, ozone, HONO, HCHO and toluene [16]. They were compared with in situ observations and with the UV-visible spectrometer made by OPSIS. The agreement with the OPSIS results is generally poor for all common detectable species.

The campaign NUAC in the frame of the EUROTRAC subproject ACE was performed in April-June 1991 on the west site of the Vosges, providing measurements of ozone, NO_2, SO_2 and HCHO. The data analysis [17] brought important information on the uncertainties and precision of these observations, and on the crucial parameters and technical issues which were later addressed during the intercomparison campaigns in 1992 and 1994.

During 1991, the Swedish groups have further developed the large scale flux method based on the observation of the zenith sky-light by an airborne DOAS instrument, made at two different flight altitudes. The method was successfully tested for NO_2, during three field campaigns.

The dual-beam extinction measurement method developed by the Swedish group was tested in the TESLAS intercomparison campaign TROLIX'91 in Bilthoven, Netherlands, June 16-27, 1991 [18]. In this method two spectra are recorded, close in time, along two paths with the same direction but with different path lengths. If these two spectra are then divided, instrument factors, such as spectral structures from the light source and spectrometer anomalies, are to a high degree cancelled. This method was used routinely to improve the detection limit of the system when measuring background concentrations and made possible the measurement of absolute atmospheric extinction spectra. The idea was to compare the theoretical LIDAR corrections for particle attenuation and Rayleigh scattering with the measured values deduced for the dual-beam measurements. DOAS measurements were made over two paths, 885 m and 2 km respectively, and covering the wavelength range 270-315 nm, an interval containing most of the wavelengths used in the LIDAR ozone measurements. During some periods two of the Lidar systems were directed horizontally giving paths nearly collinear with the DOAS paths. Thus the different systems could be inter-calibrated. During the campaign the DOAS system was also measuring ground level concentrations of O_3 and SO_2. In connection with this campaign a NO_2 intercomparison between the DOAS and the RIVM NO_2 Lidar systems over a nearly collinear path of 1116 m was performed. The comparison was made without any preceding inter-calibration and showed good agreement between the two methods.

During Spring 1993 a measurement campaign was conducted at the Swedish field site at Draget close to the TOR station at Rörvik. The aim of this campaign was a comparative study of NO_2 measurements between Draget and Rörvik as well as a first attempt to measure NO_3. The NO_2 measurements show generally good agreement, however at some of high concentrations (20 $\mu g/m^3$) the DOAS concentrations are considerably higher than the TOR station data. This may be due to local contamination, different meteorological conditions or instrumental differences. Significant concentrations of NO_3 (50 ng/cm^3) was detected during one night, the concentration being below the current detection limit of 20 ng/m^3 during the rest of the period.

A DOAS system optimised for operation in the wavelength region 215-245 nm was designed by the Swedish group for the detection of NO, NH_3 and SO_2. The system was tested in a field campaign at Lanna agricultural field station in Sweden in Summer 1993. The aim of this campaign was to intercompare different techniques for measurement of fluxes of NH_3 resulting from spreading of manure. The system was compared with a long path FTIR system as well as different gradient, denuder and field chamber techniques. The system showed good agreement with the other techniques, demonstrated linearity over three orders of magnitude, and had a detection limit for NH_3 of 15 ppb for 1 minute integration time. During the campaign the NO flux generated by microbial activity on nitrogen compounds was also measured.

Two other field campaigns at the Weybourne Atmospheric Observatory (WAO) were organised as part of the Land Ocean Interaction Study, a major UK community project funded by the Natural Environment Research Council. An important focus of LOIS is the chemistry of NO_3 in the marine boundary. This radical is a significant night-time oxidant of organic species, as well as providing a removal pathway for anthropogenic nitrogen oxides. Fig. 11.11 shows a time-resolved concentration profile of NO_3 typical of the marine boundary layer in the autumn. Note the appearance of NO_3 at detectable levels (above 1 ppt) after sunset, and the abrupt disappearance at dawn (32 hours). The NO_3 lifetime ($\tau(NO_3)$) in the marine boundary layer at Weybourne was found to be of the order of 10 minutes in the spring/summer, and appears to be controlled by dimethyl sulphide (DMS) produced in the ocean by phytoplankton. $\tau(NO_3)$ decreases to about 2 minutes in the autumn/winter. Since the DMS concentration is low at this time of the year, the most likely sink for NO_3 is the heterogeneous reaction on moist aerosol surfaces of N_2O_5 with which NO_3 is in equilibrium [19]. Another high-light of these measurements in the marine boundary layer was the observation of nitrous acid (HONO) being formed during the night at sub-ppb levels.

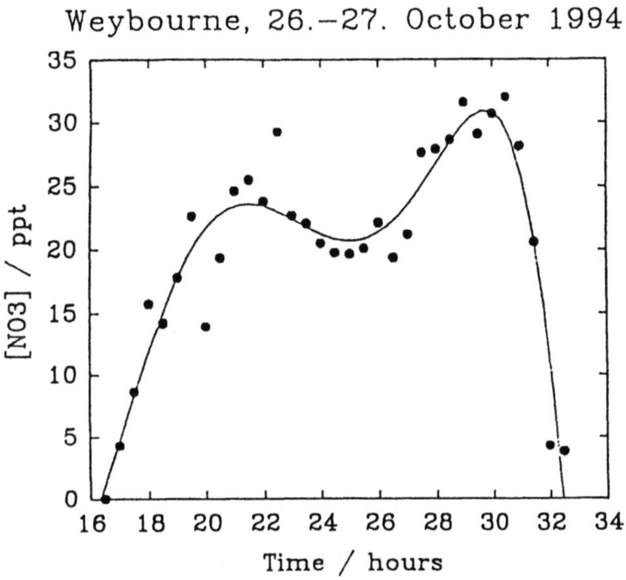

Fig. 11.11: A time-resolved profile of the NO_3 in the marine boundary layer at the Weybourne Atmospheric Observatory, north Norfolk. Note that 24 hours on the abscissa refers to midnight.

The DOAS was also employed to study NO_3 at the Izaña Observatory in Tenerife in the Canary Islands. The Meteorological Station lies on a mountain platform at an altitude of 2376 m above sea-level and 400 km west of Africa; it is a baseline station of the Background Air Pollution Monitoring Network managed by the World Meteorological Organisation. The air at the observatory is usually representative of the free troposphere, as it rests above an inversion at around 2000 m which acts as a cap for the pollution generated below. During the campaign, the concentration of NO_3, NO_2 and O_3 were measured by the DOAS with a path length of 1.98 km. The measurements of NO_2 and O_3 were compared with point sampling instruments at the Meteorological Station, yielding excellent correlation: for NO_2, a 12-hour comparison reduced a correlation plot with a slope of 0.997 and an r^2 of 0.678, despite the NO_2 concentration ranging between only 100-500 ppt; for O_3, the correlation plot produced a slope of 0.948 with an r^2 of 0.948.

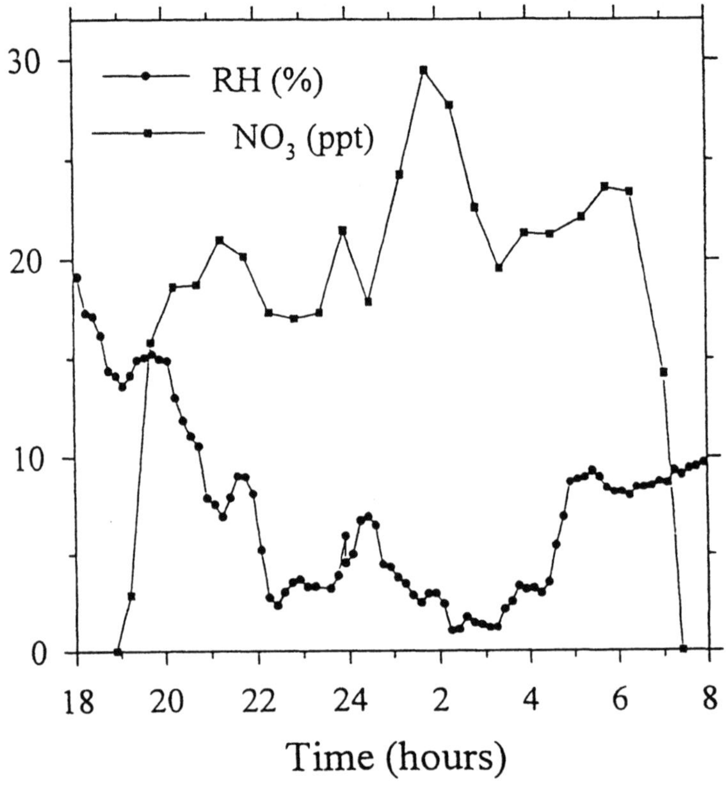

Fig. 11.12: Time-resolved profiles of the NO_3 concentration and relative humidity (RH), measured at the Izana Observatory, Tenerife.

Measurements of NO_3 were made for 20 full nights during the campaign. The NO_3 was seen to be strongly correlated with O_3, and anti-correlated with the relative humidity (RH) during the night, as shown in Fig. 11.12. The reason for the inverse relationship with RH, which has been reported previously [19, 20] is most likely that the surfaces of aerosols (and the ground) must be deliquesced in order to provide effective sinks for gas-phase NO_3 or N_2O_5. However, the RH at Izana at night-time was often very low. For instance, during the night shown in Fig. 11.12, $\tau(NO_3)$ is of the order of 4 hours when the RH is below 5 %. Since the measured concentrations of reactive hydrocarbons such as isoprene were very low, one possible mechanism for removing NO_3 is the homogeneous gas-phase reaction $N_2O_5 + H_2O \rightarrow 2HNO_3$. The field data indicate that this reaction would require a rate coefficient of about $9 \cdot 10^{-21}$ cm^3 $molecule^{-1}$ s^{-1} in order to account for $\tau(NO_3)$, which is in sensible accord with the upper limit to this reaction of $2 \cdot 10^{-21}$ cm^3 $molecule^{-1}$ s^{-1} at 298 K measured in the laboratory.

11.3.6 Laboratory measurements and wavelength range definition for DOAS

New measurements of absorption cross sections in the UV and Visible range were performed for SO_2 , NO_2 and CS_2 at different resolutions by means of the FTS at the University of Brussels. High resolution spectra (0.02 nm at 300 nm) were obtained at room temperature for SO_2, NO_2 and CS_2 and also at 250 and 273 K for NO_2. Comparison with previous measurements shows good agreement, better than 5 %. SO_2 results were published by Vandaele *et al.* [21].

One species of vital importance in research concerning photochemical oxidants is ozone. The traditional interval used a narrow range of the Huggins bands (320-330 nm) which has unfortunately relatively weak differential absorption cross sections, leading to low sensitivity in tropospheric measurements. An alternative is to measure around 250 nm where the differential absorption cross sections are much higher. Unfortunately the absolute absorption cross section is high in this region, restricting the path length to a few hundred meters. After a careful investigation of the spectrum of ozone and its interfering species, the Swedish group found the region 280-285 nm to be a good alternative and new measurements were performed in that range [22, 23]. Under normal conditions measurements over several km are feasible, and an improvement of nearly one order of magnitude in detection limit as compared to the Huggins band can be achieved.

11.4 Achievements

Due to the funding difficulties encountered by a majority of potential participants to TOPAS, the objectives of this EUROTRAC sub-project were redefined at the end of 1991. The basic objective, namely the development and improvement of instruments for the detection of tropospheric constituents by DOAS was emphasised and intercomparison campaigns were planned to evaluate their performances.

New absorption long path systems were designed and extensively used, with the new concept of the combined transmitting/receiving telescope offering several advantages with respect to the classical double-ended system. Furthermore, a multipass cell was built and tested during campaigns.

The instruments were improved by using a variety of new type of detectors. Careful characterisation was carried out during the time-frame of the subproject. The Fourier transform spectrometers were used for the first time for DOAS measurements. Several performances were improved *e.g.* in terms of measurement frequency. Dedicated softwares for instrument automation and data analysis were developed and improved, taking into account the specific problems imposed by each type of instrument and its associated detector.

The major achievements were the two intercomparison campaigns conducted in 1992 and 1994, respectively in an urban (Brussels, Belgium) and in a pristine site (Weybourne, UK).

This first intercomparison campaign of DOAS instruments operating in the visible and UV ranges brought to light several fundamental issues which have contributed to improve this technique in its aim to study the composition of the troposphere. Significant progress has been made by all the participants due to the synergy of this type of campaign. After a first comparison of results made without any constraints, it was obvious that a common set of absorption cross sections had to be used by each group and also triggered improved measurements of these [21]. The campaign also served to exchange various instrumental and software improvements between the groups.

The measurements made during the campaign by eight different instruments were carefully analysed and yielded an absolute accuracy limit for NO_2, O_3 and SO_2 measurements made by the DOAS technique using an absorption path of a few hundred meters in a relatively non-polluted urban troposphere.

The dispersion ($\pm 1\sigma$) of the simultaneous DOAS measurements on NO_2, O_3 and SO_2 performed by this large variety of UV-Visible instruments is:

$$NO_2: \pm 5 \times 10^{10} \text{ molec/cm}^3$$

$$O_3: \pm 6 \times 10^{10} \text{ molec/cm}^3$$

$$SO_2: \pm 1 \times 10^{10} \text{ molec/cm}^3$$

There remain nevertheless differences between the results obtained from the various instruments. These are most certainly linked to the various retrieval methods used although it was concluded that these retrieval methods are well optimised and well adapted to each instrument.

This campaign and its results are described in detail in Camy-Peyret *et al.* [24].

In order to improve our knowledge of the accuracy of the presently available type of instruments, further work was required. In particular it appeared interesting to compare results obtained by the DOAS technique to those obtained by other techniques. In order to do so, it was felt useful to conduct a campaign in a more stable environment, where concentrations vary less than in an urban site. This would reduce the difficulties encountered in the first campaign related to the various integration times used by the different instruments. Also, cell measurements should be performed by all instruments. Although cell measurements poorly represent the difficulties associated with real atmospheric measurements, they would probably reveal more clearly differences in results due to the retrieval methods used. Finally, the atmospheric pollution conditions encountered in the first campaign did not allow the detection limits of each experimental set-up to be reached.

All these points were addressed in a second campaign which took place in September 1994 at the Weybourne Atmospheric Observatory situated in a remote area on the coast of East Anglia (UK). The comparisons concerned SO_2, NO_2, O_3, NO_3, HCHO and HONO.

The campaign compared seven different instruments through ten days of continuous measurement under field conditions. The DOAS were compared with commercial point monitors for NO_2, O_3 and SO_2 and a home-built monitor for HONO. The preliminary results are very encouraging although the full analysis of the data has not yet been completed. Two other comparisons were also made during the campaign. Firstly the DOAS were tested using cells containing known amounts of NO_2 and SO_2. The agreement was to within about 7 %. Secondly the group's software was tested using synthetic spectra including NO_2, O_3 and HCHO. Agreement was found within about a 5 % error. This represents a considerable improvement on the results of the first comparison exercise.

In the framework of the TOPAS sub-project, tropospheric measurements of trace gases by using DOAS were significantly improved from the instrumental and retrieval method point of views. They were fully validated, mainly during the two intercomparison campaigns conducted in 1992 and 1994. This kind of exercise demonstrated the necessity for quality assurance and quality control procedures in order to obtain reliable tropospheric data for high quality scientific studies of the troposphere. It is also obvious that further improvements are required and can only be obtained by means of instrument-oriented projects in future tropospheric research programmes (*e.g.* EUROTRAC 2).

One instrument has been fully commercialised, namely the "Système d'Analyse par Observations Actives" (SANOA) developed by the CNRS French group lead by Dr. J.-P. Pommereau. For the first time, a Fourier transform spectrometer has been used in the UV-visible range for DOAS and its performances demonstrated. This instrument is commercialised by BRUKER (Germany). Furthermore, HOFFMANN MESSTECHNIK (Germany) has also developed an instrument for DOAS in the troposphere which will be used in 1996 by an Austrian research institute. It is very unfortunate that OPSIS declined to participate in the intercomparison campaign in 1992. Consequently, the performances of this commercial instrument were not evaluated.

11.5 References

1. B. Galle, H. Axelsson, P. Ragnarsson, M. Rudin; A transmitting/receiving telescope for DOAS measurements using retroreflector technique, Proceedings of OSA-conference, Optical Remote Sensing of the Atmosphere, Incline Village, Nevada, Febr. 12-15, 1990.
2. H. Axelsson, B. Galle, K. Gustavsson, P. Ragnarsson M. Rudin; A transmitting/ receiving telescope for DOAS measurements using retroreflector technique, in: P. Borrell, P.M. Borrell, W. Seiler (eds.), *Proc. EUROTRAC Symp. '90*, SPB Academic Publishing, The Hague 1991, pp. 477–478.
3. P. Laville, J.P. Pommereau, F. Goutail; Evaluation of a long path diode array spectrometer for O_3, NO_2, SO_2 and water vapour monitoring in: P. Borrell, P.M. Borrell, W. Seiler (eds.), *Proc. EUROTRAC Symp. '90*, SPB Academic Publishing, The Hague 1991, pp. 479.
4. F. Goutail, J.P. Pommereau, M. Nunes-Pinharanda; Ambient air monitoring by differential absorption spectroscopy in the ultraviolet and the visible: the SANOA instrument, *Proc. Symp. Atmos. Spectroscopy Applications*, Reims 1993.
5. J.U. White; Very long optical paths in air, *J. Opt. Soc. Am.* **66** (1976) 411-416.
6. D. Ritz, M. Hausmann, U. Platt; An improved open path multi-reflection cell for the measurement of NO_2 and NO_3, in: H.I. Schiff, U. Platt (eds.), *Proc. Int. Symp. on Environmental Sensing, in Optical Methods in Atmospheric Chemistry, Proc. SPIE* **1715** (1992) 200-211.
7. G.M.B. Dobson; A photoelectric spectrophotometer for measuring atmospheric ozone, *Proc. Phys. Soc.* **43** (1931) 324-328.
8. A.W. Brewer, C.T. McElroy, J.B. Kerr; Nitrogen dioxide concentration in the atmosphere, *Nature* **246** (1973) 129-133.
9. J.F. Noxon; Nitrogen dioxide in the stratosphere and troposphere measured by ground-based absorption spectroscopy, *Science* **189** (1975) 547-549.
10. J. Blamont, J.-P. Pommereau, G. Souchon; *CR Acad. Sciences, Serie B* (1975) 247-252.
11. U. Platt, D. Perner; Measurements of atmospheric trace gases by long path differential UV/Visible absorption spectroscopy, *Springer Ser. Opt. Sci.* **39** (1983) 95-105.
12. U. Platt; Differential optical absorption spectroscopy (DOAS), in: M.W. Sigrist (ed.), *Air Monitoring by Spectroscopic Techniques*, Chemical analysis series **127** (1994) 27-84.

13. J.M.C. Plane, N. Smith; Atmospheric monitoring by differential optical absorption spectroscopy, in: R.E Hester R.J.H. Clark (eds.), *Spectroscopy in Environmental Sciences*, Wiley, London 1995, pp. 223-262.

14. J. Stutz, U. Platt; Numerical analysis of DOAS spectra with linear and nonlinear least square fits, in: P.M. Borrell, P. Borrell, T. Cvitas, W. Seiler (eds.), *Proc. EUROTRAC Symp. '94*, SPB Academic Publishing, The Hague 1994, pp. 930-934.

15. M. Carleer R. Colin, A.C. Vandaele, P.C. Simon; Measurement of NO_2 and SO_2 absorption cross sections, in: P.M. Borrell, P. Borrell, T. Cvitas, W. Seiler (eds.), *Proc. EUROTRAC Symp. '92*, SPB Academic Publishing, The Hague 1993, pp. 419-422.

16. J.-P. Pommereau, F. Goutail, M. Nunes-Pinharanda; Application de la spectroscopie UV-Visible à la mesure de la pollution urbaine, *Coll. Pollution Atm. à l'échelle locale et régionale*, ADEME, Paris 1993.

17. F. Goutail, J.P. Pommereau, M. Nunes-Pinharanda; Surveillance de la composition de l'air par spectrométrie uv-visible, *Proc. Coll. Horizons de l'Optique*, Limoges 1993.

18. J. Bösenberg, B. Galle, J. Mellqvist *et al.*; Tropospheric ozone Lidar experiment, TROLIX '91 field phase report, *Report* Nr. 102, Max-Planck-Institut für Meteorologie, Hamburg 1993.

19. N. Smith, J.M.C. Plane, C.-F. Nien, P.A. Solomon; Night-time radical chemistry in the San Joaquin Valley, *Atmos. Environ.* in press, 1996.

20. U. Platt, A.M. Winer, H.W. Biermann, R. Atkinson, J.N. Pitts, Jr.; Measurement of nitrate radical concentrations in continental air, *Environ. Sci. Technol.* **18** (1984) 365-369.

21. A.-C. Vandaele, P.C. Simon, J.M. Guilmot, M. Carleer, R. Colin; SO_2 absorption cross section measurement in the UV using a Fourier transform spectrometer, *J. Geophys. Res.* **99** (1994) 25599-25605.

22. H. Axelsson, H. Edner, B. Galle, P. Ragnarsson, M. Rudin; Differential optical absorption spectroscopy (DOAS) measurements of ozone in the 280-290 nm wavelength region, *Appl. Spectrosc.* **44** (1990) 1654-1658.

23. H. Axelsson, H. Edner, B. Galle, P. Ragnarsson, M. Rudin; Differential absorption cross section of ozone in the 250-350 nm wavelength range, in: P. Borrell, P.M. Borrell, W. Seiler (eds.), *Proc. EUROTRAC Symp. '90*, SPB Academic Publishing, The Hague 1991, pp. 475–476.

24. C. Camy-Peyret, B. Bergquist, B. Galle, M. Carleer, C. Clerbaux, R. Colin, C. Fayt, F. Goutail, M. Nunes-Pinharanda, J.-P. Pommereau, M. Hausmann, F. Heinz, U. Platt, I. Pundt, T. Rudoph, C. Hermans, P.C. Simon, A.-C. Vandaele, J. Plane, N. Smith; Intercomparison of instruments for tropospheric measurements using differential optical absorption spectroscopy, *J. Atmos. Chem.* **23** (1996) 51-80.

Chapter 12

Individual Reports from TOPAS Contributors

12.1 The Belgian Contribution to Differential Optical Absorption Studies of the Troposphere between 1990 and 1995

R. Colin[1], M. Carleer[1], J.M. Guilmot[1], P.C. Simon[2], A.C. Vandaele[2], C. Hermans[2], P. Dufour[3] and C. Fayt[3]

[1] Laboratoire de Chimie Physique Moléculaire, Université Libre de Bruxelles, Brussels, Belgium
[2] Institut d'Aéronomie Spatiale de Belgique, Brussels, Belgium
[3] Laboratoire d'Informatique, Université de Mons-Hainaut, Mons, Belgium

Summary

The work accomplished during the period 1990 to 1995 by the three Belgian groups participating in the EUROTRAC-TOPAS project is described. It involved the construction and permanent improvement of two DOAS (Differential Optical Absorption Spectroscopy) spectrometers. One of these is a Fourier transform instrument which was applied to atmospheric measurements in the Visible-UV region for the first time. Long path absorption set-ups were constructed. Software was developped to operate and automate the instruments and specific algorithms were written for the retrieval of concentrations from the atmospheric spectra. Both instruments were validated by participating in two EUROTRAC-TOPAS intercomparison campaigns which gathered instruments from six european countries. Finaly, high resolution absorption cross sections of several atmospherically relevant species were measured in the laboratory.

Aims of the research

The principal aim of the EUROTRAC-TOPAS subproject was to develop reliable high precision equipment for visible and UV spectroscopic long path absorption measurements of the minor constituants of the troposphere, using the Differential Optical Absorption Spectroscopy technique (DOAS). The three above mentioned belgian groups have contributed to this aim by the following accomplishments during the 1990-1995 period :

- development of grating spectrometers using multi element detectors
- introduction of Fourier transform spectrometers in these type of measurements
- development of appropriate software algorithms for the retrieval of concentrations from atmospheric spectra
- participation in intercomparison campaigns aiming at the evaluation of the performance, precision and accuracy of various instruments developped in Europe during this period
- measurements in the laboratory of precise UV-Visible absorption cross sections of several atmospherically important molecules.

Scientific results

Instrument developments

Two spectrometers, one based on grating dispersion and one on Fourier transform interferometry were adapted to UV-Visible atmospheric monitoring.

– A first system based on a grating spectrometer was developed, using a 1/8 m crossed Czerny-Turner instrument from ORIEL (type Multispec) with an *f* number of 3.7. This instrument has a flat focal field adequate for use with photodiode arrays (PDA) up to 25 mm long. The combination of an entrance slit of 100 μm with a grating with 600 grooves/mm and a 1024 pixel PDA provides a measured full width at half maximum (FWHM) of 1.2 nm. A typical spectral window extends from 270 to 600 nm. In this case, the sampling defined by the size of the pixel (25 μm) and the spatial resolution corresponding to the FWHM is 4 pixels. The detector is a EG&G PDA of 1024 pixels which can be cooled to –45 °C by means of a two-stage Peltier cooler in order to reduce the dark current. Such a low temperature can be only reached by running methanol at –5 °C in a cooling loop connected to an external cryostat. The readout electronics were manufactured by Princeton Instrument Inc. and provide a dynamic range of 16 bits. The detector assembly is integrated in a vacuum-tight container evacuated by a small pump to avoid water vapor freezing on the detector at low temperature. Filters can be placed in front of the entrance slit in order to exclude second order overlap. The filter wheel has a closed position used to measure the dark count corresponding to each measurement integration time.

  The system includes a computer, associated with the detector controller and the spectrograph with a self-scanning photodiode array of 1024 pixels. The detector controller provides power, thermostating and timing signals to the detector head, coordinates data gathering with the experiment, sets exposure time, digitizes and averages data and transmits it to the computer. New software has been written in order to make instrument automation possible. Successive spectra can be taken with the filter wheel at different positions and with exposure times and numbers of scans automatically computed according to the current signal.

This system was improved by using another PDA detector (EG&G 1453, 1024 diodes), which does not require to be continuously evacuated and whose controller can be accessed through a classic IEEE input/output card. A second system was built on a 0.275 m Czerny-Turner SPECTRAPRO spectrometer (Acton Research Corporation). This spectrometer is specially adapted for use with a CCD type detector. The input light enters the spectrometer through an optical fiber. Three different gratings are mounted inside the instrument and their positions can be controlled by the same software used for the detector readout. The detector is a 512 × 512 pixels EG&G CCD detector, enhanced for use in the UV region.

- The adaptation of a Fourier transform spectrometer to field atmospheric measurements in the UV-Visible was carried out progressively throughout the period considered. In a first step, a fixed high resolution BRUKER 120HR spectrometer situated in the University of Brussels laboratory was tested in the urban site of the University campus using a long path (1 km) set-up (see below). These preliminary experiments carried out in 1990 demonstrated for the first time the possibility to use Fourier spectrometry to obtain reliable concentration measurements of tropospheric species (SO_2, O_3 and NO_2). The spectrometer was controlled by an Aspect computer which also digitizes the interferogram. The use of a high speed vector processor Fourier transforms the interferogram into a spectrum in a matter of a few seconds, depending on the required resolution. The choice of the light source (either a 1000 W high pressure "ozone-free" xenon lamp or a 250 W tungsten filament lamp) and of the detector (either a solar blind UV vacuum-diode or a silicon photodiode) depends on the spectral region investigated. Two regions are thus defined : one spans from 38000 to 27000 cm^{-1} (260-370 nm) and the other from 30 000 to 14 300 cm^{-1} (330-700 nm). During these preliminary tests, the resolution was 16 cm^{-1} (~0.15 nm) and the dispersion of the spectra is then 7.7 cm^{-1} per spectral point. Interferograms are digitized and recorded during the forward and the backward travels of the moving mirror, in a double sided mode; each spectrum is the result of the Fourier transform of the coaddition of 2000 interferograms. The time required to record a spectrum was about 45 minutes which is mainly due to the frequency response of the detector-preamplifier combination used. Software retrieval programs were also developped for the particular needs of a Fourier type spectrum.

  In 1993 a new mobile high resolution Fourier transform spectrometer (BRUKER 120M) adapted to UV-Visible atmospheric measurements was developped with the collaboration of the BRUKER SPECTROSPIN Co.

Several unfavorable features of the previous instrument were corrected, eg. measurement time were reduced from 45 to 5 min. for typical molecules such as SO_2, NO_2 and O_3 and a path length of ~500 m.

The main advantages of Fourier transform spectrometry compared to grating instruments are (i) the large spectral coverage achieved in one sampling, allowing to study many molecules simultaneously, (ii) an internal high precision wavenumber calibration device (He-Ne laser line), (iii) a constant, linear dispersion wavenumber scale, (iv) a simple, well-known instrumental line shape function when perfectly tuned and (v) an available high resolution (0.01 cm^{-1} in the UV-Visible) when required by sharp structured absorption features.

Long path absorption devices

A long path (784 m) absorption system was set up on the campus of the University of Brussels. It served for the preliminary Fourier experiments described above and was used by almost all participants in the 1992 Intercomparison campaign to be described below. The light source consists either of a 800 watt "ozone free" xenon lamp or of a Tungsten filament, depending on the spectral range studied. The xenon lamp, used for the UV/Visible region is enclosed in a envelope preventing transmission of ultraviolet light which could possibly create ozone within the lamp housing, thereby affecting the quality of the ozone measurements. The Tungsten filament is used when working in the visible range.

The light is collimated onto a 30 cm Cassegrain-type telescope, which projects the beam onto a slightly parabolic mirror, situated 394 m away. This mirror sends the beam back into a second 30 cm Cassegrain telescope, which collimates the light onto the entrance aperture or slit of the spectrometer.

For use with the mobile FTS spectrometer, a single path mobile set-up using a Xe lamp source was developped for use over distances ranging from 150 to 400 m.

Software developments

In contrast to laboratory spectroscopy, in atmospheric spectroscopy, the true intensity $I_0(\lambda)$, as would be received from the light source in the absence of any atmospheric absorption, is usually difficult to determine. It would involve in principle removing the air from the absorption path. The solution to this apparent dilemna lies in measuring the so called "differential" absorption. This quantity can be defined as the part of the total absorption which is due to a given molecular species and which varies "rapidly" with wavelength, in contrast to the background absorption which varies "slowly" with wavelength. Accordingly the total intensity can be split into two parts $I(\lambda) = I_B(\lambda).\delta I(\lambda)$ where $I_B(\lambda)$ is due to the background absorption varying slowly with λ and describing a general "slope" and $\delta I(\lambda)$ shows rapid variations due to absorption lines or bands. The meaning of "rapid" and "slow" variation of the intensity as a function of wavelength is, of course, related to the observed wavelength interval and to the width of the absorption bands or lines to be detected. The extinction due to Rayleigh and Mie scattering can usually be assumed to be slowly varying with λ. This slow variation with λ is removed by the application of various filtering techniques, in order to leave the differential absorption peaks of the atmospheric trace gases. This filtering process is an important part of the DOAS (Differential Optical Absorption Spectroscopy) method; various methods are employed according to the types of spectrograph and detector used. Several methods such as polynomial interpolation, Fourier transform filtering and filtering after removal of residual emission peaks from the source were carefully compared in order to optimize the procedure. No unique method is adaptable to both Fourier and grating instruments and this called for an overall intercomparison of instruments and algorithms used by TOPAS groups which were carried out during the two campaigns to be described below.

Much software progress was accomplished also in order to make the instruments independent of expert personal assistance. Several automation programs were developped.

Intercomparison field campaigns

The Brussels campaign

In order to assess the accuracy and advantages of several recently built European DOAS instruments which use a wide variety of solutions to the above mentioned difficulties, a first international intercomparison TOPAS campaign was organised. It took place during two weeks (7-18 Sept. 1992) at an urban site in Brussels and brought together eight different instruments from six different groups (one from the UK, Germany, France, Sweden and two from Belgium).

Prior to the campaign, two preparatory meetings were held. At these meetings, a number of general principles for the conduct of the campaign were adopted :

(i) Apart from predetermined coordinated periods of measurements during which all instruments measured simultaneously, each group was free to make measurements within the constraints imposed on those instruments sharing a common light source. Before and after each coordinated period, calibration runs of the light sources were performed.

(ii) After each coordinated period, measurements were processed within 24 hours and given to the Referee without knowing the results of the other groups. These were then compared at briefings which followed each period.

(iii) Several of the instruments can measure many atmospheric species. However in order to maximise the efficiency of intercomparing the measurements, three target molecules (NO_2, O_3 and SO_2) were chosen in spite of the fact that some instruments could not cover all three species.

(iv) The spectrum of a low pressure Hg lamp was recorded by each instrument, configured in an identical manner to that used to acquire atmospheric spectra (*i.e.* same slit size). This provided the "working" resolution of each instrument.

Due to the fact that this campaign was held in an urban site, in which rapid fluctuations of the pollutants can occur, and also because the various instruments used different optical paths and different observation times, a very careful analysis of the data was needed. After a first comparison of results made without any constraints, it was obvious that a common set of absorption cross sections had to be used by each group. This led to improved cross section measurements (see below). The campaign also served to exchange various instrumental and software improvements between the groups.

The comparison showed that rapid fluctuations in concentrations were not responsible for the discrepancies between instruments using almost the same paths, because the overall accuracy for each molecule did not improve with respect to the comparison of the fast (non-integrated) measurements. The absolute accuracy did not vary in spite of a fairly large concentration variation of the three species investigated during specific coordonated periods (by factors of ~7, ~8 and ~6 for NO_2, O_3 and SO_2 respectively). It may therefore be stated that the dispersion of the results is mainly due to the instruments themselves including the use of individual algorithms for the retrieval of the data. The dispersion ($\pm 1\sigma$) of the DOAS measurements on NO_2, O_3 and SO_2 performed by this large variety of UV-Visible instruments using a path length of a few hundred meters was:

NO_2	:	$\pm 5 \times 10^{10}$	molec.cm^{-3}	or	2.0 ppb,
O_3	:	$\pm 6 \times 10^{10}$	molec.cm^{-3}	or	2.4 ppb,
SO_2	:	$\pm 1 \times 10^{10}$	molec.cm^{-3}	or	0.4 ppb.

The lower value for SO_2 was expected because the signature (cross section) of this molecule is stronger than that of the two others.

From the various comparisons made during the campaign it is clear that the small discrepancies between the instruments are not entirely random. It is difficult to attribute these differences to particular aspects of the hardware and software of the instruments, which have such different designs. The large number of variables involved, such as the technique for adapting the resolution of the cross sections to that of the instrument, the number of molecules measured simultaneously, the type of regression used, the need for a dark current spectrum and the retrieval procedure used, are all examples of the different approaches to the treatment of the raw data. After close examination of all these aspects, it appears that the differences in results between the various instruments are most likely to be linked to the various retrieval methods used, although it was concluded that these retrieval methods are well optimized and well adapted to each instrument.

The Weybourne campaign

In order to improve our knowledge of the accuracy of the presently available type of instruments, further work was required. In particular it appeared interesting to compare results obtained by the DOAS technique to those obtained by other techniques. In order to do so, it was felt useful to conduct a campaign in a more stable environment, where concentrations vary less than in an urban site. This would reduce the difficulties encountered in the first campaign related to the various integration times used by the different instruments. Also, cell measurements should be performed by all instruments. Although cell measurements poorly represent the difficulties associated with real atmospheric measurements, they would probably reveal more clearly differences in results due to the retrieval methods used. Finally, the atmospheric pollution conditions encountered in the first campaign did not allow the detecton limits of each experimental setup to be reached.

All these points were addressed in a second campaign which took place in September 1994 at the Weybourne Atmospheric Observatory situated in a remote area on the coast of East Anglia (UK). The comparisons concerned SO_2, NO_2, O_3, NO_3, HCHO and HONO. The results of this campaign, in which the same groups (except the Swedish group) took part, are now being examined.

Laboratory absorption cross section measurements

The choice of the absorption cross section values for atmospherically important molecules has a direct impact on the accuracy of the DOAS measurements. This technique requires cross sections at the resolution of the instrument used, and preferably differential cross sections measured by that particular instrument in order to eliminate systematic errors. Because the spectral resolution of the different available DOAS instruments varies between 0.15 and 1.2 nm, it became necessary to dispose of a new set of high resolution asorption cross sections and to convole them with the actual spectral line shape of each instrument in order to obtain an appropriate "common set" of cross sections. Such measurements were undertaken in the Laboratoire de Chimie Physique Moléculaire using the high resolution BRUKER Fourier transform spectrometer (120HR). A carefully designed temperature controlled absorption cell was constructed. Precise absorption cross section in the appropriate visible and UV regions were measured for SO_2, CS_2, NO_2 and N_2O_4.

Conclusions

During the course of the period 1990 to 1995 much progress has been made in the development of UV-Visible methods of determing the concentrations of minor constituants of the troposphere. The Belgian group have contributed to this. New instruments were constructed and older ones greatly improved, basic data such as high resolution cross sections have been measured, various retrieval algorithms were tested and a huge step forward was accomplished in validating the DOAS method as a whole by participating in intercomparison campaigns.

Some aspects have yet to be improved. One for example is the difficulty still encountered in obtaining measurements of the important OH radical. Considering its low concentration this could probably be achieved by the high resolution Fourier spectrometer using still longer path lengths. Results on other important tropospheric molecules (NO, CS_2, NH_3, benzene, toluene,...) having UV-Visible absorption features have been obtained by other TOPAS groups.

All these improvements have led the UV-Visible DOAS method of monitoring to be a highly performing, automated, precise and reliable technique in the study of atmospheric chemical processes.

Acknowledgments

This work was supported by the Belgian Federal Office for Scientific, Technical and Cultural Affairs (EUROTRAC/TOPAS programmes). Financial support for the acquisition of the instruments from the Fonds National de la Recherche Scientifique (Belgium) and National Lotery (Belgium) is gratefully acknowledged.

12.2 Development of DOAS for Atmospheric Trace Species Monitoring

Bo Galle, H. Axelsson, B. Bergqvist, A. Eilard, J. Mellqvist
and L. Zetterberg

Swedish Environmental Research Institute, P.O. Box 47086,
S-40258 Gothenburg, Sweden

Summary

The DOAS (Differential Optical Absorption Spectroscopy) technique has been developed for and applied in applications relevant to EUROTRAC. The development comprises a new single ended design with a transmitting/receiving telescope at one end and a retroreflector at the remote end. The design facilitates the recording of lamp spectra and the use of multiple paths. Software improvements involves algorithms for generation of calibration spectra from literature data and for optimisation of the evaluation algorithms. A previously not used spectral region for ozone evaluation is investigated. Use of this region considerably improves the detection limit of ozone when measured over long paths. Dual path spectroscopy is exploited as a means to improve detection limits as well as make possible absolute extinction measurements. A fully automatic system operating on two paths have been installed close to the TOR-station Rörvik on the Swedish west coast. Besides this the systems and methods developed have participated in a large number of field campaigns measuring NO_2, NO, NO_3, SO_2, CH_2O, HNO_2, NH_3 and various aromatic hydrocarbons, and thereby proved to be a valuable and reliable method for use in atmospheric research.

Aims of the research

The aim of the research has been to develop and apply the DOAS technique for background monitoring of trace gases of relevance for EUROTRAC.

Principal scientific results

The development can be divided into development of hardware, development of software and development of measurement strategies.

Development of hardware

DOAS systems

Three different DOAS systems have been developed, each of them optimised for different tasks:

SERI1 is a fully automated system optimised for background trace gas monitoring [1]. This system has a fibre-coupled 30 cm diameter transmitting/receiving telescope, specially designed to operate against retroreflectors on different paths. Computer controlled mirror alignment makes it possible to switch automatically between different optical paths as well as helps keeping the system aligned on long paths. A lamp shutter makes it possible to block the outgoing beam and record a background spectrum for subtraction of scattered sunlight. The dispersive element is a 0.5 m Czerny Turner spectrograph with computer controlled grating setting, and an effective resolution of 0.5 nm. Spectra are recorded at a repetition rate of 100 Hz using a slotted disc arrangement. By moving the grating the instrument is sequentially tuned to different wavelength regions depending on which species is studied. This optimises detection limit on behalf of simultaneousity. In background monitoring this is usually no problem, however, since the temporal variability is small.

SERI2 is a system optimised for urban air monitoring. In its normal operation this system is fibre-coupled to a 20 cm diameter transmitting/receiving telescope similar in design to the SERI1 telescope. The dispersive element is a home built spectrograph [2] with a curved holographic flat field grating. The detector is a cooled photodiode array of 1024 elements with an effective resolution of 0.1 nm. This system uses a fixed wavelength region, 270-370 nm, in which a number of interesting species (O_3, SO_2, NO_2, CH_2O, HNO_2) can be measured. The absorption bands of the various species may not always be optimal what regards detection limit, in the used wavelength region, but makes possible simultaneous measurements of the different species. This quality is, due to the high temporal variability and relatively high concentrations in urban air, regarded more important than optimal detection limit.

SERI3 is a system specially designed for the wavelength region 215-235 nm, in which a number of molecular species of interest for atmospheric chemistry (NO, NH_3 SO_2) exhibit absorption features which are suitable for DOAS measurement. At these short wavelengths all optical components (optical fibres, lamps) has to be carefully selected to minimise intensity losses. This system is capable of operation over up to 100 m [3, 4] and the length of the light paths is primarily limited by the absorption of atmospheric oxygen. In this system UV light from a 150 W xenon arc lamp (Hamamatsu L2273) is transmitted over the measurement path by means of two 20 cm diameter telescopes (f = 80 cm). Light is collected in an optical fibre and fed to a Czerny-Turner spectrometer (Thermo Jarrel-Ash, focal length 275 mm, *f*/3.8). The spectrum is scanned by means of an slotted disc arrangement, digitised in a 12 bit A/D converter and stored in a home built MCA plug-in card. Spectra are analysed on a IBM-compatible AT-computer using traditional DOAS evaluation algorithms.

A transmitting/receiving telescope

DOAS-measurements were originally performed using a *double-ended* system with a transmitting and a receiving telescope positioned some km apart. In a number of applications however it is more convenient, and sometimes even necessary, to use *single-ended* systems. For this purpose a combined transmitting/receiving system was designed [5, 6]. The system is designed to use different parts of the parabolic main mirror for transmitting and receiving, and uses retroreflectors at the remote end of the path. The design is simple, easy to align, has good light economy and is compact. With this system a number of advantages can be gained. As the lamp is located close to the spectrometer, lamp reference spectra can easily be obtained and the lamp can be intermittently blocked out by an automatic shutter to record background noise and light levels. Multiple path measurements can easily be performed by realignment of the main mirror towards different retroreflectors, opening up interesting measurement strategies as described below. Finally, measurements can be made also on remote locations with the need for electricity and a stable platform only at one end.

Development of software

Synthetic calibration

It is usually advantageous to utilise literature absorption cross sections in the retrieval of total columns from recorded DOAS spectra. Some species like O_3, NO_3 and NH_3 are because of their reactivity difficult to calibrate by laboratory methods, while for other species like NO, NO_2 and SO_2 the use of literature absorption cross sections makes calibration easy and consistent.

In order to be able to use literature absorption cross sections for calculation of concentrations from atmospheric spectra it is necessary to degrade the high resolution literature data in accordance with the specific instrument function of each instrument. To do this the instrument line shape (ILS) function is determined experimentally by means of a low pressure mercury lamp, which emits narrow spectral lines in the wavelength regions normally used in DOAS measurements. By convoluting the high resolution literature data with this measured ILS-function, a calibration spectrum with a given concentration can be synthesised, for a specific instrument. A system used in field operation may get misaligned or defocused making a new calibration necessary. By using literature absorption cross sections the system may easily be recalibrated by recording the ILS-function.

Algorithm optimisation

A computer program has been developed in which different parameters (polynom degree, fitting interval) in the DOAS evaluation procedure are varied and optimised using statistical methods [7]. By applying this program to a number of atmospheric spectra recorded under different conditions the evaluation parameters can be objectively optimised. This program also makes it possible to compare different algorithms and subroutines (*i.e.* polynom fitting contra Fourier filtering or smoothing for short pass filtering).

Ozone absorption cross sections

One species of vital importance in research concerning photochemical oxidants is ozone. The traditionally used Huggins bands (320-330 nm) has unfortunately relatively weak differential absorption cross sections leading to bad sensitivity in tropospheric measurements. An alternative is to measure around 250 nm where the differential absorption cross sections are much higher. Unfortunately also the absolute absorption cross section is high in this region, restricting the pathlength to a few hundred meters. Thus after a careful investigation of the spectrum of ozone and its interfering species, we found the region 280-285 nm to be a good alternative [8, 9]. Under normal conditions measurements over several km are feasible, and an improvement of nearly one order of magnitude in detection limit as compared to the Huggins band can be achieved.

Development of measurement strategies

Dual beam measurements

In this method two spectra are recorded, close in time, along two paths with the same direction but with different pathlengths. If these two spectra are then divided, instrument factors, such as spectral structures from the light source and spectrometer anomalies, are to a high degree cancelled. This method has been used routinely to improve the detection limit of the system when measuring background concentrations

Extinction measurements

The dual beam technique also makes possible the measurement of absolute atmospheric absorption spectra. Based on this, a method for measurement of spectrally resolved long path averages of particle extinction has been developed [10-12].

In tropospheric research there is a need to determine the attenuation due to particles and molecules. The need may be direct, as in studies of spectral balance or photochemical reactions, or indirect as in corrections for range resolved LIDAR measurements of ozone.

In the method two spectra are recorded over two different, nearly collinear, paths with lengths A and B. The use of a retroreflector based system with stepper motor controlled main telescope mirror, makes it possible to perform the measurements at short intervals and using a common light source. Thus by dividing the spectra, instrument factors are cancelled and an atmospheric extinction spectrum over the path B-A is obtained.

The extinction in the atmosphere is due to absorption and scattering of particles and molecules. With differential absorption spectroscopy (DOAS) the concentration of the absorbing gases can be derived. If the absolute absorption spectra of these gases have been previously recorded, then the absorption contribution from these gases to the extinction in the measured atmospheric spectrum can be removed by using the Beer-Lambert law. The remaining attenuation in the spectrum is then due only to particle absorption and scattering of particles and molecules. Knowing the atmospheric density (pressure and temperature) the molecule scattering, Rayleigh-scattering, can removed, leaving only the extinction due to particles. This method was successfully tested during the EUROTRAC ozone-LIDAR intercomparison TROLIX'91 [13].

Passive DOAS

Passive DOAS, using zenith scattered sky light has become an important technique for analysing total columns of stratospheric molecules. The technique may however also be used for flux measurements in tropospheric applications. A method based on this principle was developed in an attempt to measure emission of NO_2 from starting aircraft under realistic conditions [14, 15]. By measuring the zenith sky total column of NO_2 when the exhaust plume from starting aircraft is drifting over the instrument, a vertically integrated concentration of NO_2 is obtained. If the wind is perpendicular to the runway, and spectra are collected during the entire passage of the plume, the emission of NO_2 in g/s can be calculated using the wind speed at the proper height. In this application the instrument was static while the plume was moving with a known speed. Alternatively the instrument may move and thus the total emission of different species from various activities may be determined.

Applications

TOR station Rörvik

At Draget, a few km south of the EUROTRAC TOR station Rörvik, a test site for the DOAS research has been established. The system is located in a small house in a remote area of the coast with access to three different paths; 1 × 9 km, 2 × 2.3 km and 2 × 0.8 km. When not participating in field campaigns, the SERI1 system is permanently installed at the site for automatic routine measurements of NO_2, SO_2 and O_3. During campaigns also NO_3 have been measured.

TOPAS intercomparison, Brussels

A DOAS intercomparison campaign involving 7 different groups and 9 different systems was conducted in Brussels Sept. 9-18, 1992. The aim of this campaign was to compare different instruments and evaluation algorithms as well as demonstrate to the EUROTRAC community the operability of the technique. The instruments were all located at *Laboratoire de Chimie Physique Moleculaire at Universite Libre de Bruxelles*. We participated successfully with the two fundamentally different systems *SERI1* and *SERI2*, applied over three different optical paths.

Ozone-LIDAR intercomparison TROLIX'91

The above mentioned extinction measurement method was tested in the TESLAS intercomparison campaign TROLIX'91 in Bilthoven, Netherlands, June 16-27,

1991 [13]. The idea was to compare the theoretical LIDAR corrections for particle attenuation and Rayleigh scattering with our measured values. DOAS measurements were made over two paths, 885 m. and 2 km respectively, and covering the wavelength range 270-315 nm, an interval containing most of the wavelengths used in the LIDAR ozone measurements. During some periods two of the LIDAR systems were directed horizontally giving paths nearly collinear with our DOAS paths. Thus the different systems could be intercalibrated. During the campaign the DOAS system was also measuring ground level concentrations of O_3 and SO_2. In connection with the campaign we also performed an NO_2 intercomparison between our DOAS and the RIVM NO_2 LIDAR over a nearly collinear path of 1116 m. The comparison was made without any preceding intercalibration and showed good agreement between the two methods.

Other field experiments

Besides the above mentioned EUROTRAC activities the described DOAS systems (*SERI1, SERI2, SERI3*) have been applied in a large number of field campaigns. These include measurements of fluxes of NH_3 and NO from spreading of manure [3, 4], measurements of hydrocarbon emissions from industry [16, 17], studies of HNO_2 formation downwind a city and urban air measurements.

Conclusions

The DOAS technique has, through a number of different field activities, proven to be a valuable and reliable tool for atmospheric research.

Acknowledgements

The financial support of the Swedish Environmental Protection Agency as well as the Swedish Environmental Research Institute is greatly appreciated.

References

1. B. Galle, H. Axelsson, J. Mellqvist, L. Zetterberg; A DOAS System for background Air pollution monitoring, in: P.M. Borrell, P. Borrell, T. Cvitas, W. Seiler (eds.), *Proc. EUROTRAC Symp. '92*, SPB Academic Publishing, The Hague 1993, p. 214.
2. J. Pettersson; Spektrometer för UV-ljus, *IVL-Report* 1990 (in Swedish).
3. J. Mellqvist, B. Galle; Application of open path DOAS and FTIR for a static chamber approach to measure emissions of NH_3, N_2O and NO from a fertilized field, to be published.
4. S.G. Sommer, H. Mikkelsen, J. Mellqvist; Evaluation of meteorological techniques for measurements of ammonia loss from pig slurry, *Agric. For. Meteorol.* in press.
5. B. Galle, H. Axelsson, P. Ragnarsson, M. Rudin; A transmitting/receiving telescope for DOAS measurements using retroreflector technique, *Proc. OSA Conf. Optical Remote Sensing of the Atmopshere*, Incline Village, Nv. 1990.

6. H. Axelsson, B. Galle, K. Gustavsson, P. Ragnarsson, M. Rudin; A transmitting/receiving telescope for DOAS measurements using retroreflector technique, in: P. Borrell, P.M. Borrell, W. Seiler (eds.), *Proc. EUROTRAC Symp. '90*, SPB Academic Publishing, The Hague 1991, pp. 477-478.

7. C. Bjur; Datoroptimering av utvärderingsalgoritmer för DOAS spektroskopi, *IVL Report* 1993 (in Swedish).

8. H. Axelsson, H. Edner, B. Galle, P. Ragnarsson, M. Rudin; Differential optical absorption spectroscopy (DOAS) measurements of ozone in the 280-290 nm wavelength region, *Appl. Spectrosc.* **44** (1990) 1654-1658.

9. H. Axelsson, H. Edner, B. Galle, P. Ragnarsson, M. Rudin; Differential absorption cross section of ozone in the 250-350 nm wavelength range, in: P. Borrell, P.M. Borrell, W. Seiler (eds.), *Proc. EUROTRAC Symp. '90*, SPB Academic Publishing, The Hague 1991, pp. 475-476.

10. B. Galle, J. Mellqvist; A long path method for measurement of spectrally resolved particle attenuation, *Proc. SPIE Conf. Optical Remote Sensing for Environmental Monitoring*, Berlin 1992.

11. B. Galle, A. Apituley, J. Bösenberg, H. Edner, J. Mellqvist, A Long Path Optical Absorption Method for Aerosol Absorption Measurements, in: P.M. Borrell, P. Borrell, T. Cvitas, W. Seiler (eds.), *Proc. EUROTRAC Symp. '92*, SPB Academic Publishing, The Hague 1993, p. 213.

12. B. Galle, J. Mellqvist; A long path method for measurement of spectrally resolved particle attenuation". *Proc. SPIE Conf. Optical Remote Sensing for Environmental Monitoring*, Berlin 1992.

13. J. Bösenberg, B. Galle, J. Mellqvist *et al*,; Tropospheric ozone lidar experiment, TROLIX '91 field phase report, *Report* No. 102, Max-Planck-Institut für Meteorologie, Hamburg 1993.

14. B. Galle, A. Eilard, J. Mellqvist; Measurements of emissions of NO_2 from military aircraft takeoff, *IVL-Report* 1991 (in Swedish).

15. B. Galle, H. Axelsson, J. Mellqvist, M. Rudin; A method for measurement of large scale fluxes of tropospheric trace gases, in: P.M. Borrell, P. Borrell, T. Cvitas, W. Seiler (eds.), *Proc. EUROTRAC Symp. '92*, SPB Academic Publishing, The Hague 1993, p. 215.

16. H. Axelsson, H. Kloo, H. Edner, P. Ragnarsson; Long path DOAS measurements of light aromatic hydrocarbons, in: *Optical Remote Sensing and Applications to Environmental and Industrial Safety Problems*, A&WMA, Houston 1992.

17. H. Axelsson, H. Edner, A. Eilard, A. Emanuelsson, B. Galle, H. Kloo, P. Ragnarsson; Measurement of aromatic hydrocarbons with the DOAS technique, *Appl. Opt.* in press.

12.3 Differential Optical Absorption Studies at East Anglia

Nicola Smith, Hugh Coe, Beverley Allan and John Plane

School of Environmental Sciences, University of East Anglia,
Norwich NR4 7TJ, United Kingdom.

Summary

A new DOAS instrument was developed at UEA between 1992 and 1994, and has been deployed in major field campaigns at the Weybourne Atmospheric Observatory in north Norfolk, UK, and at the Izaña Observatory, Tenerife. The instrument also took part in TOPAS intercomparison exercises in Brussels and at Weybourne, where the target molecules included NO_2, O_3, SO_2, HONO, CH_2O and NO_3. The UEA DOAS has primarily been used to study the chemistry of the nitrate radical. In the marine boundary layer the NO_3 lifetime was observed to be less than 10 minutes, due to reaction with biogenic organic species such as dimethyl sulphide and the heterogeneous loss of N_2O_5 (and perhaps NO_3) on moist aerosol surfaces. In the free troposphere at very low relative humidities, the lifetime was found to be several hours.

Aims of the research

- to redesign and improve the original DOAS described by Plane and Nien [1],
- to install this instrument in the new Weybourne Atmospheric Observatory, operated by the University of East Anglia on the coast of north Norfolk
- to improve the performance of the instrument by taking part in the intercomparison exercises organised by TOPAS.
- to measure NO_3 in the marine boundary layer as part of a major campaign to study the night-time chemistry of this important radical
- to carry out a comparative study of NO_3 in the dry, relatively clean free troposphere over Tenerife

Principal experimental results

Fig. 12.1 illustrates the new DOAS design, which is significantly modified from our earlier configuration [1]. The instrument employs coaxial transmitting and receiving telescopes which use different sections of the same primary mirror in a Newtonian configuration. This is achieved with an additional secondary mirror. The transmitter, powered by a 450 W xenon arc lamp, uses the front secondary and the outside of the primary mirror to create an outgoing beam in the shape of an annulus. The retro-reflector then folds the light beam and displaces part of it to the inside of the annulus; this part of the return beam then arrives at the inner portion of the primary mirror and is directed by the rear secondary mirror into a 5 m quartz optical fibre. The received light is then dispersed by a 0.5 m Czerny-Turner spectrometer onto a 1024-element diode array [1]. There are several advantages to this optical design. First, the transmitter and receiver are in one location, and the need for a second telescope [1] is dispensed with. Second, a very accurate reflector can be used to reflect a large fraction of the light back into the telescope. Third, light losses due to beam divergence and atmospheric scintillations are reduced [2]. At the Weybourne Atmospheric Observatory (WAO) in North Norfolk, the light beam is folded back to the receiver by the retro-reflector situated 2.75 km from the observatory, providing an overall light path of 5.5 km.

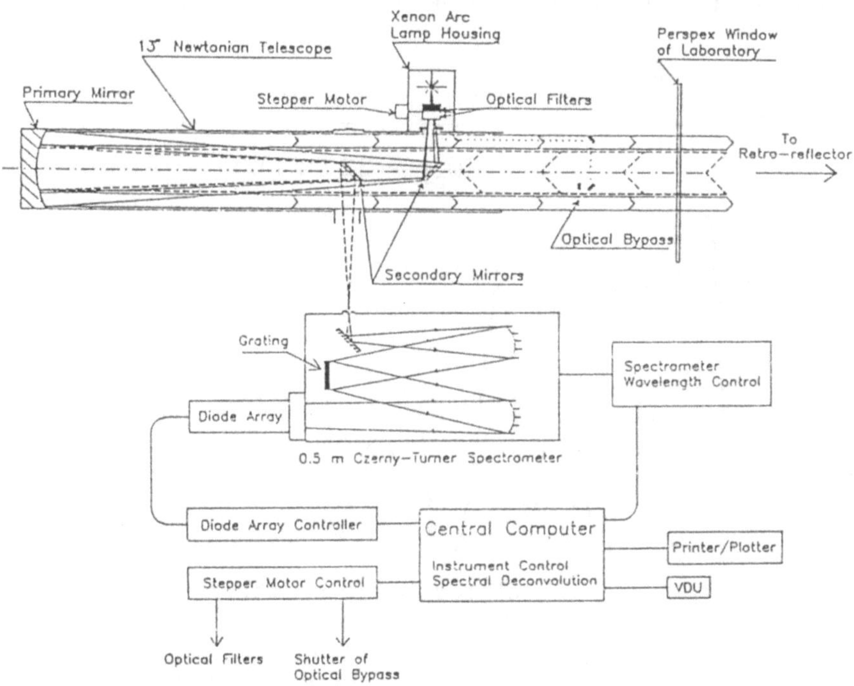

Fig. 12.1: Schematic diagram of the new coaxial DOAS system developed during this project.

The DOAS is completely automated for unattended deployment in the field. The instrument is controlled by a microcomputer which scans the spectrometer wavelength, interfaces to the diode array controller for initiating and recording spectra, and operates two optical shutters by means of stepper motors. One shutter opens the optical bypass to take spectra of the source lamp. The second shutter system inserts a 620 nm long-pass filter into the transmitter beam when making NO_3 measurements at 662 nm. This avoids the photolysis of NO_3 in the beam path by light below 620 nm, and excludes second-order light from reaching the spectrometer grating. The second shutter system can also insert a disc in front of the lamp to block the transmitted beam while a spectrum of the scattered light from the atmosphere is being recorded.

The spectral deconvolution routine has been described in detail by Plane and Nien [1] and Plane and Smith [2]. The atmospheric absorption spectrum is first converted into a differential optical density spectrum. A set of appropriate library spectra are then fitted to this spectrum by using a least-squares routine employing singular value decomposition. An example of such a fit is shown in Fig. 12.2, where an NO_2 reference spectrum has been fitted to an atmospheric spectrum in the 345-380 nm region. Fig. 12.3 illustrates excellent correlation between NO_2 measured by the DOAS and the concentration determined by a chemiluminescence point sampling instrument (Cranox) at the Weybourne Atmospheric Observatory.

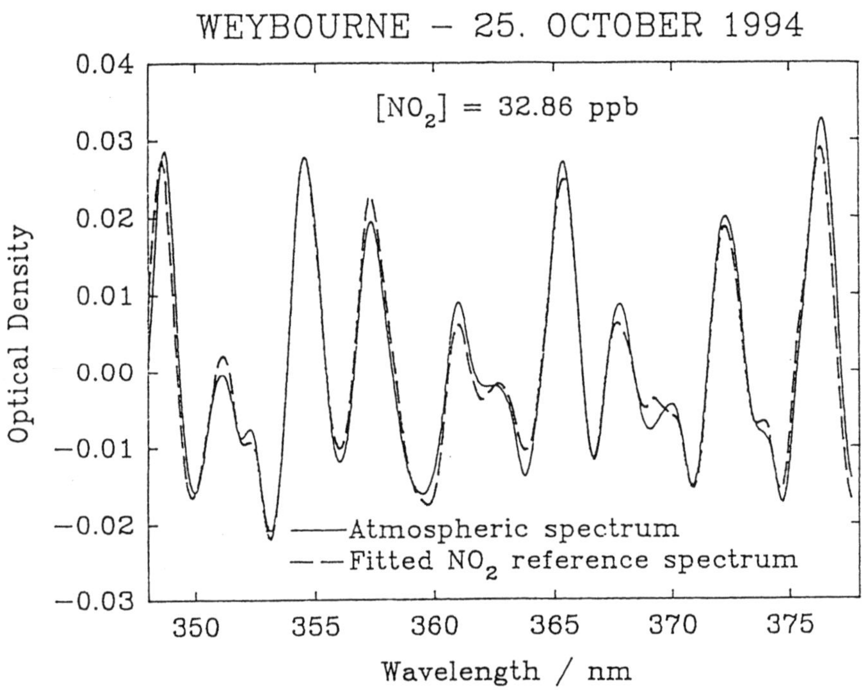

Fig. 12.2: Fit of an NO_2 reference spectrum to an atmospheric spectrum in the 345-380 nm region. The standard error of the fitted concentration is ± 0.4 ppb.

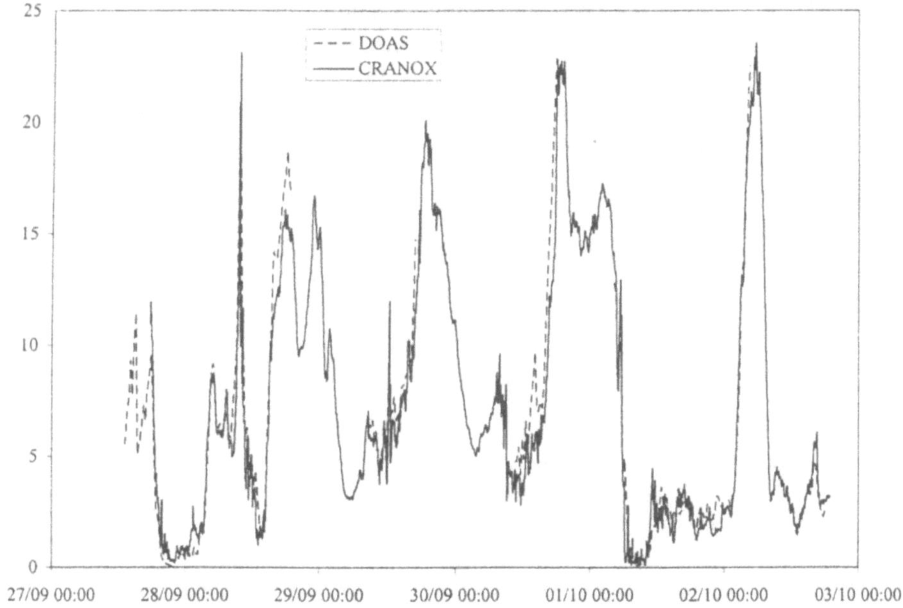

Fig. 12.3: Comparison of the NO_2 concentration measured by the DOAS against that measured by a chemiluminescence point sampling instrument (CRANOX) (courtesy of K. Clemitshaw and S.A. Penkett).

During this project the DOAS was employed in several field campaigns. The DOAS intercomparison exercises held in Brussels (September 1992) and at the WAO (September 1994) are described separately in the TOPAS subproject report. Two other field campaigns at the WAO were organised as part of the *Land Ocean Interaction Study*, a major U.K. community project funded by the Natural Environment Research Council. An important focus of LOIS is the chemistry of NO_3 in the marine boundary layer. This radical is a significant night-time oxidant of organic species, as well as providing a removal pathway for anthropogenic nitrogen oxides. DOAS measurements of the NO_3 radical utilise the strong absorption peak at 662 nm and are made by the UEA DOAS in the 645-680 nm wavelength region. There is additional absorption in this region due to the (311) vibrational overtone of water [1, 2], and this must first be removed from the night-time spectra. A daytime reference spectrum is therefore taken when photolysis reduces the NO_3 concentration to below the detection limit. This spectrum is then divided into a night-time spectrum in order to yield the absorption structure due to NO_3, as shown in Fig. 12.4. Fig. 12.5 shows a time-resolved concentration profile of NO_3 typical of the marine boundary layer in the autumn. Note the appearance of NO_3 at detectable levels (above 1 ppt) after sunset, and the abrupt disappearance at dawn (32 hours). The NO_3 lifetime ($t(NO_3)$) in the marine boundary layer at Weybourne has been found to be of the order of 10 minutes in the spring/summer, and appears to be controlled by dimethyl sulphide produced in the ocean by phytoplankton. $t(NO_3)$ decreases to about 2 minutes in the autumn/winter. Since the DMS concentration is low at this time of the year, the most likely sink for NO_3 is the heterogeneous reaction on moist aerosol surfaces of N_2O_5 (with which NO_3 is in equilibrium) [3]. Another highlight of these measurements in the marine boundary layer has been the observation of nitrous acid (HONO) being formed during the night at sub-ppb levels.

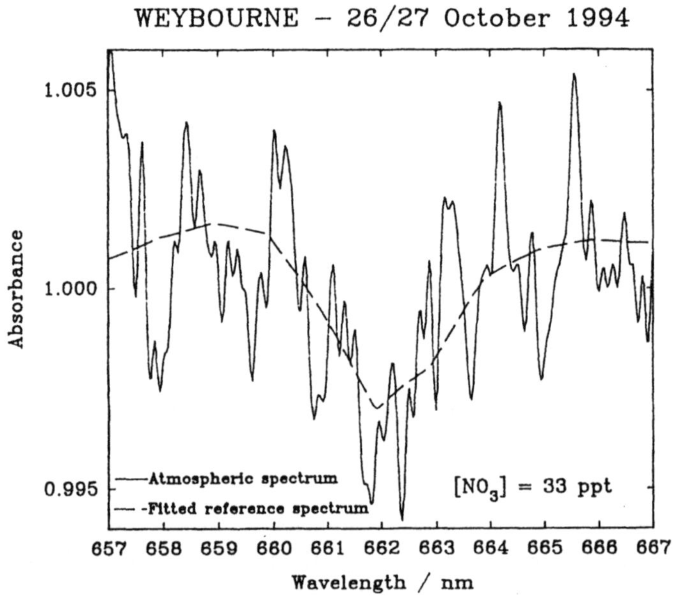

Fig. 12.4: Fit of an NO_3 reference spectrum to an atmospheric spectrum in the 657-667 nm region, obtained by dividing the night-time spectrum by a daytime reference spectrum taken the previous afternoon.

The DOAS was also employed to study NO_3 at the Izaña Observatory in Tenerife in the Canary Islands. The Meteorological Station lies on a mountain platform at an altitude of 2376 m above sea-level and 400 km west of Africa; it is a baseline station of the Background Air Pollution Monitoring Network managed by the World Meteorological Organisation. The air at the observatory is usually representative of the free troposphere, as it rests above an inversion at around 2000 m which acts as a cap for the pollution generated below. During the campaign, the concentrations of NO_3, NO_2 and O_3 were measured by the DOAS with a pathlength of 1.98 km. The measurements of NO_2 and O_3 were compared with point sampling instruments at the Meteorological Station, yielding excellent correlations: for NO_2, a 12-hour comparison produced a correlation plot with a slope of 0.997 and an r^2 of 0.678, despite the NO_2 concentration ranging between only 100 - 500 ppt; for O_3, the correlation plot produced a slope of 0.948 with an r^2 of 0.948.

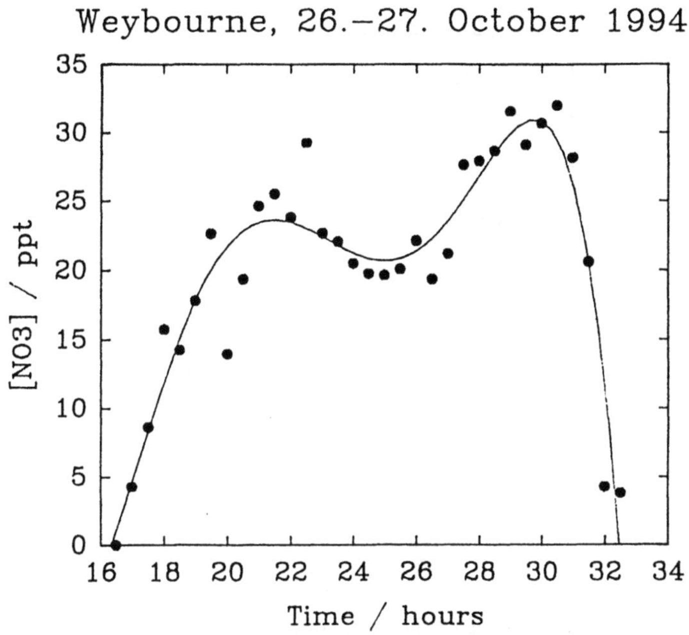

Fig. 12.5: A time-resolved profile of the NO_3 in the marine boundary layer at the Weybourne Atmospheric Observatory, North Norfolk. Note that 24 hours on the abscissa refers to midnight.

Measurements of NO_3 were made for 20 full nights during the campaign. The NO_3 was seen to be strongly correlated with O_3, and anti-correlated with the relative humidity (RH) during the night, as shown in Fig. 12.6. The reason for the inverse relationship with RH, which has been noted previously [3, 4], is most likely that the surfaces of aerosols (and the ground) must be deliquesced in order to provide effective sinks for gas-phase NO_3 or N_2O_5. However, the RH at Izaña at night-time was often very low. For instance, during the night shown in Fig. 12.6, $t(NO_3)$ is of the order of 4 hours when the RH is below 5 %. Since the measured concentrations of reactive hydrocarbons such as isoprene were very low, one possible mechanism for removing NO_3 is the homogeneous gas-phase reaction $N_2O_5 + H_2O \rightarrow 2\ HNO_3$. The field data indicate that this reaction would require a rate coefficient of about 9×10^{-21} cm^3 $molecule^{-1}$ s^{-1} in order to account for $t(NO_3)$, which is in sensible accord with the upper limit to this reaction of 2×10^{-21} cm^3 $molecule^{-1}$ s^{-1} at 298 K measured in the laboratory [5].

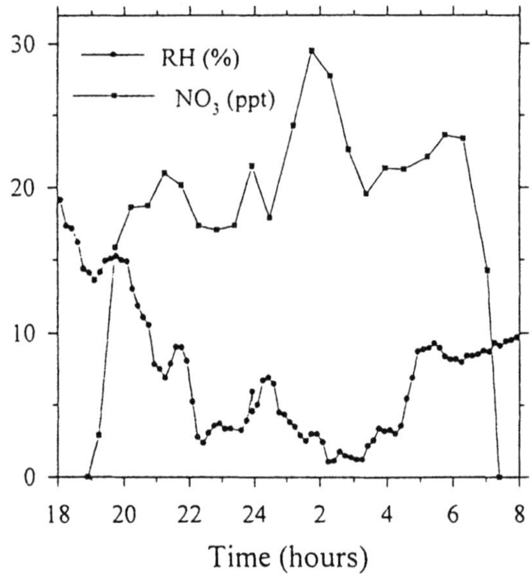

Fig. 12.6: Time-resolved profiles of the NO_3 concentration and relative humidity (RH), measured at the Izaña Observatory, Tenerife.

Conclusions

A new and much enhanced DOAS instrument has been successfully developed. Significant modifications to both the hardware and software resulted from participation in the two TOPAS intercomparison exercises. The instrument has been used in major field campaigns at two very different locations to study the chemistry of the nitrate radical and other trace species that play an important role in the oxidising capacity of the troposphere.

Acknowledgments

This work was supported by the Natural Environment Research Council (U.K.) and the Commission of the European Community.

References

1. J.M.C. Plane, C.-F. Nien; Differential optical absorption spectrometer for measuring atmospheric trace gases, *Rev. Sci. Instrum.* **63** (1992) 1867-1876.
2. J.M.C. Plane, N. Smith; Atmospheric monitoring by differential optical absorption spectroscopy, in: R.E. Hester, R.J.H. Clark (eds.), *Spectroscopy in Environmental Sciences*, John Wiley, London 1995, pp. 223-262.
3. N. Smith, J.M.C. Plane, C.-F. Nien, P.A. Solomon; Nighttime radical chemistry in the San Joaquin Valley, *Atmos. Environ.* in press.

4. U. Platt, A.M. Winer, H.W. Biermann, R. Atkinson, J.N. Pitts, Jr.; Measurement of nitrate radical concentrations in continental air, *Environ. Sci. Technol.* **18** (1984) 365-369.

5. R. Atkinson, D.L. Baulch, R.A. Cox, R.F. Hampson Jr., J.A. Kerr, J. Troe; Evaluated kinetic and photochemical data for atmospheric chemistry: Supplement III, *J. Phys. Chem. Ref. Data* **18** (1989) 881-1097.

12.4 A New Generation of DOAS Instruments

J. Stutz and U. Platt

Institut für Umweltphysik, Im Neuenheimer Feld 366, D-69120 Heidelberg

Summary

An improved design of a long path DOAS system based on a photodiode array detector (PDA) was developed. The known high residual structures, caused by a changing illumination of the spectrograph and detector could be overcome by the use of a quartz fibre mode mixer (FMM) in combination with the scanning multichannel technique (SMT). The system improves the detection limits of the older type instruments with 'slotted disk' detectors, which was determined by photon noise. The detection limit of the new instruments is given by residual structures in the spectra. As the photon noise, which is approximately a factor of ten lower with an PDA, is smaller than residual structures, the light path length in the atmosphere can be increased, thus further decreasing the detection limit.

A new analysis method, modelling atmospheric absorption spectra with the reference absorption spectra of the pure trace gases, enhances the detection limit by automatically correcting the spectral misalignment of these spectra. A new method to calculate the error on the concentration was developed. The error can now be calculated within a few percent of the true value. Monte-Carlo experiments showed, that the routine is able to identify trace gas absorptions even if the absorption structures are smaller than the 1σ noise in the spectrum.

Investigations of the behaviour of a newly built multireflection cell, based on the design of White 1976, showed that this systems are able to perform automatic long term DOAS measurements in the atmosphere. Problems with the lamp stray light in the system could be solved by a mathematical correction of measurements with an interrupted light path. The correction of this stray light is crucial to derive the correct trace gas concentrations.

Both instruments and the analysis routine were used in two TOPAS intercomparison campaigns (Brussels 1992 and Weybourne 1994). The trace gas concentrations were in good agreement with *in-situ* measurements.

Introduction

Measurements of tropospheric trace gas concentrations by differential optical absorption spectroscopy (DOAS) have been performed since 1979. With an instrumental setup developed by Platt *et al.* [1] it was possible to identify highly reactive species, *i.e.* OH [2], NO_3 [3], HONO [4], in the open atmosphere for the first time. The instruments were based on a 'slotted disk' detector (SD), which scanned the spectra by moving a slit over the spectrum and detecting the light by a photomultiplier. Typical measurement times of this systems are 20 minutes on a

5 km light path. After evaluating atmospheric spectra by removing all trace gas absorptions, residual structures of typically 2×10^{-4} (1σ) remained, thus defining the measurement limit (Table 12.1). The residual structure of the instrument were caused by the statistical photon noise, thus indicating that the method was limited by the measured light intensity.

The use of a photodiode array detector (PDA) [5] promised to improve the detection limit. The PDA records all wavelength simultaneously, thus collecting 100-1000 times more photons, compared to the sequential scanning of the SD. As this multiplex advantage of the PDA reduces photon noise by more than a factor of ten, an improvement of the detection limit was expected. Unfortunately, new problems were introduced using this detector in DOAS instruments. Random structures of 2×10^{-3} (1σ) were found in atmospheric spectra. This is much higher than the expected photon noise and even higher than the residual structures found with the SD instruments. The detection limit of 'naive' PDA systems, limited by the size of the residual structures, is therefore a factor of ten higher than for the SD instruments. The aim of this project was to identify the problems with PDA detectors and to find measures against the residuals structures occuring in atmospheric spectra to achieve better detection limits.

The use of a multireflection cell promised another improvement in DOAS measurements. The system is less sensitive to atmospheric turbulence due to the folded light path and the special optical setup. Due to the ability to change the absorption path length automatically, the multireflection cell can adapt better to changing atmospheric visibilty conditions, which often disturbs long path measurements. Another important advantage is that the system allows better comparison to *in-situ* measurements, as only a small volume of air is observed, in contrast to the long path systems where the measured concentrations are averaged over several kilometres. A system based on the design by White [6] was supposed to be built and its behaviour in the open atmosphere investigated.

Another important part of DOAS is the numerical analysis of atmospheric absorption spectra. Methods used in the past were based on linear least square procedures, often removing the different trace gas absorptions iteratively [7]. These method did not include the possibility to correct the spectral misalignment of the trace gas absorption reference spectra. A new analysis method based on the numerical modelling of atmospheric absorption spectra with the trace gas reference spectra, able to correct spectral misalignment, promised to improve the detection limit of the measurements. The estimation of the detection limit is based on a correct estimation of the measurement error by the analysis procedure. This estimation has to be included in the analysis routine.

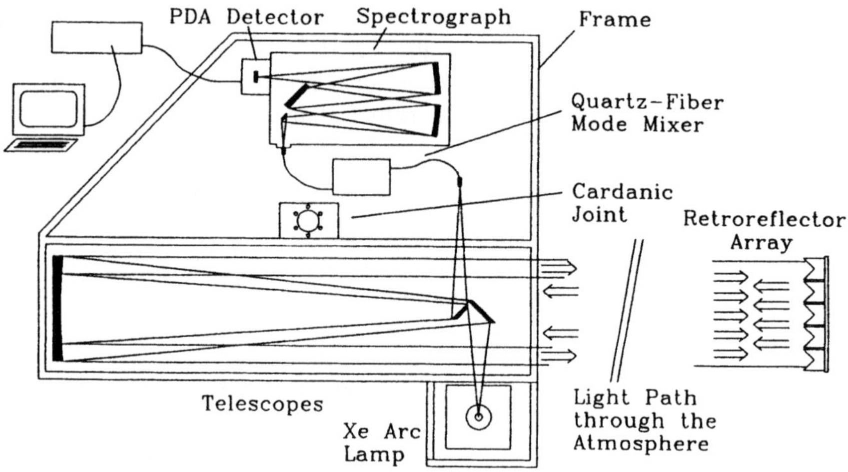

Fig. 12.7: Long path DOAS system with PDA. Light of a xenon arc lamp is collimated by a Newtonian telescope and transmitted through the open atmosphere. The light is reflected back by a retroreflector array in a few kilometres distance. The light is then collected by a second Newtonian telescope and focused on a quartz fibre. In the quartz fibre mode mixer the light looses its information of direction and is fed into the spectrograph-detector system. The telescope and spectrograph are fixed on a frame mounted an a cardanic joint. The hole instrument can be turned by stepper motors.

Table 12.1: Detection limit of long path DOAS systems with 'slotted disk' detector (SD) and photodiode array detector (PDA) The detection limits for the PDA and 5 km light path was derived during the TOPAS campaign in Weybourne 1994. The longer paths were calculated using the detection limits of the 5 km path length

	O_3 [ppb]	SO_2 [ppt]	NO_2 [ppt]	HCHO [ppt]	HONO [ppt]	NO_3 [ppt]
SD 5km	2	100	200	500	60	2
PDA 5 km	0.7 - 1.4	50-100	100 - 200	200 - 500	30 - 60	1 - 3
PDA 10 km	0.35 - 0.7	25 - 50	50 - 100	100 - 250	15 - 30	0.5 - 1.5
PDA 15 km	0.2 - 0.45	16 - 33	33 - 66	66 - 166	10 - 20	0.33 - 1
PDA 20 km	0.17 - 0.35	12 - 25	25 - 50	50 - 125	7 - 15	0.25 - 0.75

Results

Hardware improvements of the long path system

A long path DOAS system based on the coaxial design by Axelson [8, 9] was built, replacing the setup by Platt *et al.* [1]. A schematic drawing of the new setup is given in Fig. 12.7.

First measurements in the atmosphere, without the quartz fibre in Fig. 12.7, showed residual structures after the evaluation of the atmospheric spectra of 0.2 % (1σ). Further investigations revealed, that changing illumination of the spectrograph and detector are responsible for this structures. The change of illumination can be caused by a moving focus on the entrance slit due to turbulence and temperature fluctuations in the atmosphere. Different measures to reduce this effect were tested in the laboratory. A more uniform illumination of the spectrograph could be achieved by using diffusers in front of the entrance slit. The lowest residuals were found using a quartz fibre mode mixer, which reduced the structures by a factor of ten and more.

Another problem of the PDA is the correction of the pixel sensitivity. First a mirror-lens-system was integrated in the coaxial telescope to measure the lamp spectrum directly. Unfortunately it is impossible to measure the lamp spectrum without strongly disturbing the illumination geometry compared to atmospheric measurements. To measure the pixel sensitivity with the same illumination geometry the scanning multichannel method (SMT) proposed by Knoll *et al.* [10] was adapted to the DOAS method [11]. The pixel sensitivity can be derived by measuring $(2\,n + 1)$ atmospheric spectra S_i at different wavelengths $\lambda = \lambda_0 + i\,\Delta\lambda$ with $i = -n, -(n-1),..., -1, 0, 1,..., n-1, n$. The spectra are then added pixel by pixel. This sum ΣS_i, contains the pixel sensitivity and low pass filtered structures of the trace gas absorptions. The spectra $S'_i = S_i/\Sigma S_i$ are then aligned numerically to the wavelength λ. The sum of the aligned spectra S'_i is a high pass filtered atmospheric spectrum without the pixel sensitivity. Different to the approach of Brauers *et al.* [11] the spectral alignment is performed by a fitting routine calculating the optimal shift. This is necessary as the wavelength reproducibility of the spectrograph is only 0.01nm, whereas $\Delta\lambda$ was chosen to 0.4 nm, thus the actual wavelength shift $\Delta\lambda$ varies from its intended value. If not aligned correctly, the uncertainty of the wavelength leads to random variations in the spectral resolution of the instrument, making it impossible to remove strong absorption structures in the spectra which often overlay small absorption features. Therefore the exact wavelength alignment of the spectra S'_m is crucial for the use of SMT in DOAS measurements.

The use of the quartz fibre mode mixer (Fig. 12.7) and the SMT reduced the residual structures in atmospheric spectra by a factor of 20 down to 1×10^{-4} (1σ). This is comparable or better than the SD instruments described by Platt *et al.* [1, 7]. As the residual structures are not still larger than the photon noise, it is possible to reduce the measurement time to five minutes or less, depending on the wavelength region used. Alternatively it is possible to increase the light path length through the atmosphere and thus further improve the detection limit.

The detection limit of the instrument was verified during the TOPAS intercomparison campaign in Weybourne, England, with a light path of 5 km. During this campaign the instrument measured O_3, NO_2, SO_2, HCHO, HONO and NO_3 automatically. The experimental detection limits for this trace gases are listed in Table 12.1 together with calculated data to be reached with light path length larger than 5 km.

The multi-reflection system

A multireflection cell based on the design of White 1976 was built [12]. The baselength of the cell was chosen to 15 m allowing light paths as long as 2 km with 144 reflections in the cell. This high number of reflection could only be achieved with dielectric coated mirrors, which limits the spectral region that can be observed. With the chosen mirror coating only NO_2 and NO_3 were measured with the cell. The mechanical stability of the mirror mounting proved to be a

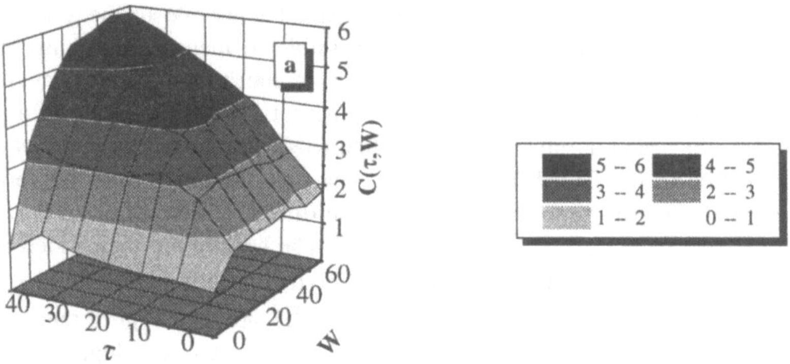

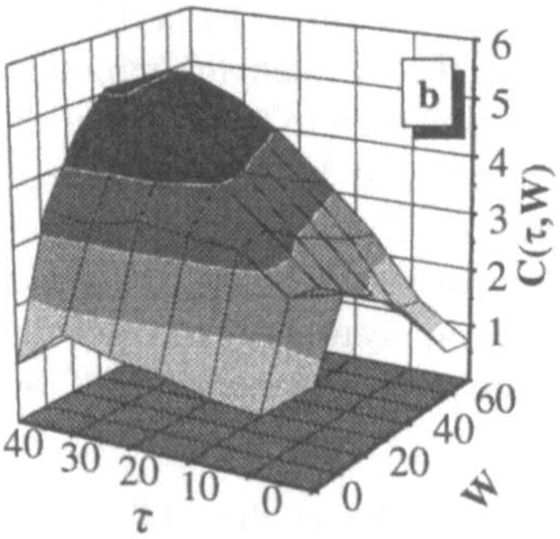

Fig. 12.8: Correction factor for the linear fit (a) and the nonlinear fit (b) in the analysis routine. The factor $C(\tau, W)$ depends on the half width of the absorption structure τ (in pixel) and the width W of the running mean (in pixel) of a noise spectrum that is similar to the residual spectrum

major problem, as changes in the adjustment reduced the observed light intensity. The mechanical setup was therefore rebuilt for maximum stability and an computer controlled adjustment routine was developed. The time between two manual readjustments was prolonged to 3 days, thus enabling the cell to perform automatic long term measurements. Another problem with multireflection cells is the stray light in the cell. Two different sources could be identified: Solar light scattered in the cell can easily be measured by taking spectra without the light of the lamp. Also the lamp light itself produces stray light in the cell. This is difficult to measure as the measured stray light intensity depends on the path length illuminated by the lamp and the path length observed by the spectrograph. By measuring the stray light interrupting the light path in the middle and then applying a mathematical correction this stray light contribution could be removed from the spectra.

The detection limits determined in field measurements in Heidelberg and Norwich are 1ppb NO_2 and 10 ppt NO_3 on a 2160 m light path respectively.

Software improvements

To derive concentrations of atmospheric trace gases, the atmospheric absorption spectra measured by DOAS instruments have to be analysed. A new routine was developed modelling the atmospheric absorption spectra with the reference spectra of the pure trace gases. The routine is a combination of a linear least square procedure, calculating the size of the absorptions in the atmospheric spectrum, and a nonlinear least square procedure, calculating the alignment correction for the reference spectra. The correction is necessary, as even small differences in the spectral alignment, due to instrument imperfections, can cause large systematic errors in the trace gas concentrations.

An important feature of the analysis routine is the estimation of the error of the concentrations. To calculated the correct error of the concentrations, the errors of the optical densities of the absorption bands and the error of the alignment have to be considered. If the alignment uncertainty is not considered the total error of the concentration can be underestimated by up to 50 %. A new method to calculate this total error was included in the analysis routine [13]. Monte-Carlo tests showed, that the error can now be calculated within 10 % of the statistical fluctuations due to noise in the spectra.

Unfortunately the described routine can only calculate the correct error if the errors of the pixel intensities are independent from each other. This is not the case if unknown absorption structures are found in the residual spectrum. In the presence of these structures the total error can be underestimated by the analysis routine by up to a factor of five. To derive the correct error the linear least square

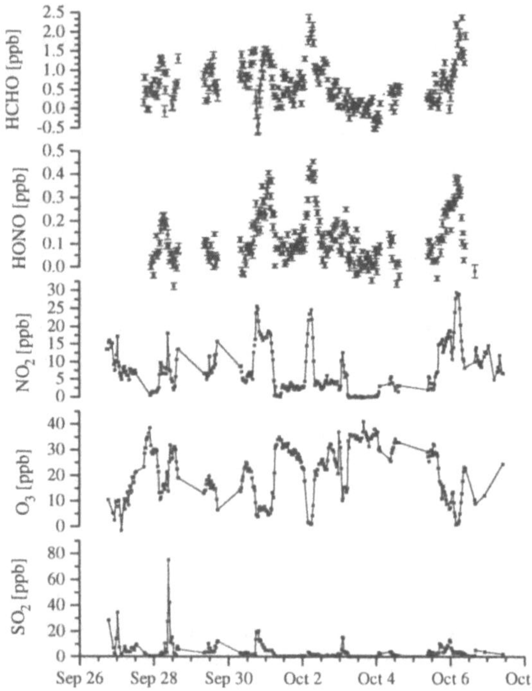

Fig. 12.9: Trace gas concentrations measured by the long path system during the TOPAS intercomparison in Weybourne 1994.

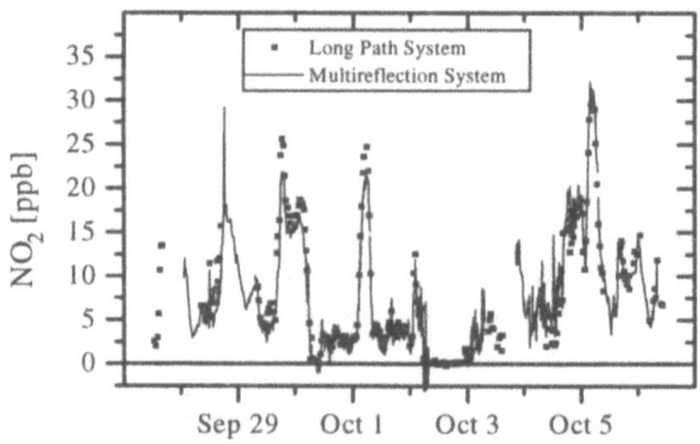

Fig. 12.10: Comparison of the NO_2 concentration measuresd by the long path system and the White cell during the TOPAS intercomparison in Weybourne 1994.

procedure has to be extended to a correlated least square [14]. This method requires the covariance matrix of the errors of the pixel intensities. This matrix is difficult to obtain. An approximation can be built using the description of the residuals by smoothed noise spectra (usually by a running mean of width *W*). As the method requires the calculation of the matrix, it is very time consuming and can only be used for the linear procedure. To correct the results of the normal least square routine, described above, Monte-Carlo tests were used to derive correction factors $C(\tau, W)$ depending on the spectral width of the absorption spectrum τ and the width of the running mean W describing the shape of the residual (Fig. 12.8). The errors calculated by normal least square procedures must be multiplied by $C(\tau, W)$ to obtain the correct error in the presence of residual structures.

With the factors and the new routine it is now possible to estimate the error of DOAS measurements correctly. The Monte-Carlo tests also showed, that trace gas concentrations can be detected even if their absorption structures are much smaller than the actual noise level.

Both instruments participated on the TOPAS intercomparison campaigns in Brussels and Weybourne. The results of Brussels will be published in 1995. The concentrations measured in Weybourne are shown in Fig. 12.9 and Fig. 12.10. The NO_2 concentrations measured by the long path system and the White cell are in good agreement especially for the last days of the campaign. The differences in the first part of the campaign can be explained by a drift of the resolution of the spectrograph of the multireflection cell.

Conclusions

The use of photodiode array detectors in long path DOAS instruments improved the detection limits of atmospheric trace gas concentrations compared to the older SD instruments, while at the same time allowing much higher (~5 fold) time resolution. The expected detection limits, given by photon noise and thus intensity, can be reached by inceasing the path length until photon noise is equal to the residual structures of the instrument. The campaign in Weybourne showed, that the concentrations of the measured trace gases are in good agreement to *in-situ* measurements. The campaign also showed that the instrument is capable of automatic long term measurements.

The analysis of atmospheric absorption spectra was improved with a new routine, which is able to correct spectral misalignment of the reference absorption spectra of the trace gases. A correct estimation of the errors of the trace gas concentrations further lowered the detection limit, as the least square methods are able to detect absorption structures smaller than the noise in the spectra.

The investigation of the behavior of multireflection cell proved that it is possible to perform automatic long term DOAS measurements with White cells. Supposing that the residual structures actually found in the spectra measured by this system can also be reduced by using a quartz fibre mode mixer and the scanning multichannel technique, an improvement of the detection limit to 100 ppt NO_2 and 1ppt NO_3 respectively, can be expected.

Acknowledgements

This study was funded by the BMFT (Grant No. 07EU737)

References

1. U. Platt, D. Perner, H. Pätz; Simultaneous measurements of atmospheric CH_2O, O_3 and NO_2 by differential optical absorption, *J. Geophys. Res.* **84** (1979) 6329-6335.
2. D. Perner, D.H. Ehalt, H.W. Paetz, U. Platt, E.P. Roeth, A. Volz; OH-radicals in the lower troposphere, *Geophys. Res. Lett.* **3** (1976) 466-468.
3. U. Platt P. Perner, G.W. Harris, A.M. Winer, J.N. Pitts; Detection of NO_3 in the polluted troposphere by differential optical absorption, *Geophys. Res. Lett.* **7** (1980) 89-92.
4. D. Perner, U. Platt; Detection of nirous acid in the atmosphere by differential optical absorption, *Geophys. Res. Lett.* **6** (1979) 917-920.
5. J. Stutz, U. Platt; Problems in using diode arrays for open path DOAS measurements of atmospheric species, Procedings international Symposium on Environmental Sensing, vol.1715, Berlin, 1992.
6. J.U. White; Very long optical paths in air; *J. Opt. Soc. Amer.* **66** (1976) 411-416.
7. U. Platt; Differential optical absorption spectroscopy (DOAS), in: M.W. Sigrist (ed.), *Air Monitoring by Spectroscopic Techniques*, Chemical Analysis Series, Vol. 127, 1994.
8. H. Axelson, B. Galle, K. Gustavsson, P. Ragnarsson, M. Rudin; A transmitting / receiving telescope for DOAS-measurements using retroreflector technique, *Technol. Dig. Series* **4** (1990) 641 - 644.
9. M. Hausmann, T. Rudolf, U. Platt; A new coaxial "Long-Path-DOAS"-system: First application to BrO measurements in the arctic troposphere, *Proc. Int. Symp. Environmental Sensing*, Vol. 1715, Berlin 1992.
10. P. Knoll, R. Singer W. Kiefer; Improving spectroscopic techniques by a scanning multichannel method, *Appl. Spectrosc.* **44** (1990) 776-782.
11. T. Brauers, M. Hausmann, U. Brandenburger, H-P. Dorn; Improvement of differntial optical absorption spectroscopy using multi-channel-scanning-technique, *Appl. Opt.* in press.
12. D. Ritz, M. Hausmann, U. Platt; An improved open path multi-reflection cell for the measurement of NO_2 and NO_3, *Proc. Int. Symp. Environmental Sensing*, Vol. 1715, Berlin 1992.
13. J. Stutz, U. Platt; Numerical analysis of DOAS spectra with linear and nonlinear least square fits, in: P.M. Borrell, P. Borrell, T. Cvitas, W. Seiler (eds.), *Proc. EUROTRAC Symp. '94*, SPB Academic Publishing, The Hague 1994, pp. 930-934.
14. D.L. Albritton, A.L. Schmeltekopf, R.N. Zare; An introduction to the least-square fitting of spectroscopic data, in: K. Narahari Rao, M. W. Weldon (eds.), *Molecular Spectroscopy: Modern Research*, Academic Press, Orlando 1976.

12.5 Development of a Long Path UV-visible Spectrometer for Atmospheric Composition Monitoring

Jean-Pierre Pommereau, Florence Goutail, Patricia Laville and Manuel Nunes-Pinharanda

Service d'Aéronomie du CNRS, 91371 Verrières le Buisson, France

Summary

The project was to develop a long path UV-visible spectrometer for the simultaneous measurement of a number of species by the DOAS technique in polluted urban and industrial areas. An instrument, the SANOA, has been built, with which several aspects of the use of this approach in the field were investigated. The measurement of the well known uv absorbing compounds SO_2, NO_2 and ozone, on a short distance with a relatively simple arrangement, has been validated. But in addition it was demonstrated that the method allows also the measurement of several other compounds present in polluted areas which cannot be monitored easily by other methods. The instrument is now commercialised for use in Air Quality monitoring networks as well as for scientific investigations anticipated for the near future.

Aims of Research

The general objective of the project was to explore the potential of UV-visible long path spectrometry in the measurement of a variety of atmospheric trace gases and to develop an instrument which could be used in air quality monitoring stations as well as during specific scientific campaigns aiming at the study of air pollution chemistry.

More specifically, the objectives included the design and the testing of a diode array instrument; the evaluation of its performances in the field as well as in the laboratory; the evaluation of its limitations; and the investigation of atmospheric absorbents present in the urban or industrial areas which could be measured using the method.

Principal Scientific Results

A long path instrument, the SANOA, has been designed at the beginning of the project in 1990-91, which was progressively improved following the results of the testing in the laboratory and in the field. Its performances for the measurement of NO_2, SO_2 and ozone in urban situation, have been first evaluated by comparison to those of other long path analysers during the first TOPAS campaign held at Brussels in 1992. As the result, this campaign demonstrated a reasonable agreement between the various instruments but highlighted also some limitations of the approach, particularly the offset of zero difficult to compensate for. This first campaign was followed in 1993-94, by a series of qualification tests for six absorbents (SO_2, NO_2, O_3, CHOH, C_7H_8 and C_6H_6) at the industrial laboratory of INERIS in France. The results demonstrated the specificity, the linearity and stability of the SANOA regarding the 6 species, but also showed some differences in sensitivity compared to conventional chemical methods, which required investigation. In parallel, a second model of the SANOA has been tested in a large variety of polluted situations in urban and industrial areas, including Athens during the EC supported MEDCAPHOT-Trace campaign in 1994. Several absorbents not taken previously into account in the analysis (naphthalene, *m* and *p*-xylene, phenol, CS_2, HNO_2, *p* and *o*-cresol, ethylbenzene), sometimes present at very high concentration, were progressively identified. Their cross-sections, often not available in the literature, have been established in the laboratory and the systematic measurement of these species was introduced progressively in the spectral analysis software. In 1994, two SANOA participated also in the second TOPAS campaign held at Weybourne in UK, the results of which are still under interpretation.

At present time, the SANOA is commercialised under a CNRS licence, by a French company, ENVIRONNEMENT SA. A demonstration test of use of the SANOA in an Air Quality Monitoring station is currently conducted in cooperation with the above company.

The SANOA instrument

The instrument consists of a projector equipped with a 150 watt xenon lamp and a measuring unit located at a distance of 200 to 400 m. The measuring unit is connected to a computer which controls the operation of the instrument, the acquisition of the spectra, the data processing, the visualisation of the results and the transmission to a data centre when needed.

The measuring unit includes an optical head for collecting the light, a broad band spectrometer (200-370 nm) of 0.3 nm resolution equipped with a flat field holographic grating (360 grooves/mm), and a 512 elements NMOS diode array detector. The duration of exposure adjusted automatically varies between 1 and a few seconds depending on the atmospheric transmission.

The spectral analysis is based on a least squares fit between atmospheric spectra and absorption cross-sections specific of each constituent, measured in the laboratory.

First TOPAS intercomparison at Brussels in 1992

The SANOA participated in the TOPAS intercomparison campaign at Brussels in 1992 by measuring SO_2, NO_2 and O_3 (plus other constituents not selected for the exercise), every 5 minutes without interruption during the two weeks of the campaign. Although working on a shorter distance (230 m) than that of the other instruments, its results were found comparable since the dispersion between the concentrations observed by the various instruments during the comparison periods did not exceeded 1.5 ppb for NO_2, 3.2 ppb for O_3 and 0.8 ppb for SO_2. The results of this campaign will not be discussed here since they are presented in the general TOPAS report. However, one of its results was the demonstration of the existence of some systematic small differences between the instruments which could be due to differences in the spectral analysis software's, to the spectral ranges selected for the measurement of each species, to the cross-sections in use, to the method for determining the zero of the instrument *etc.* which required further attention.

Evaluation at the industrial laboratory of INERIS

Although having its own limits, one of the method for investigating the performances of a DOAS instrument, is a comparison of its measurements with those of conventional chemical analysers in the laboratory. This has been done for the SANOA by the industrial laboratory of INERIS (Institut National de l'Environnement Industriel et des Risques) in France, by introducing a permanent flow of a mixture of nitrogen and calibrated gases into a 2 m long cell placed in the optical path of the instrument, while the concentration of the gas was also measured at the output of the cell. These tests performed in 1993-1994 for 6 gases (SO_2, NO_2, CHOH, O_3, C_7H_8 and C_6H_6) concluded at the linearity, the long term stability of zero as well as fidelity of the measurements, but at also at a systematic lower sensitivity of the SANOA measurements of 9 to 28 % depending on the constituent. Further investigation in the laboratory, has shown that the systematic underestimation was partly due to the low resolution of the absorption cross-sections used in the SANOA. After replacing them by high resolution cross-sections convolved with an instrument function measured with a mercury lamp, the discrepancy reduced, but not totally. A disagreement remained of 2 to 15 % depending on the constituent, which may be related to the limits of the cell approach: decomposition on the walls, partial photolysis of the constituent in the intense uv light beam, dimerisation in the case of NO_2.

In addition thermal testing for –50 to +50 °C, have shown a temperature dependence of the measurements partly due to a deformation of the optics in front of the spectrometer causing an increase of straylight at low temperature and partly also to a variation of the resolution of the spectrometer, which needed to be compensated.

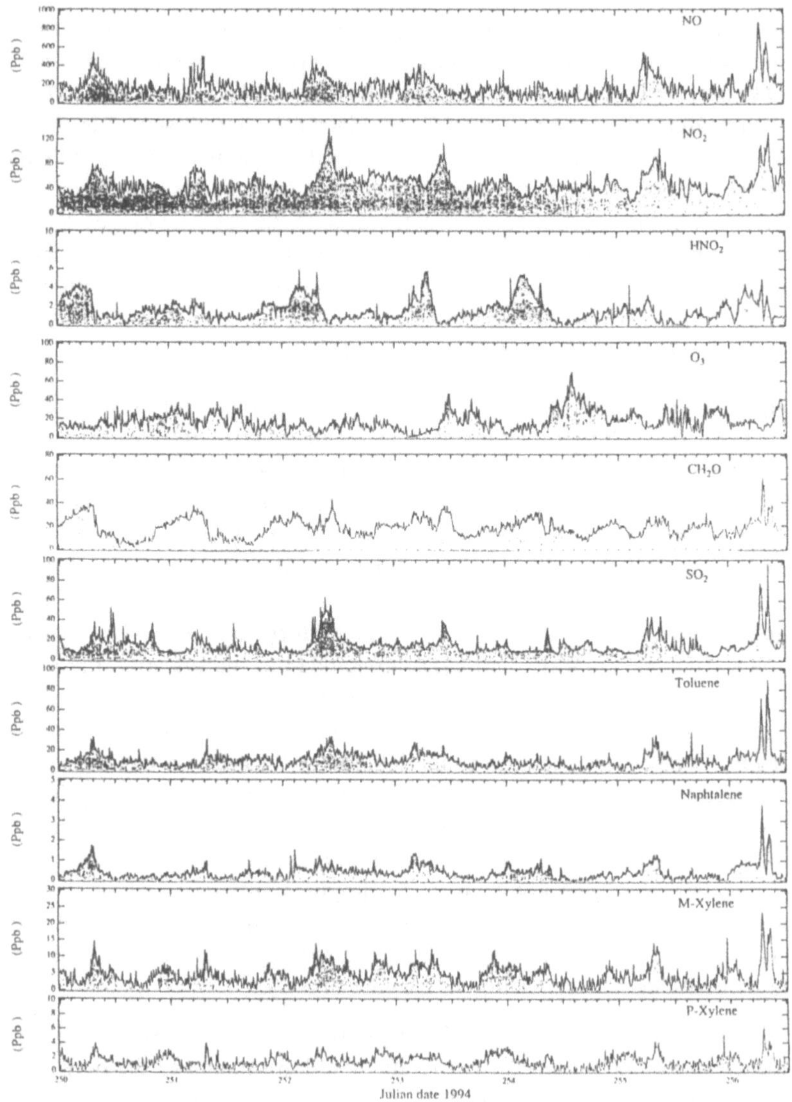

Fig. 12.11: Results of one week measurements in the city of Athens during the EC campaign MEDCAPHOT-TRACE in August 1994.

The detection limits following the Norm X 20-300 (approx. 5 standard deviations) measured by the INERIS and converted for a distance of 400 m, were: 0.6 μg/m^3 for SO_2, 2.3 μg/m^3 for NO_2, 1.7 μg/m^3 for CH_2O, 2.8 μg/m^3 for O_3, 5 μg/m^3 for C_7H_8 and, 4.1 μg/m^3 for C_6H_6.

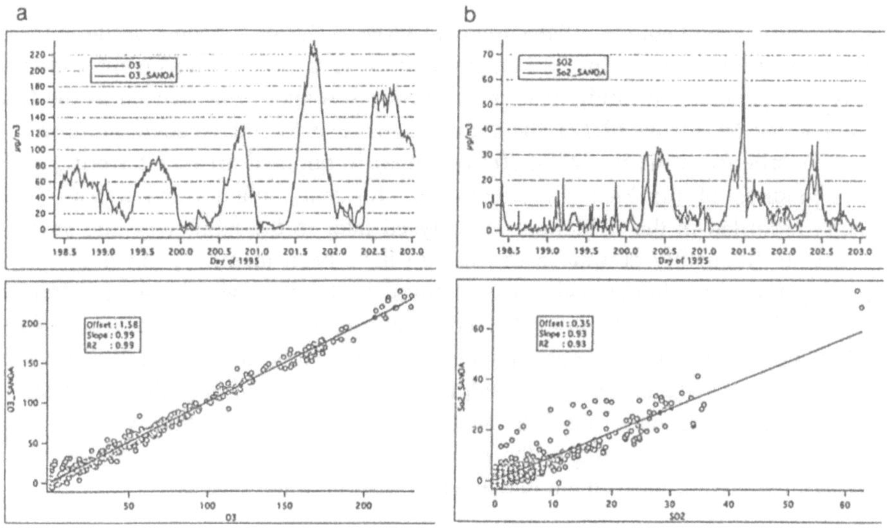

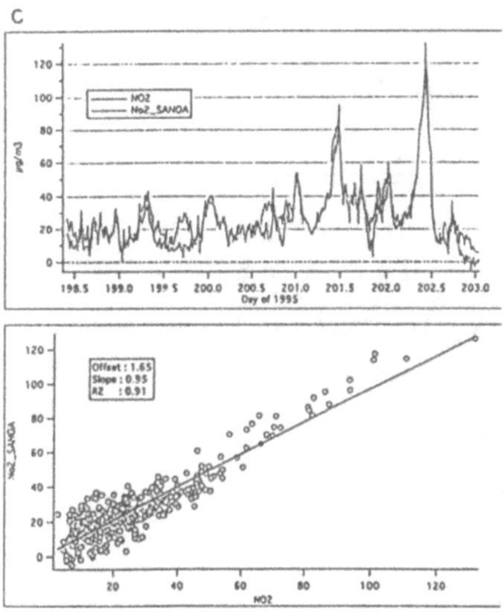

Fig. 12.12: Comparison of SANOA's measurements and those of traditionnal chemical analysers in the suburbs of Paris:
a) O_3 by UV photometry;
b) SO_2 by fluorescence;
c) NO_2 by chemiluminescence.

Investigations in the field

In parallel to the above laboratory testing, a second SANOA has been used to investigate its performances in the field. A variety of demonstration and exploration tests have been conducted in cooperation with the manufacturer at a number of sites in France as well as in major cities or industrial suburbs in Europe (Paris, Geneva, Bilbao, Brussels, Milan *etc.*). In 1994, a SANOA participated in the MEDCAPHOT-TRACE campaign supported by the European Commission for investigating the chemistry of the frequent large pollution episodes in the city of Athens in the summer (Fig. 12.11). In addition to a general agreement with the many measurements of traditional analysers for SO_2 and ozone as well as the few available for NO_2 (Fig. 12.12), these campaigns revealed the frequent presence of large absorbents in the uv (240-300 nm range) in polluted areas, interfering with the signatures of the known aromatic HC: benzene, toluene and xylene. The spectra recorded in the field were re-analysed later in the laboratory and several absorbents frequently present in the spectra have been identified unambiguously by comparison with home made cross-sections: naphthalene which can vary from 1 to 5 ppb in urban area to several hundred ppb close to a tar manufacture (Fig. 12.13), bursts of hundreds ppb of phenol and *m*-xylene in the vicinity of a car manufacture, ethylbenzene, carbon sulphur (CS_2) and cresol in polluted industrial suburbs. Although their absolute accuracy still requires further work in the laboratory, the measurements of the above absorbents introduced into the analysis, resulted in a great improvement in the determination of the BTX.

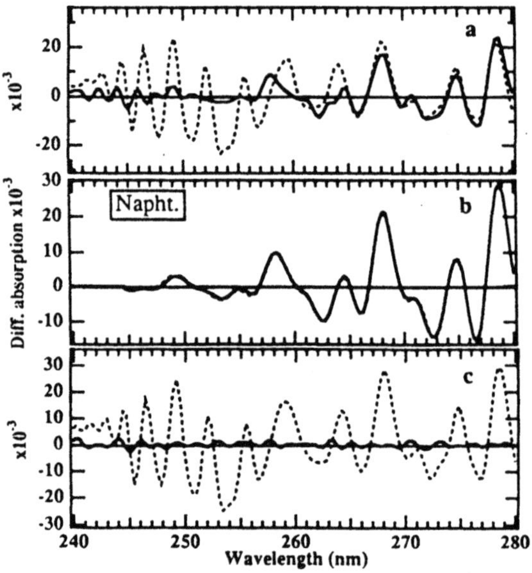

Fig. 12.13: Identification of the presence of naphtalene in the suburbs of Bilbao (Spain): a) atmospheric differential spectrum (dotted) and initial residual (full); b) cross-sections of naphtalene measured in the laboratory; c) atmospheric differential spectrum (dotted) and final residual (full) after introducing naphthalene in the analysis.

Second TOPAS intercomparison

Two SANOA participated to the second TOPAS campaign at Weybourne (UK) in October 1994 aiming at an intercomparison of long path and conventional point analysers at a remote site, in presence of low concentration of pollutants. Since the

SANOA is not designed for extremely long distances as 1600 m or 5000 m used by some of the other instruments, the two SANOAs as well as the Fourier interferometer of IASB were run on short path lengths, 155 m and 292 m (folded beam) respectively, within the limits of the station. In such conditions only ozone, NO_2 and SO_2 were measured. The results of the campaign still under interpretation, should be available shortly.

Improvements of measurements

Several limitations of the current SANOA were identified during the many tests and comparison campaigns, regarding particularly the sensitivity of the instrument, its offset of zero and the presence of interfering absorbers in the atmosphere. They have been investigated thoroughly in the laboratory. The absorption cross-sections of a number of constituents including that of O_2 and O_4, have been measured at a spectral resolution higher than that of the SANOA, convolved with the instrument's function and tested by comparison to atmospheric spectra for identifying the absorbers and removing at best their signatures in the spectra. The spectrum of the xenon lamps and its changes in temperature has been studied as well as the influence of enhanced solar straylight in presence of fog during daytime. The thermal drift of the instrument in wavelength and resolution, was investigated and compensated for. The various mathematical functions (numerical filters, interpolation *etc.*) in use in the spectral analysis software were tested in detail and improved when possible. Overall, this resulted in a reduction of the residual spectral optical thickness, which in a polluted urban site after removal of all known absorbents, has been reduced to 2×10^{-4} peak to peak.

At present time, the list of pollutants measured by the SANOA and their detection limit for a path of 300 m, is the following: NO (1.9 μg/m^3), SO_2 (0.8 μg/m^3), NO_2 (3 μg/m^3), HONO (3 μg/m^3), CH_2O (2.3 μg/m^3), ozone (3.8 μg/m^3), toluene (5 μg/m^3), *m*-xylene (4.4 μg/m^3), benzene (5 μg/m^3), naphthalene (5.3 μg/m^3).

Industrialisation

In parallel to the above research activities, the SANOA has been industrialised in cooperation between the research team and a manufacturer. All hardware and software aspects have been revisited and when needed the design has been modified for easier manufacturing or improved reliability. The SANOA is now commercialised, under licence of CNRS, by the French company ENVIRONNEMENT SA. Long term demonstration and qualification tests are currently conducted at Air Quality monitoring stations for defining better the methodology of the use of a long path monitor in the middle of a set of conventional chemical analysers.

Conclusions

The initial objectives of the project have been fulfilled. A relatively simple instrument for air pollution monitoring has been built and tested in the field. It allows reliable and simultaneous measurements of SO_2, NO_2 and ozone on a short distance, but also that of a number of compounds present in polluted areas which cannot be monitored easily by other methods. The instrument, currently commercialised for use in Air Quality monitoring networks, allows also to investigate the chemistry of polluted urban and industrial areas within the frame of several on-going scientific projects which in turn should allow to improve again its performances and the list of detectable compounds.

Acknowledgements

This work was supported by the CNRS (Programme PACB), the Ministère de l'Environnement (SRETIE) and the air quality agency (ADEME) in France, which are gratefully acknowledged.

Chapter 13

TOPAS Publications 1991 - 1995

1991

Armerding, W., A. Herbert, T. Schindler, M. Spiekermann , F.J. Comes; A long-path absorption method for monitoring OH and other tropospheric trace gases with a rapidly tunable UV-laser, in: P. Borrell, P.M. Borrell, W. Seiler (eds.), *Proc. EUROTRAC Symp. '90*, SPB Academic Publishing bv, The Hague 1991, pp. 469-473.

Axelsson, H., B. Galle, K. Gustavsson, P. Ragnarsson, M. Rudin; A transmitting/receiving telescope for DOAS-measurements using retroreflector technique, in: P. Borrell, P.M. Borrell, W. Seiler (eds.), *Proc. EUROTRAC Symp. '90*, SPB Academic Publishing bv, The Hague 1991, pp. 477-478.

Axelsson, H., H. Edner, B. Galle, P. Ragnarsson , M. Rudin; Differential absorption cross section of ozone in 250-350 nm wavelength range, in: P. Borrell, P.M. Borrell, W. Seiler (eds.), *Proc. EUROTRAC Symp. '90*, SPB Academic Publishing bv, The Hague 1991, pp. 475-476.

Carleer, M., R. Colin, A.C. Vandaele, P.C. Simon; Detection of minor tropospheric constituents using Fourier transform spectroscopy, in: *Proc. Opt. Soc. Amer. Topical Meeting on Optical Remote Sensing of the Atmosphere*, Williamsburg 1991, *Opt. Soc. Amer. Technical Digest Series* **18** (1991) 278-280.

Galle, B., A. Eilard, J. Mellquist; Mätning av NO_2 emissionen vid start med flygplanet Viggen, *IVL-Report,* Gothenburg 1991.

Laville, P., J.P. Pommereau , F. Goutail ; Evaluation of a long path diode-array spectrometer for O_3, NO_2, SO_2 and water vapor monitoring, in: P. Borrell, P.M. Borrell, W. Seiler (eds.), *Proc. EUROTRAC Symp. '90*, SPB Academic Publishing bv, The Hague 1991, pp. 479-482.

Pommereau, J.P.; Introduction to TOPAS, in: P. Borrell, P.M. Borrell, W. Seiler (eds.), *Proc. EUROTRAC Symp. '90*, SPB Academic Publishing bv, The Hague 1991, pp. 467-468.

Zetzsch, C., C.Leonard; Development of a Frequency-Modulated CW-Laser-System for the Measurement of Tropospheric OH and Other Trace Gases, in: P. Borrell, P.M. Borrell, W. Seiler (eds.), *Proc. EUROTRAC Symp. '90*, SPB Academic Publishing bv, The Hague 1991, pp. 483-485.

1993

Bösenberg, J., G. Ancellet, A. Apituley, H. Bergwerff, G. V. Cossart, H. Edner, J. Fiedler, B. Galle, C.N. de Jonge, J. Mellquist, V. Mitev, T. Schaberl, G. Sonnemann, J. Spaakman, D.J.P. Swart, E. Wallinder; Tropospheric ozone lidar experiment, TROLIX '91: Field phase report, *Report* Nr. 102, Max-Planck-Institut für Meteorologie, Hamburg 1993, pp. 1-239.

Brauers, T., H.-P. Dorn, R. Neuroth, U. Platt; Measurements of NO_3 in the marine troposphere, in: P.M. Borrell, P. Borrell, T. Cvitaš, W. Seiler (eds.), *Proc. EUROTRAC Symp. '92*, SPB Academic Publishing bv, The Hague 1993, pp. 783-787.

Carleer, M., R. Colin, A. C. Vandaele, P. C. Simon; Measurement of NO_2 and SO_2 absorption cross sections, in: P.M. Borrell, P. Borrell, T. Cvitaš, W. Seiler (eds.), *Proc. EUROTRAC Symp. '92*, SPB Academic Publishing bv, The Hague 1993, pp. 419-422.

Carleer, M., A.C. Vandaele, R. Colin, P.C. Simon; Urban Pollution Monitoring by DOAS Fourier Transform Spectroscopy, in: A. Barbe, L. Rothman (eds.), *Proc. Atmospheric Spectroscopy Applications Meeting*, Reims 1993, pp. 220-223.

Edner, H., P. Ragnarson, S. Svanberg; A multi-path DOAS system for large area pollution monitoring, in: P.M. Borrell, P. Borrell, T. Cvitaš, W. Seiler (eds.), *Proc. EUROTRAC Symp. '92*, SPB Academic Publishing bv, The Hague 1993, pp. 220-223.

Fayt, C., P. Dufour, C. Hermans, M. Van Roozendael, P.C. Simon; Instrument and software development for DOAS measurements of atmospheric constituents, in: P.M. Borrell, P. Borrell, T. Cvitaš, W. Seiler (eds.), *Proc. EUROTRAC Symp. '92*, SPB Academic Publishing bv, The Hague 1993, pp. 231-233.

Galle, B., A. Apituley, J. Bösenberg, H. Edner, J. Mellquist; A long path optical absorption method for aerosol absorption measurements, in: P.M. Borrell, P. Borrell, T. Cvitaš, W. Seiler (eds.), *Proc. EUROTRAC Symp. '92*, SPB Academic Publishing bv, The Hague 1993, p. 213.

Galle, B., H. Axelsson, J. Mellquist, L. Zetterberg; A DOAS system for background air pollution monitoring, in: P.M. Borrell, P. Borrell, T. Cvitaš, W. Seiler (eds.), *Proc. EUROTRAC Symp. '92*, SPB Academic Publishing bv, The Hague 1993, p. 214.

Galle, B., H. Axelsson, J. Mellqvist, M. Rudin; A method for measurement of large scale fluxes of tropospheric trace gases, in: P.M. Borrell, P. Borrell, T. Cvitaš, W. Seiler (eds.), *Proc. EUROTRAC Symp. '92*, SPB Academic Publishing bv, The Hague 1993, p. 215.

Galle, B., J. Mellqvist; A long path method for measurement of spectrally resolved particle attenuation, in: H. Schiff, U. Platt (eds.), *Optical Methods in Atmospheric Chemistry, Proc SPIE* **1715** (1993) 258-266.

Goutail, F., J.-P. Pommereau, M. Nunes-Pinharanda; Ambient air monitoring by differential absorption spectroscopy in the ultraviolet and the visible: the SANOA instrument, in: A. Barbe, L. Rothman (eds), *Proc. Symp. Atmos. Spectroscopy Applications*, Reims 1993, pp. 55-58.

Goutail, F., J.P. Pommereau, M. Nunes-Pinharanda, Surveillance de la composition de l'air par spectrométrie uv-visible, *Proc. Coll. Horizons de l'Optique*, Limoges, 1993.

Pommereau, J.-P., F. Goutail, M. Nunes-Pinharanda, Application de la spectroscopie UV-Visible à la mesure de la pollution urbaine, *Coll. Pollution Atm. à l'échelle locale et régionale*, ADEME, Paris 1993.

Goutail, F., J.P. Pommereau, M. Nunes-Pinharanda; Surveillance de la composition de l'air par spectrometrie UV-Visible, in: *Proc. Coll. Horizons de l'Optique,* Limoges 1993, pp. ???.

Kraft, M., D. Perner; Measurement of tropospheric OH by laser long-path absorption spectroscopy, in: P.M. Borrell, P. Borrell, T. Cvitaš, W. Seiler (eds.), *Proc. EUROTRAC Symp. '92,* SPB Academic Publishing bv, The Hague 1993, p. 239.

Pommereau, J.P., F. Goutail, M. Nunes-Pinharanda; Application de la spectroscopie UV-Visible ê la mesure de la pollution urbaine, in: *Coll. Pollution Atm. a l'echelle locale et régionale,* ADEME, Paris 1993, pp. ???.

Ritz, D., M. Hausmann, U. Platt; An improved open path multi-reflection cell for the measurement of NO_2 and NO_3, in: H. Schiff, U. Platt (eds.), *Proc. Int. Symp. on Environmental Sensing, Optical Methods in Atmospheric Chemistry, Proc SPIE* **1715** (1993) 200-211.

Vandaele A.C., M. Carleer, R. Colin, P.C. Simon; Urban Air Pollution Monitoring by DOAS Transform Spectroscopy, in: A. Barbe, L. Rothman (eds), *Proc. Workshop Atmospheric Spectroscopy Application,* Reims 1993, pp.220-223.

Vandaele, A. C., P. C. Simon, M. Carleer, R. Colin; Detection of minor tropospheric constituents in the near UV by Fourier transform spectroscopy, in: P.M. Borrell, P. Borrell, T. Cvitaš, W. Seiler (eds.), *Proc. EUROTRAC Symp. '92,* SPB Academic Publishing bv, The Hague 1993, pp. 234-238.

Vandaele, A.C., M. Carleer, R. Colin, P.C. Simon; Detection of urban O_3, NO_2, H_2CO and SO_2 using Fourier transform spectroscopy, in: H. Schiff, U. Platt (eds.), *Optical Methods in Atmospheric Chemistry, Proc SPIE* **1715** (1993) 288-292.

1994

Dorn, H.-P., U. Brandenburger, T. Brauers, M. Hausmann; A new *in situ* laser long-path absorption instrument for the measurement of tropospheric OH, in: P.M. Borrell, P. Borrell, T. Cvitaš, W. Seiler (eds.), *Proc. EUROTRAC Symp. '94,* SPB Academic Publishing bv, The Hague 1994, pp. 425-429.

Stutz, J., U. Platt; Numerical analysis of DOAS spectra with linear and nonlinear least square fits, in: P.M. Borrell, P. Borrell, T. Cvitaš, W. Seiler (eds.), *Proc. EUROTRAC Symp. '94,* SPB Academic Publishing bv, The Hague 1994, pp. 930-934.

Camy-Peyret, C., M. Carleer, C. Clerbaux, R. Colin, A.C. Vandaele, C. Hermans, P.S. Simon, B. Bergqvist, B. Galle, C. Fayt, F. Goutail, M. Nunes-Pinharanda, J.P. Pommereau, M. Hausmann, F. Heintz, U. Platt, I. Pundt, T. Rudolph, J. Plane, N. Smith; Intercomparison of instruments for tropospheric measurements using differential optical absorption spectroscopy, in: P.M. Borrell, P. Borrell, T. Cvitaš, W. Seiler (eds.), *Proc. EUROTRAC Symp. '94,* SPB Academic Publishing bv, The Hague 1994, pp. 935-939.

Carleer, M., R. Colin, J.M. Guilmot, P. Simon, A.C. Vandaele; UV - visible absorption cross-sections of relevant atmospheric species, in: P.M. Borrell, P. Borrell, T. Cvitaš, W. Seiler (eds.), *Proc. EUROTRAC Symp. '94,* SPB Academic Publishing bv, The Hague 1994, pp. 940-943.

Vandaele, A.C., M. Carleer, R. Colin, P.C. Simon; Long path monitoring of tropospheric O_3, NO_2, H_2CO and SO_2, in: R.D. Hudson (ed.), *Ozone in the Troposphere and Stratosphere* (Part 1), *Proc. Quadrennial Ozone Symp.* 1992, *NASA Conference Publication* **3266** (1994) 166-169.

Vandaele, A.C., P.C. Simon, J.M. Guilmot, M. Carleer, R. Colin; SO_2 absorption cross section measurement in the UV using a Fourier transform spectrometer, *J. Geophys. Res.* **99** (1994) 25 599-25 605.

1996

Camy-Peyret, C., B. Bergqvist, B. Galle, M. Carleer, C. Clerbaux, R. Colin, C. Fayt, F. Goutail, M. Nunes-Pinharanda, J.P. Pommereau, M. Hausmann, F. Heinz, U. Platt, I. Pundt, T. Rudolph, C. Hermans, P.C. Simon, A.C. Vandaele, J. Plane, N. Smith; Intercomparison of instruments for tropospheric measurements using differential optical absorption spectroscopy, *J. Atmos. Chem.* **23** (1996) 51-80.

Theses

M. Sc. / Diploma

Pundt, I.; Charakterisierung eines Vielfachreflexionssystems zu Messung der NO_2- und NO_3 Konzentration in der Atmosphäre, University of Heidelberg, 1993.

Ph. D.

Laville, P.; Mesure de composés gazeux troposphériques par spectrométrie d'absorption active uv et visible, University of Paris VII, 1992.

Subject Index

A

B

C

D

N

O

P

R

S

T

MIX
Papier aus verantwortungsvollen Quellen
Paper from responsible sources
FSC® C105338

If you have any concerns about our products,
you can contact us on
ProductSafety@springernature.com

In case Publisher is established outside the EU,
the EU authorized representative is:
Springer Nature Customer Service Center GmbH
Europaplatz 3, 69115 Heidelberg, Germany

Printed by Libri Plureos GmbH
in Hamburg, Germany